Progress in Mathematical Physics
Volume 59

Gerardo F. Torres del Castillo

Spinors in Four-Dimensional Spaces

Birkhäuser

Gerardo F. Torres del Castillo
Instituto de Ciencias
Universidad Autónoma de Puebla
Ciudad Universitaria
72570 Puebla, Puebla, México
gtorres@fcfm.buap.mx

ISBN 978-0-8176-4983-8 e-ISBN 978-0-8176-4984-5
DOI 10.1007/978-0-8176-4984-5
Springer New York Dordrecht Heidelberg London

Library of Congress Control Number: 2010930605

Mathematics Subject Classification (2010): 15A66, 15B10, 22E43, 53B30, 81Q05, 81R20, 83C50, 83C60

Printed on acid-free paper

www.birkhauser-science.com

Preface

The aim of this book is to present in an elementary manner the spinor formalism applicable to four-dimensional spaces of any signature, without assuming previous knowledge about spinors, nor an advanced knowledge about Lie groups. The approach followed here is not based on the Clifford algebras and does not make use of the language of fiber bundles.

This book should provide the basic notions for graduate students interested in the foundations of the two-component spinor formalism or its applications in general relativity or relativistic quantum mechanics. It is also intended for use as a reference book, since it contains several results scattered in research papers, as well as some previously unpublished results.

Throughout the book I have tried to stress the advantages of the two-component spinor formalism over other formalisms encountered in the literature, and the examples considered here have been selected with this purpose. Whenever possible, I have tried to derive the relations given in the book making use of the spinor formalism, not simply translating into the spinor notation some relation obtained by means of another formalism.

My first encounter with the two-component spinor formalism was through Professor Jerzy Plebański's monograph entitled *Spinors, Tetrads and Forms* (Plebański 1974). This monograph circulated in handwritten form in the 1970s, and when Professor Plebański decided to update his work and get it published as a book, he invited me to collaborate in the task. After several not very successful attempts to complete Professor Plebański's manuscript, I decided to start from scratch, rewriting everything. Unfortunately, owing to his much deteriorated health, Professor Plebański was unable to give his opinion on the first chapter, which was almost finished before his death in 2005. To a great extent, I have tried to follow the conventions employed in the works of Professor Plebański and collaborators.

This book differs from the books about spinors already published and also from Professor Plebański's monograph in many details. It develops the two-component spinor formalism for four-dimensional spaces of any signature, giving a detailed study of the orthogonal groups based on spinors. It also gives a detailed account of the relationship with the standard treatment of the Dirac bispinors.

The concept of the mate of a spinor (which can be defined only when the signature is Euclidean or ultrahyperbolic) is introduced and applied. Many worked examples are included in the manuscript; in some of them, the bases of the self-dual and the anti-self-dual two-forms are employed to find the connection and curvature of the manifold. The self-dual electromagnetic fields are studied to find the solutions of the source-free Maxwell equations, presenting results and derivations not previously given in book form. The $D(k,0)$ Killing spinors and their applications are discussed in some detail, and the formalism of the $\mathscr{H}\mathscr{H}$ spaces is also included with complete derivations. The Killing bispinors are studied making use of the two-component spinor formalism.

In Chapter 1, which deals with the algebraic aspects of the spinor formalism and contains a fairly complete discussion about the orthogonal groups, it is assumed that the reader has some familiarity with linear algebra, elementary group theory, and the use of tensor indices. Chapter 2 deals with the elementary applications of the spinor formalism to Riemannian manifolds, assuming some basic knowledge about differentiable manifolds, Riemannian connections, vector fields, and differential forms. For the last two chapters, it is convenient to have also some knowledge of general relativity. Most of the examples considered in the book are taken from general relativity and differential geometry.

I would like to thank the reviewers of the original version of the manuscript for their very helpful comments. I am also grateful to Gerald Kaiser and Ann Kostant at Birkhäuser for their support.

Puebla, Puebla, México *Gerardo F. Torres del Castillo*
October 2009

Contents

Chapter 1
Spinor Algebra

In this chapter the algebraic background of the two-component spinor formalism is developed, showing that a four-dimensional real vector space with an inner product can be considered as a subspace of the tensor product of two two-dimensional complex vector spaces (the spin spaces), and several examples of the usefulness of this identification are given, deriving various properties of the orthogonal transformations as well as several tensor relations.

Most existing applications of the two-component spinor formalism are related to the space-time of special or general relativity, and it is sometimes asserted that the two-component spinor formalism is tied to the signature of the space-time metric. As shown below in detail, the two-component spinor formalism can be applied to four-dimensional spaces of any signature.

In the traditional tensor formalism, one is acquainted with the fact that it is possible to construct tensors with any number of indices by means of the tensor product of vectors. As we shall see, in turn, the vectors of a four-dimensional vector space can be constructed from the tensor product of simpler objects, which are the two-component spinors. In this sense, it is often said that the spinors are square roots of vectors and that the spinors are more fundamental objects than the vectors themselves. Any vector or tensor has a spinor equivalent, and all the operations between vectors and tensors have a straightforward counterpart in the spinor algebra. As shown throughout this book, in many cases, the expression of the spinor equivalent of a tensor is much simpler than that of the corresponding tensor, and therefore it is easier to derive many relations employing spinors instead of tensors. Some of the advantages of the spinor formalism come from the fact that each spinor index takes two values only.

Apart from giving the basic rules of the spinor algebra, this chapter contains a detailed study of the orthogonal groups, as an example of the simplifications that can be achieved by making use of the spinor formalism. The connection between the two-component spinors and the Dirac spinors, for all signatures of the metric, is developed here, and the algebraic classification of totally symmetric spinors is also discussed.

G.F.T. del Castillo, *Spinors in Four-Dimensional Spaces*, Progress in Mathematical Physics 59, DOI 10.1007/978-0-8176-4984-5_1,

1.1 Orthogonal Groups

Let V be a four-dimensional real vector space with a metric tensor g (i.e., g is a nondegenerate, symmetric, not necessarily definite bilinear form). (Some authors, e.g., Hall 2004, reserve the name *metric tensor* for the tensor field defining a Riemannian manifold, reserving *inner product* for the nondegenerate bilinear form of a vector space.) It is always possible to find an orthogonal basis of V, $\{\mathbf{e}_1, \mathbf{e}_2, \mathbf{e}_3, \mathbf{e}_4\}$, such that for $a = 1,2,3,4$, $g(\mathbf{e}_a, \mathbf{e}_a) = 1$ or -1; in other words, the 4×4 matrix (g_{ab}) $(a, b = 1,2,3,4)$ that represents the metric tensor g with respect to the basis $\{\mathbf{e}_1, \mathbf{e}_2, \mathbf{e}_3, \mathbf{e}_4\}$, defined by $g_{ab} = g(\mathbf{e}_a, \mathbf{e}_b)$, is diagonal with 1's or (-1)'s along the diagonal. Such a basis will be called *orthonormal*. We can assume that the basis vectors $\mathbf{e}_1, \mathbf{e}_2, \mathbf{e}_3, \mathbf{e}_4$ are ordered in such a way that the first p entries of the diagonal are equal to 1 and the last $q = 4 - p$ entries are equal to (-1).

The numbers, p and q, of 1's and (-1)'s appearing in the diagonal matrix (g_{ab}) do not depend on the orthonormal basis chosen, but are fixed by the metric tensor. The pair of numbers (p, q) defines the *signature* of g. Since $p + q$ must be equal to 4, the knowledge of p or q defines the signature of g; some authors define the signature as $p - q$. Thus, (g_{ab}) must be the matrix $\operatorname{diag}(1,1,1,1)$, $\operatorname{diag}(1,1,1,-1)$, $\operatorname{diag}(1,1,-1,-1)$, $\operatorname{diag}(1,-1,-1,-1)$, or $\operatorname{diag}(-1,-1,-1,-1)$. Since the last two cases are obtained from the first two by reversing the sign of g and relabeling the basis vectors, we will consider only the first three cases

$$(g_{ab}) = \begin{cases} \operatorname{diag}(1,1,1,1) & \text{(Euclidean signature)}, \\ \operatorname{diag}(1,1,1,-1) & \text{(Lorentzian or hyperbolic signature)}, \\ \operatorname{diag}(1,1,-1,-1) & \text{(Kleinian or ultrahyperbolic signature)}. \end{cases} \tag{1.1}$$

In what follows it will be assumed that the metric tensor of V has one of the forms (1.1).

For a given signature, there are infinitely many bases with respect to which (g_{ab}) takes one of the forms (1.1). If $\{\mathbf{e}_1, \mathbf{e}_2, \mathbf{e}_3, \mathbf{e}_4\}$ and $\{\mathbf{e}'_1, \mathbf{e}'_2, \mathbf{e}'_3, \mathbf{e}'_4\}$ are two orthonormal bases of V with respect to which g is represented by the same matrix (g_{ab}), i.e.,

$$g(\mathbf{e}'_a, \mathbf{e}'_b) = g(\mathbf{e}_a, \mathbf{e}_b), \tag{1.2}$$

then the fact that $\{\mathbf{e}'_1, \mathbf{e}'_2, \mathbf{e}'_3, \mathbf{e}'_4\}$ is a basis of V implies the existence of a 4×4 real matrix $(L^a{}_b)$ such that

$$\mathbf{e}_a = L^b{}_a \mathbf{e}'_b \tag{1.3}$$

(with a sum over repeated indices; here and in what follows, lowercase Latin indices $a, b, \ldots$ take values from 1 to 4). Substituting (1.3) into (1.2), using the bilinearity of g, one obtains

$$g_{ab} = L^c{}_a L^d{}_b g_{cd}. \tag{1.4}$$

This condition implies that $\det(L^a{}_b) = 1$ or -1 (and, hence, $(L^a{}_b)$ is invertible). The matrices $(L^a{}_b)$ satisfying (1.4) are called *orthogonal*, and as can readily be seen, the set of orthogonal matrices forms a group with the usual matrix multiplication as the group operation, which is denoted by $\mathrm{O}(p,q)$, where (p,q) is the signature of g. The matrices satisfying (1.4) with $\det(L^a{}_b) = 1$ form a subgroup of $O(p,q)$, denoted by $SO(p,q)$. (When q is equal to zero, we simply write O(4) or SO(4) instead of $O(4,0)$ or $SO(4,0)$, respectively.)

If (g^{ab}) is the inverse of the matrix (g_{ab}), i.e., $g_{ab}g^{bc} = \delta^c_a$, from (1.4) we obtain $\delta^e_b = g^{ea}g_{ab} = g^{ea}L^c{}_aL^d{}_bg_{cd} = g^{ea}L^c{}_ag_{cd}L^d{}_b$, which means that the inverse of $(L^a{}_b)$, with entries $(L^{-1})^a{}_b$, is given by

$$(L^{-1})^e{}_d = g^{ea}L^c{}_ag_{cd}, \tag{1.5}$$

or, following the standard rules for raising and lowering tensor indices by means of the metric tensor (g_{ab}) and its inverse (g^{ab}) (e.g., $t^a = g^{ab}t_b$, $t_a = g_{ab}t^b$), we can write (1.5) as

$$(L^{-1})^e{}_d = L_d{}^e. \tag{1.6}$$

The inverse of a matrix satisfying (1.4) also belongs to $O(p,q)$ and therefore also satisfies condition (1.4), i.e., $g_{ab} = (L^{-1})^c{}_a(L^{-1})^d{}_bg_{cd}$, and according to (1.6) we conclude that

$$g_{ab} = L_a{}^cL_b{}^dg_{cd}. \tag{1.7}$$

In order to motivate the spinor notation employed in what follows, it is convenient to consider the space $\mathbb{R}^{2,2}$, which is the vector space $\mathbb{R}^4$ with the ultrahyperbolic metric tensor $(g_{ab}) = \mathrm{diag}(1,1,-1,-1)$ with respect to the canonical basis. The mapping

$$(x,y,z,w) \mapsto \frac{1}{\sqrt{2}}\begin{pmatrix} -x-z & -y-w \\ w-y & x-z \end{pmatrix} \tag{1.8}$$

is a one-to-one correspondence between $\mathbb{R}^{2,2}$ and the vector space formed by the 2×2 real matrices (the factor $1/\sqrt{2}$ is introduced in order to get agreement with the conventions adopted in the following sections). Denoting by P the matrix on the right-hand side of (1.8), we see that

$$\det P = -\frac{1}{2}(x^2+y^2-z^2-w^2),$$

i.e., apart from the factor $-1/2$, $\det P$ is the inner product of the vector (x,y,z,w) with itself.

If K and M are 2×2 matrices, both real or both pure imaginary, the transformation

$$P \mapsto P' \equiv KPM \tag{1.9}$$

is linear and, by means of the correspondence (1.8), is equivalent to some linear transformation of $\mathbb{R}^{2,2}$ into itself. Since $\det P' = \det(KPM) = (\det K)(\det P)(\det M)$, it follows that if

$$(\det K)(\det M) = 1, \tag{1.10}$$

then $\det P = \det P'$; that is, denoting by (x',y',z',w') the vector corresponding to P' according to (1.8), the inner product of (x,y,z,w) with itself is equal to the inner product of (x',y',z',w') with itself. This implies that the mapping (1.9) corresponds to an orthogonal transformation of $\mathbb{R}^{2,2}$, i.e., an element of $O(2,2)$.

Assuming that the matrices K, M satisfy the condition (1.10), letting $\widetilde{K} \equiv (\det K)^{-1/2} K$, $\widetilde{M} \equiv (\det K)^{1/2} M$, we have

$$P' = KPM = \widetilde{K} P \widetilde{M},$$

and $\det \widetilde{K} = 1 = \det \widetilde{M}$; hence, in order for (1.9) to correspond to an orthogonal transformation, we can assume that the determinants of K and M are equal to 1. (Note that since K and M are both real or both pure imaginary, $(\det K)^{1/2}$ is real or pure imaginary and the matrices $\widetilde{K}$ and $\widetilde{M}$ are also both real or both pure imaginary.) Hence, by representing the points of $\mathbb{R}^{2,2}$ by 2×2 matrices, as in (1.8), the simple algebraic conditions $\det K = 1 = \det M$, allow us to find orthogonal transformations.

If we denote the entries of P by $P^i{}_j$, where the superscript labels the row and the subscript labels the column, and similarly for the other 2×2 matrices, the transformation (1.9) is equivalent to

$$P'^i{}_j = K^i{}_k P^k{}_l M^l{}_j = K^i{}_k M^l{}_j P^k{}_l. \tag{1.11}$$

This last equation is similar to the transformation law for the components of a rank-two tensor, with the difference that in the latter case, the matrix K would be the inverse of M, while in (1.11), K and M can be two arbitrary unimodular 2×2 real or pure imaginary matrices.

Hence, it is convenient to distinguish the two indices labeling the entries of the matrix P in a way that explicitly indicates which one of the matrices K and M appearing in (1.9) is employed in the transformation. Specifically, the entries of K will be labeled with undotted indices $A, B, \ldots$ that take two values only, $K = (K^A{}_B)$ $(A, B, \ldots = 1, 2)$, while the entries of M will be labeled with dotted indices $\dot{A}, \dot{B}, \ldots$, $M = (M^{\dot{A}}{}_{\dot{B}})$ $(\dot{A}, \dot{B}, \ldots = \dot{1}, \dot{2})$, and the entries of P and P' will be labeled with one undotted index and one dotted index, $P = (P^A{}_{\dot{B}})$, so that (1.9) amounts to

$$P'^A{}_{\dot{B}} = K^A{}_C P^C{}_{\dot{D}} M^{\dot{D}}{}_{\dot{B}} = K^A{}_C M^{\dot{D}}{}_{\dot{B}} P^C{}_{\dot{D}}. \tag{1.12}$$

As we shall show in the following sections, *all* the $SO(p,q)$ transformations can be expressed in a form analogous to (1.12), though in some cases it will be necessary or convenient to make use of linear combinations of vectors of V with complex scalars.

1.2 Null Tetrads and the Spinor Equivalent of a Tensor

Apart from the orthonormal bases considered in the foregoing section, it will be useful to consider bases, $\{\mathbf{E}_1, \mathbf{E}_2, \mathbf{E}_3, \mathbf{E}_4\}$, with respect to which the metric tensor of V is represented by the matrix

$$\begin{pmatrix} 0 & 1 & 0 & 0 \\ 1 & 0 & 0 & 0 \\ 0 & 0 & 0 & 1 \\ 0 & 0 & 1 & 0 \end{pmatrix}, \tag{1.13}$$

i.e., the only nonvanishing inner products among the vectors $\mathbf{E}_a$ are given by $g(\mathbf{E}_1, \mathbf{E}_2) = 1 = g(\mathbf{E}_3, \mathbf{E}_4)$. Such a basis will be called a *null tetrad*, since each basis vector is null ($g(\mathbf{E}_a, \mathbf{E}_a) = 0$, without summation on a). Only when the signature of the metric of V is ultrahyperbolic will it be possible to find a null tetrad formed by real vectors (see (1.16) below), and therefore, in most cases we will have to assume that the vectors $\mathbf{E}_a$ belong to the *complexification* of V (see, e.g., Hirsch and Smale 1974). For instance, if the signature of V is Euclidean and $\{\mathbf{e}_1, \mathbf{e}_2, \mathbf{e}_3, \mathbf{e}_4\}$ is an orthonormal basis of V, we can take

$$\begin{aligned} \mathbf{E}_1 &= \frac{1}{\sqrt{2}}(\mathbf{e}_1 + \mathrm{i}\mathbf{e}_2), & \mathbf{E}_2 &= \frac{1}{\sqrt{2}}(\mathbf{e}_1 - \mathrm{i}\mathbf{e}_2), \\ \mathbf{E}_3 &= \frac{1}{\sqrt{2}}(\mathbf{e}_3 - \mathrm{i}\mathbf{e}_4), & \mathbf{E}_4 &= \frac{1}{\sqrt{2}}(\mathbf{e}_3 + \mathrm{i}\mathbf{e}_4). \end{aligned} \tag{1.14}$$

Similarly, it can be readily verified that if $(g_{ab}) = \mathrm{diag}\,(1,1,1,-1)$, we can take

$$\begin{aligned} \mathbf{E}_1 &= \frac{1}{\sqrt{2}}(\mathbf{e}_1 + \mathrm{i}\mathbf{e}_2), & \mathbf{E}_2 &= \frac{1}{\sqrt{2}}(\mathbf{e}_1 - \mathrm{i}\mathbf{e}_2), \\ \mathbf{E}_3 &= \frac{1}{\sqrt{2}}(\mathbf{e}_3 + \mathbf{e}_4), & \mathbf{E}_4 &= \frac{1}{\sqrt{2}}(\mathbf{e}_3 - \mathbf{e}_4), \end{aligned} \tag{1.15}$$

while if $(g_{ab}) = \mathrm{diag}\,(1,1,-1,-1)$, a convenient choice is

$$\begin{aligned} \mathbf{E}_1 &= \frac{1}{\sqrt{2}}(\mathbf{e}_1 - \mathbf{e}_3), & \mathbf{E}_2 &= \frac{1}{\sqrt{2}}(\mathbf{e}_1 + \mathbf{e}_3), \\ \mathbf{E}_3 &= \frac{1}{\sqrt{2}}(-\mathbf{e}_2 + \mathbf{e}_4), & \mathbf{E}_4 &= \frac{1}{\sqrt{2}}(-\mathbf{e}_2 - \mathbf{e}_4). \end{aligned} \tag{1.16}$$

Conversely, if one assumes that the only nonzero inner products among the vectors $\mathbf{E}_a$ are given by $g(\mathbf{E}_1, \mathbf{E}_2) = 1 = g(\mathbf{E}_3, \mathbf{E}_4)$, then the vectors $\mathbf{e}_a$ given by (1.14)–(1.16) form an orthonormal basis with respect to which the metric tensor is represented by one of the forms (1.1), depending on the signature of the metric tensor.

Whereas the components of the metric tensor with respect to an orthonormal basis $\{\mathbf{e}_1, \mathbf{e}_2, \mathbf{e}_3, \mathbf{e}_4\}$ depend on the signature of the metric tensor [see (1.1)], the components of the metric tensor with respect to a null tetrad $\{\mathbf{E}_1, \mathbf{E}_2, \mathbf{E}_3, \mathbf{E}_4\}$ will be given *in all cases* by (1.13); the signature of the metric tensor is determined by the behavior of the vectors $\mathbf{E}_a$ under complex conjugation. (For instance, the relations $\overline{\mathbf{E}_1} = \mathbf{E}_2$, $\overline{\mathbf{E}_3} = \mathbf{E}_4$, where the bar denotes complex conjugation, satisfied by the null tetrad (1.14) imply that the metric has Euclidean signature.)

Instead of a single subscript, $a = 1,2,3,4$, labeling the vectors, $\mathbf{E}_a$, of a null tetrad, it is convenient to make use of a pair of subscripts that take two values each. Letting

$$(\mathbf{e}_{A\dot{B}}) \equiv \begin{pmatrix} \mathbf{E}_4 & \mathbf{E}_2 \\ \mathbf{E}_1 & -\mathbf{E}_3 \end{pmatrix} \tag{1.17}$$

$(A,B=1,2,\dot{A},\dot{B}=\dot{1},\dot{2})$, one finds that the definition of the null tetrad is equivalent to

$$g(\mathbf{e}_{A\dot{B}},\mathbf{e}_{C\dot{D}}) = -\varepsilon_{AC}\varepsilon_{\dot{B}\dot{D}}, \tag{1.18}$$

where

$$(\varepsilon_{AB}) = (\varepsilon^{AB}) = \begin{pmatrix} 0 & 1 \\ -1 & 0 \end{pmatrix} = (\varepsilon_{\dot{A}\dot{B}}) = (\varepsilon^{\dot{A}\dot{B}}). \tag{1.19}$$

(Some works make use of primed indices instead of dotted ones (e.g., Penrose 1960, Penrose and Rindler 1984, Stewart 1990).)

Since the vectors $\mathbf{e}_{A\dot{B}}$ belong to the complexification of V, there exist complex scalars, $\sigma^a{}_{A\dot{B}}$, such that

$$\mathbf{e}_{A\dot{B}} = \frac{1}{\sqrt{2}}\sigma^a{}_{A\dot{B}}\mathbf{e}_a, \tag{1.20}$$

and as a consequence of (1.18), they obey

$$\sigma^a{}_{A\dot{B}}\sigma^b{}_{C\dot{D}}g_{ab} = -2\varepsilon_{AC}\varepsilon_{\dot{B}\dot{D}}. \tag{1.21}$$

The scalars $\sigma^a{}_{A\dot{B}}$ will be called *connection symbols* or *Infeld–van der Waerden symbols.*

The Levi-Civita symbols (1.19) will be employed to lower or raise the *spinor indices* A,B. We shall follow the convention (e.g., Plebański 1974, 1975)

$$\psi_A = \varepsilon_{AB}\psi^B, \qquad \psi^A = \psi_B\varepsilon^{BA} \tag{1.22}$$

(thus, $\psi^2 = \psi_1$, $\psi^1 = -\psi_2$), and similarly for dotted indices. It should be remarked that other authors (e.g., Penrose 1960, Penrose and Rindler 1984, Wald 1984, Stewart 1990) follow the convention $\psi^A = \varepsilon^{AB}\psi_B$. The antisymmetry of ε_{AB} implies that

$$\psi^A\phi_A = \psi^A\varepsilon_{AB}\phi^B = -\psi^A\varepsilon_{BA}\phi^B = -\psi_B\phi^B = -\psi_A\phi^A; \tag{1.23}$$

thus,

$$\psi^A\psi_A = 0. \tag{1.24}$$

Similar results hold for objects with more than one index (dotted or undotted), e.g., $\alpha_{AB}\beta^{AB} = -\alpha^A{}_B\beta_A{}^B = \alpha^{AB}\beta_{AB}$, $\mu_{A\dot{B}C}\nu^{A\dot{B}} = -\mu^A{}_{\dot{B}C}\nu_A{}^{\dot{B}} = \mu^{A\dot{B}}{}_C\nu_{A\dot{B}}$. Furthermore, from (1.22) it follows that

$$\varepsilon^A{}_B = \delta^A_B \qquad \text{and} \qquad \varepsilon_A{}^B = -\delta^B_A. \tag{1.25}$$

The definitions (1.19) are consistent with the rules (1.22) in the sense that, for instance, lowering the indices of ε^{AB}, one obtains ε_{AB}, that is, $\varepsilon_{AC}\varepsilon_{BD}\varepsilon^{CD} = \varepsilon_{AB}$.

The Levi-Civita symbols (1.19) will appear very often in what follows, in many cases as a consequence of the fact that any object antisymmetric in two indices must be proportional to one of the symbols (1.19).

Proposition 1.1. *If* $\psi_{AB} = -\psi_{BA}$, *then*

$$\psi_{AB} = \tfrac{1}{2}\psi^R{}_R\varepsilon_{AB}. \tag{1.26}$$

Proof. Any antisymmetric 2×2 matrix is of the form $\left(\begin{smallmatrix} 0 & a \\ -a & 0 \end{smallmatrix}\right)$, which can be written as $a(\varepsilon_{AB})$; therefore, if ψ_{AB} is antisymmetric, it must be a multiple of ε_{AB}: $\psi_{AB} = a\varepsilon_{AB}$, for some a. Then, by contracting both sides of this equality with ε^{AB}; one obtains $\psi^A{}_A = 2a$, which leads to (1.26). □

Similarly, $\psi^{AB} = \frac{1}{2}\psi^R{}_R\varepsilon^{AB}$, if ψ^{AB} is antisymmetric in A, B, with analogous identities for objects with dotted indices. Furthermore, for an object with more than two indices, e.g., $h_A{}^{\dot{B}}{}_C{}^D$, such that $h_A{}^{\dot{B}}{}_C{}^D = -h_C{}^{\dot{B}}{}_A{}^D$, we have $h_A{}^{\dot{B}}{}_C{}^D = \frac{1}{2}h^{R\dot{B}}{}_R{}^D\varepsilon_{AC}$. As a corollary, noting that (1.26) can be written as $\psi_{AB} = \frac{1}{2}(\varepsilon^{RS}\psi_{RS})\varepsilon_{AB}$, we see that if $\psi_{AB\ldots\dot{C}\dot{D}\ldots}$ is any object with at least two indices of the same type (dotted or undotted), a difference of the form

$$\psi_{A\ldots R\ldots S\ldots\dot{C}\dot{D}\ldots} - \psi_{A\ldots S\ldots R\ldots\dot{C}\dot{D}\ldots},$$

where only two indices of the same type interchange places, is antisymmetric in R, S, and therefore it is equal to $\frac{1}{2}\varepsilon^{PQ}(\psi_{\ldots P\ldots Q\ldots} - \psi_{\ldots Q\ldots P\ldots})\,\varepsilon_{RS} = \frac{1}{2}(\psi_{\ldots}{}^{Q}{}_{\ldots Q\ldots} - \psi_{\ldots Q\ldots}{}^{Q}{}_{\ldots})\,\varepsilon_{RS} = \psi_{\ldots}{}^{Q}{}_{\ldots Q\ldots}\,\varepsilon_{RS}$. Thus,

$$\psi_{\ldots R\ldots S\ldots} - \psi_{\ldots S\ldots R\ldots} = \psi_{\ldots}{}^{Q}{}_{\ldots Q\ldots}\,\varepsilon_{RS}, \tag{1.27}$$

with analogous relations for the cases where the interchanged indices are dotted or appear as superscripts (dotted or undotted). By raising, e.g., the index R on both sides of (1.27), one obtains

$$\psi_{\ldots}{}^{R}{}_{\ldots S\ldots} - \psi_{\ldots S\ldots}{}^{R}{}_{\ldots} = \psi_{\ldots}{}^{Q}{}_{\ldots Q\ldots}\,\delta^R_S. \tag{1.28}$$

Note also that (1.27) implies that

$$\psi_{\ldots R\ldots S\ldots} = \psi_{\ldots S\ldots R\ldots} \qquad \Leftrightarrow \qquad \psi_{\ldots}{}^{Q}{}_{\ldots Q\ldots} = 0. \tag{1.29}$$

If $(M^R{}_S)$ is a (real or complex) 2×2 matrix, then

$$\varepsilon_{AC}M^A{}_BM^C{}_D = \det(M^R{}_S)\varepsilon_{BD}. \tag{1.30}$$

In fact, letting $\psi_{BD} \equiv \varepsilon_{AC}M^A{}_BM^C{}_D$, it follows that

$$\psi_{DB} = \varepsilon_{AC}M^A{}_DM^C{}_B = -\varepsilon_{CA}M^C{}_BM^A{}_D = -\psi_{BD},$$

and since $\psi^R{}_R = \psi^1{}_1 + \psi^2{}_2 = -\psi_{21} + \psi_{12} = 2\psi_{12} = 2\varepsilon_{AC}M^A{}_1M^C{}_2 = 2(M^1{}_1M^2{}_2 - M^2{}_1M^1{}_2) = 2\det(M^R{}_S)$, making use of (1.26) one obtains (1.30).

Given a set of connection symbols $\sigma^a{}_{A\dot{B}}$, the scalars

$$\tilde{\sigma}^a{}_{A\dot{B}} = K^C{}_AM^{\dot{D}}{}_{\dot{B}}\sigma^a{}_{C\dot{D}} \tag{1.31}$$

also satisfy (1.21) (with the same g_{ab}) if and only if $\det(K^A{}_B)\det(M^{\dot{A}}{}_{\dot{B}}) = 1$. In fact, from (1.31) and (1.30) we have

$$\begin{aligned}
\tilde{\sigma}^a{}_{A\dot{B}}\tilde{\sigma}^b{}_{C\dot{D}}g_{ab} &= K^E{}_AM^{\dot{F}}{}_{\dot{B}}\sigma^a{}_{E\dot{F}}K^G{}_CM^{\dot{H}}{}_{\dot{D}}\sigma^b{}_{G\dot{H}}g_{ab} \\
&= K^E{}_AM^{\dot{F}}{}_{\dot{B}}K^G{}_CM^{\dot{H}}{}_{\dot{D}}(-2\varepsilon_{EG}\varepsilon_{\dot{F}\dot{H}}) \\
&= -2\det(K^A{}_B)\,\varepsilon_{AC}\,\det(M^{\dot{A}}{}_{\dot{B}})\,\varepsilon_{\dot{B}\dot{D}} \\
&= -2\varepsilon_{AC}\varepsilon_{\dot{B}\dot{D}}.
\end{aligned}$$

Similarly,

$$\widetilde{\sigma}^a{}_{A\dot{B}} = K^{\dot{C}}{}_A M^D{}_{\dot{B}} \sigma^a{}_{D\dot{C}} \tag{1.32}$$

satisfy (1.21) if and only if $\det(K^{\dot{A}}{}_B)\det(M^A{}_{\dot{B}}) = 1$. As we shall show in the following section, given two sets of Infeld–van der Waerden symbols $\sigma^a{}_{A\dot{B}}$ and $\widetilde{\sigma}^a{}_{A\dot{B}}$ (satisfying (1.21) with the *same* metric tensor g_{ab}), there exist two unimodular 2×2 matrices such that (1.31) or (1.32) holds.

A Convenient Choice for the Infeld–van der Waerden Symbols

Comparing (1.14) with (1.20), one finds that when $(g_{ab}) = \mathrm{diag}(1,1,1,1)$, the Infeld–van der Waerden symbols can be chosen as

$$\begin{aligned} (\sigma^1{}_{A\dot{B}}) &= \begin{pmatrix} 0 & 1 \\ 1 & 0 \end{pmatrix}, & (\sigma^2{}_{A\dot{B}}) &= \begin{pmatrix} 0 & -\mathrm{i} \\ \mathrm{i} & 0 \end{pmatrix}, \\ (\sigma^3{}_{A\dot{B}}) &= \begin{pmatrix} 1 & 0 \\ 0 & -1 \end{pmatrix}, & (\sigma^4{}_{A\dot{B}}) &= \begin{pmatrix} \mathrm{i} & 0 \\ 0 & \mathrm{i} \end{pmatrix}. \end{aligned} \tag{1.33}$$

(Note that the first three matrices in (1.33) are the usual Pauli matrices.)

Similarly, when $(g_{ab}) = \mathrm{diag}(1,1,1,-1)$, from (1.15), one finds that we can take

$$\begin{aligned} (\sigma^1{}_{A\dot{B}}) &= \begin{pmatrix} 0 & 1 \\ 1 & 0 \end{pmatrix}, & (\sigma^2{}_{A\dot{B}}) &= \begin{pmatrix} 0 & -\mathrm{i} \\ \mathrm{i} & 0 \end{pmatrix}, \\ (\sigma^3{}_{A\dot{B}}) &= \begin{pmatrix} 1 & 0 \\ 0 & -1 \end{pmatrix}, & (\sigma^4{}_{A\dot{B}}) &= \begin{pmatrix} -1 & 0 \\ 0 & -1 \end{pmatrix}, \end{aligned} \tag{1.34}$$

and if $(g_{ab}) = \mathrm{diag}(1,1,-1,-1)$, then

$$\begin{aligned} (\sigma^1{}_{A\dot{B}}) &= \begin{pmatrix} 0 & 1 \\ 1 & 0 \end{pmatrix}, & (\sigma^2{}_{A\dot{B}}) &= \begin{pmatrix} -1 & 0 \\ 0 & 1 \end{pmatrix}, \\ (\sigma^3{}_{A\dot{B}}) &= \begin{pmatrix} 0 & 1 \\ -1 & 0 \end{pmatrix}, & (\sigma^4{}_{A\dot{B}}) &= \begin{pmatrix} -1 & 0 \\ 0 & -1 \end{pmatrix}, \end{aligned} \tag{1.35}$$

which are all real [see (1.16)]. By means of the relation (1.31), with the $\sigma^a{}_{A\dot{B}}$ given by (1.35) and the unimodular matrices

$$(\overset{0}{K}{}^A{}_B) \equiv \frac{1}{2}\begin{pmatrix} 1-\mathrm{i} & -1+\mathrm{i} \\ 1+\mathrm{i} & 1+\mathrm{i} \end{pmatrix}, \qquad (\overset{0}{M}{}^{\dot{A}}{}_{\dot{B}}) \equiv \frac{1}{2}\begin{pmatrix} 1+\mathrm{i} & -1-\mathrm{i} \\ 1-\mathrm{i} & 1-\mathrm{i} \end{pmatrix}, \tag{1.36}$$

one obtains another set of connection symbols for the same signature given by (dropping the tilde)

$$
\begin{aligned}
(\sigma^1{}_{A\dot{B}}) &= \begin{pmatrix} 0 & -\mathrm{i} \\ \mathrm{i} & 0 \end{pmatrix}, \qquad (\sigma^2{}_{A\dot{B}}) = \begin{pmatrix} 0 & 1 \\ 1 & 0 \end{pmatrix}, \\
(\sigma^3{}_{A\dot{B}}) &= \begin{pmatrix} -\mathrm{i} & 0 \\ 0 & \mathrm{i} \end{pmatrix}, \qquad (\sigma^4{}_{A\dot{B}}) = \begin{pmatrix} -1 & 0 \\ 0 & -1 \end{pmatrix}.
\end{aligned}
\tag{1.37}
$$

The choices given in (1.33)–(1.35) and (1.37) are convenient because they satisfy the relations

$$
\overline{\sigma^a{}_{A\dot{B}}} = \begin{cases} -\sigma^{aA\dot{B}} & \text{if the } \sigma^a{}_{A\dot{B}} \text{ are given by (1.33)}, \\ \sigma^a{}_{B\dot{A}} & \text{if the } \sigma^a{}_{A\dot{B}} \text{ are given by (1.34)}, \\ \sigma^a{}_{A\dot{B}} & \text{if the } \sigma^a{}_{A\dot{B}} \text{ are given by (1.35)}, \\ \eta_{AC}\eta_{\dot{B}\dot{D}}\sigma^{aC\dot{D}} & \text{if the } \sigma^a{}_{A\dot{B}} \text{ are given by (1.37)}, \end{cases}
\tag{1.38}
$$

where

$$
(\eta_{AB}) \equiv \begin{pmatrix} 1 & 0 \\ 0 & -1 \end{pmatrix} \equiv (\eta_{\dot{A}\dot{B}}).
\tag{1.39}
$$

By contracting both sides of (1.21) with $\sigma_c{}^{C\dot{D}}$, one obtains

$$
\sigma^a{}_{A\dot{B}}\sigma^b{}_{C\dot{D}}\sigma_c{}^{C\dot{D}}g_{ab} = -2\varepsilon_{AC}\varepsilon_{\dot{B}\dot{D}}\sigma_c{}^{C\dot{D}} = -2\sigma_{cA\dot{B}},
$$

which is equivalent to

$$
\sigma^a{}_{A\dot{B}}(\sigma_{aC\dot{D}}\sigma_c{}^{C\dot{D}} + 2g_{ac}) = 0.
$$

Since the Infeld–van der Waerden symbols represent a change of basis, they correspond to an invertible relation, and therefore, from the last equation it follows that

$$
\sigma_{aA\dot{B}}\sigma_b{}^{A\dot{B}} = -2g_{ab}.
\tag{1.40}
$$

The minus sign appearing in the right-hand sides of and (1.21) and (1.40), as well as in very many equations in the rest of the book, is necessary in order to have the simple relation $\overline{\sigma^a{}_{A\dot{B}}} = \sigma^a{}_{B\dot{A}}$ when $(g_{ab}) = \operatorname{diag}(1,1,1,-1)$; this sign can be avoided, maintaining the relation $\overline{\sigma^a{}_{A\dot{B}}} = \sigma^a{}_{B\dot{A}}$, if one chooses $(g_{ab}) = \operatorname{diag}(1,-1,-1,-1)$, as in Penrose (1960), Penrose and Rindler (1984), and Stewart (1990).

If $t_{ab\ldots d}$ are the components of a tensor with respect to the orthonormal basis $\{\mathbf{e}_1, \mathbf{e}_2, \mathbf{e}_3, \mathbf{e}_4\}$, then the components of its *spinor equivalent* are defined by

$$
t_{A\dot{B}C\dot{D}\ldots G\dot{H}} \equiv \frac{1}{\sqrt{2}}\sigma^a{}_{A\dot{B}}\frac{1}{\sqrt{2}}\sigma^b{}_{C\dot{D}}\cdots\frac{1}{\sqrt{2}}\sigma^d{}_{G\dot{H}}\,t_{ab\ldots d}.
\tag{1.41}
$$

It is often convenient to write (1.41) in the form

$$
t_{A\dot{A}B\dot{B}\ldots D\dot{D}} \equiv \frac{1}{\sqrt{2}}\sigma^a{}_{A\dot{A}}\frac{1}{\sqrt{2}}\sigma^b{}_{B\dot{B}}\cdots\frac{1}{\sqrt{2}}\sigma^d{}_{D\dot{D}}\,t_{ab\ldots d},
\tag{1.42}
$$

where it is understood that the indices A and $\dot{A}$ are independent of each other, and so on; in this manner, the tensor index a is replaced by the pair of indices $A\dot{A}$.

Sometimes all the undotted indices are written to the left of all the dotted indices, maintaining the order in which the indices of each type appear, that is,

$$t_{AB\ldots D\dot{A}\dot{B}\ldots\dot{D}} \equiv \frac{1}{\sqrt{2}}\sigma^a{}_{A\dot{A}}\frac{1}{\sqrt{2}}\sigma^b{}_{B\dot{B}}\cdots\frac{1}{\sqrt{2}}\sigma^d{}_{D\dot{D}}t_{ab\ldots d}. \tag{1.43}$$

Then, according to (1.40),

$$t_{ab\ldots d} = \left(-\frac{1}{\sqrt{2}}\sigma_a{}^{A\dot{A}}\right)\left(-\frac{1}{\sqrt{2}}\sigma_b{}^{B\dot{B}}\right)\cdots\left(-\frac{1}{\sqrt{2}}\sigma_d{}^{D\dot{D}}\right)t_{A\dot{A}B\dot{B}\ldots D\dot{D}}. \tag{1.44}$$

Owing to (1.20), the spinor equivalent of a tensor gives the components of that tensor with respect to the basis $\{\mathbf{e}_{1\dot{1}}, \mathbf{e}_{1\dot{2}}, \mathbf{e}_{2\dot{1}}, \mathbf{e}_{2\dot{2}}\}$. For instance, an arbitrary element of V can be expressed as a linear combination $v^a\mathbf{e}_a$, and by virtue of (1.44) and (1.20),

$$v^a\mathbf{e}_a = -\frac{1}{\sqrt{2}}\sigma^a{}_{A\dot{A}}v^{A\dot{A}}\,\mathbf{e}_a = -v^{A\dot{A}}\,\mathbf{e}_{A\dot{A}}. \tag{1.45}$$

From the definition (1.42) it follows that the spinor equivalent of a sum [resp. product] of tensors is the sum [resp. product] of their spinor equivalents; however, the spinor equivalent of the contraction of a tensor is not always equal to the contraction of its spinor equivalent. From (1.21) we obtain, for instance,

$$t_a s^a = -t_{A\dot{A}}s^{A\dot{A}}. \tag{1.46}$$

Equation (1.45) allows us to consider the complexification of V as the tensor product of two complex two-dimensional vector spaces called *spin spaces*. We identify each null vector $\mathbf{e}_{A\dot{B}}$ with the tensor product $\mathbf{e}_A \otimes \mathbf{e}_{\dot{B}}$, where $\{\mathbf{e}_1, \mathbf{e}_2\}$ and $\{\mathbf{e}_{\dot{1}}, \mathbf{e}_{\dot{2}}\}$ are bases of the spin spaces. The elements of the spin spaces will be called *one-index spinors*, since with respect to the bases $\{\mathbf{e}_1, \mathbf{e}_2\}$ and $\{\mathbf{e}_{\dot{1}}, \mathbf{e}_{\dot{2}}\}$ they are represented by complex components of the form ϕ^A and $\phi^{\dot{A}}$, respectively (i.e., an arbitrary element of each spin space has the form $\phi^A\mathbf{e}_A$ or $\phi^{\dot{A}}\mathbf{e}_{\dot{A}}$).

By forming tensor products of one-index spinors we obtain spinors with any number of dotted or undotted indices (the spinor equivalent of a tensor is a special case for which the number of dotted indices coincides with the number of undotted indices).

From (1.38) and (1.41) it follows that the components of an n-index tensor $t_{ab\ldots d}$ with respect to the orthonormal basis $\{\mathbf{e}_1, \mathbf{e}_2, \mathbf{e}_3, \mathbf{e}_4\}$ are real if and only if

$$\begin{aligned}
&\overline{t_{A\dot{B}C\dot{D}\ldots G\dot{H}}} \\
&= \begin{cases} (-1)^n t^{A\dot{B}C\dot{D}\ldots G\dot{H}} & \text{if the } \sigma^a{}_{A\dot{B}} \text{ are given by (1.33)}, \\ t_{B\dot{A}D\dot{C}\ldots H\dot{G}} & \text{if the } \sigma^a{}_{A\dot{B}} \text{ are given by (1.34)}, \\ t_{A\dot{B}C\dot{D}\ldots G\dot{H}} & \text{if the } \sigma^a{}_{A\dot{B}} \text{ are given by (1.35)}, \\ \eta_{AR}\eta_{\dot{B}\dot{S}}\eta_{CT}\eta_{\dot{D}\dot{U}}\cdots\eta_{GV}\eta_{\dot{H}\dot{W}}t^{R\dot{S}T\dot{U}\ldots V\dot{W}} & \text{if the } \sigma^a{}_{A\dot{B}} \text{ are given by (1.37)}. \end{cases}
\end{aligned} \tag{1.47}$$

Each of the two sets of Infeld–van der Waerden symbols given above for the case of the ultrahyperbolic signature, (1.35) and (1.37), has a certain advantage over the other. The set of real Infeld–van der Waerden symbols may seem more natural in the sense that the complexification of V is not necessary and the spinor equivalent of a real tensor is real. On the other hand, the use of the complex Infeld–van der Waerden symbols allows us to reduce the number of independent components or equations to consider.

The Spinor Equivalents of Symmetric or Antisymmetric Tensors

The symmetries of a tensor are inherited by its spinor equivalent. For example, if t_{ab} is a two-index symmetric tensor, then its spinor equivalent has the symmetry

$$t_{A\dot{A}B\dot{B}} = t_{B\dot{B}A\dot{A}}, \tag{1.48}$$

but $t_{A\dot{A}B\dot{B}}$ will not necessarily coincide with $t_{B\dot{A}A\dot{B}}$ (which, according to (1.48), is equal to $t_{A\dot{B}B\dot{A}}$). In fact, using (1.27), we have

$$t_{A\dot{A}B\dot{B}} - t_{B\dot{A}A\dot{B}} = \varepsilon_{AB} t^R{}_{\dot{A}R\dot{B}},$$

but owing to (1.48), $t^R{}_{\dot{A}R\dot{B}}$ is antisymmetric,

$$t^R{}_{\dot{A}R\dot{B}} = t_{R\dot{B}}{}^R{}_{\dot{A}} = -t^R{}_{\dot{B}R\dot{A}},$$

and therefore $t^R{}_{\dot{A}R\dot{B}}$ is proportional to $\varepsilon_{\dot{A}\dot{B}}$; thus

$$t_{A\dot{A}B\dot{B}} - t_{B\dot{A}A\dot{B}} = \varepsilon_{AB}\, \tfrac{1}{2} t^{R\dot{S}}{}_{R\dot{S}} \varepsilon_{\dot{A}\dot{B}} = -\tfrac{1}{2} t^s{}_s\, \varepsilon_{AB} \varepsilon_{\dot{A}\dot{B}}, \tag{1.49}$$

which means that $t_{A\dot{A}B\dot{B}} = t_{B\dot{A}A\dot{B}} = t_{A\dot{B}B\dot{A}}$ if and only if $t_{A\dot{A}B\dot{B}}$ is the spinor equivalent of a traceless symmetric tensor. Similarly, one finds that $t_{A\dot{A}B\dot{B}\ldots C\dot{C}}$ is completely symmetric separately on the undotted and on the dotted indices if and only if $t_{A\dot{A}B\dot{B}\ldots C\dot{C}}$ is the spinor equivalent of a traceless totally symmetric tensor.

In the case of a completely symmetric object $M_{AB\ldots L}$ with n spinor indices, the $n+1$ components $M_{111\ldots1}, M_{211\ldots1}, M_{221\ldots1}, \ldots, M_{222\ldots2}$, are independent; hence an n-index traceless totally symmetric tensor has $(n+1)^2$ independent components. (Note that the components $t_{A\dot{A}B\dot{B}\ldots C\dot{C}}$ may be complex, but also in that case, taking into account the conditions (1.47), one concludes that a real n-index traceless totally symmetric tensor has $(n+1)^2$ real independent components.)

The spinor equivalent of an antisymmetric two-index tensor $t_{ab} = -t_{ba}$ satisfies $t_{A\dot{A}B\dot{B}} = -t_{B\dot{B}A\dot{A}}$; hence,

$$\begin{aligned} t_{A\dot{A}B\dot{B}} &= \tfrac{1}{2}(t_{A\dot{A}B\dot{B}} + t_{B\dot{A}A\dot{B}}) + \tfrac{1}{2}(t_{A\dot{A}B\dot{B}} - t_{B\dot{A}A\dot{B}}) \\ &= \tfrac{1}{2}(t_{A\dot{A}B\dot{B}} - t_{A\dot{B}B\dot{A}}) + \tfrac{1}{2}(t_{A\dot{A}B\dot{B}} - t_{B\dot{A}A\dot{B}}) \\ &= \tfrac{1}{2} t_A{}^{\dot{R}}{}_{B\dot{R}} \varepsilon_{\dot{A}\dot{B}} + \tfrac{1}{2} t^R{}_{\dot{A}R\dot{B}} \varepsilon_{AB} \\ &= \tau_{AB} \varepsilon_{\dot{A}\dot{B}} + \tau_{\dot{A}\dot{B}} \varepsilon_{AB}, \end{aligned} \tag{1.50}$$

where

$$\tau_{AB} \equiv \tfrac{1}{2} t_A{}^{\dot{R}}{}_{B\dot{R}}, \qquad \tau_{\dot{A}\dot{B}} \equiv \tfrac{1}{2} t^R{}_{\dot{A}R\dot{B}}. \tag{1.51}$$

Owing to $t_{A\dot{A}B\dot{B}} = -t_{B\dot{B}A\dot{A}}$ and (1.23), τ_{AB} and $\tau_{\dot{A}\dot{B}}$ are symmetric in their two indices; e.g., $t_B{}^{\dot{R}}{}_{A\dot{R}} = -t_{A\dot{R}B}{}^{\dot{R}} = t_A{}^{\dot{R}}{}_{B\dot{R}}$.

Note that if t_{ab} is real, the components τ_{AB} and $\tau_{\dot{A}\dot{B}}$ may be complex or related to one another. For instance, if the metric of V has Lorentzian signature, then from (1.47) we have $\overline{t_{A\dot{B}C\dot{D}}} = t_{B\dot{A}D\dot{C}}$ and therefore

$$\overline{\tau_{AB}} = \tfrac{1}{2}\overline{t_A{}^{\dot{R}}{}_{B\dot{R}}} = \tfrac{1}{2} t^R{}_{\dot{A}R\dot{B}} = \tau_{\dot{A}\dot{B}}$$

(for the other two signatures see Section 1.6).

In particular, from (1.50) one finds that the spinor equivalent of the antisymmetrized product $v_a w_b - v_b w_a$ of a pair of vectors is

$$\begin{aligned}
v_{A\dot{A}} w_{B\dot{B}} - v_{B\dot{B}} w_{A\dot{A}} &= \tfrac{1}{2}(v_A{}^{\dot{R}} w_{B\dot{R}} - v_{B\dot{R}} w_A{}^{\dot{R}})\varepsilon_{\dot{A}\dot{B}} + \tfrac{1}{2}(v^R{}_{\dot{A}} w_{R\dot{B}} - v_{R\dot{B}} w^R{}_{\dot{A}})\varepsilon_{AB} \\
&= \tfrac{1}{2}(v_A{}^{\dot{R}} w_{B\dot{R}} + v_B{}^{\dot{R}} w_{A\dot{R}})\varepsilon_{\dot{A}\dot{B}} + \tfrac{1}{2}(v^R{}_{\dot{A}} w_{R\dot{B}} + v^R{}_{\dot{B}} w_{R\dot{A}})\varepsilon_{AB} \\
&= v_{(A}{}^{\dot{R}} w_{B)\dot{R}} \varepsilon_{\dot{A}\dot{B}} + v^R{}_{(\dot{A}} w_{|R|\dot{B})} \varepsilon_{AB},
\end{aligned} \tag{1.52}$$

where the parentheses denote symmetrization on the indices enclosed [e.g., $\xi_{(AB)} = \tfrac{1}{2}(\xi_{AB} + \xi_{BA})$] and the indices between bars are excluded from the symmetrization.

If t_{abc} is totally antisymmetric (that is, it changes sign under any transposition of a pair of its indices), then in particular, it is antisymmetric in the last pair of indices; therefore, according to the preceding results, the spinor equivalent of t_{abc} must be of the form

$$t_{A\dot{A}B\dot{B}C\dot{C}} = \mu_{A\dot{A}BC} \varepsilon_{\dot{B}\dot{C}} + \mu_{A\dot{A}\dot{B}\dot{C}} \varepsilon_{BC}, \tag{1.53}$$

with $\mu_{A\dot{A}BC} = \mu_{A\dot{A}(BC)}$, $\mu_{A\dot{A}\dot{B}\dot{C}} = \mu_{A\dot{A}(\dot{B}\dot{C})}$. The antisymmetry in the first pair of indices of t_{abc} is then equivalent to

$$\mu_{A\dot{A}BC} \varepsilon_{\dot{B}\dot{C}} + \mu_{A\dot{A}\dot{B}\dot{C}} \varepsilon_{BC} = -\mu_{B\dot{B}AC} \varepsilon_{\dot{A}\dot{C}} - \mu_{B\dot{B}\dot{A}\dot{C}} \varepsilon_{AC}. \tag{1.54}$$

Contracting the last equation with $\varepsilon^{\dot{B}\dot{C}}$, we obtain [see (1.29)]

$$\begin{aligned}
2\mu_{A\dot{A}BC} &= -\mu_B{}^{\dot{B}}{}_{AC} \varepsilon_{\dot{A}\dot{B}} - \mu_B{}^{\dot{B}}{}_{\dot{A}\dot{B}} \varepsilon_{AC} \\
&= -\mu_{B\dot{A}AC} - \mu_B{}^{\dot{B}}{}_{\dot{A}\dot{B}} \varepsilon_{AC} \\
&= -\mu_{A\dot{A}BC} + \mu_{A\dot{A}BC} - \mu_{B\dot{A}AC} - \mu_B{}^{\dot{R}}{}_{\dot{A}\dot{R}} \varepsilon_{AC},
\end{aligned}$$

and hence, making use of (1.27),

$$3\mu_{A\dot{A}BC} = \mu^R{}_{\dot{A}RC} \varepsilon_{AB} - \mu_B{}^{\dot{R}}{}_{\dot{A}\dot{R}} \varepsilon_{AC}. \tag{1.55}$$

Contracting now this last equation with ε^{BC}, we obtain

$$0 = -\mu^R{}_{\dot{A}RA} - \mu_A{}^{\dot{R}}{}_{\dot{A}\dot{R}},$$

and therefore, defining $t_{B\dot{A}} \equiv \frac{2}{3}\mu^{R}{}_{\dot{A}RB}$, from (1.55) we obtain

$$2\mu_{A\dot{A}BC} = t_{C\dot{A}}\varepsilon_{AB} + t_{B\dot{A}}\varepsilon_{AC}. \tag{1.56}$$

In an entirely similar way, from (1.54) one arrives at

$$2\mu_{A\dot{A}\dot{B}\dot{C}} = -t_{A\dot{C}}\varepsilon_{\dot{A}\dot{B}} - t_{A\dot{B}}\varepsilon_{\dot{A}\dot{C}}. \tag{1.57}$$

Substituting (1.56) and (1.57) into (1.53), it follows that the spinor equivalent of a three-index antisymmetric tensor is of the form

$$\begin{aligned} t_{A\dot{A}B\dot{B}C\dot{C}} &= \tfrac{1}{2}[(t_{C\dot{A}}\varepsilon_{AB} + t_{B\dot{A}}\varepsilon_{AC})\varepsilon_{\dot{B}\dot{C}} - (t_{A\dot{C}}\varepsilon_{\dot{A}\dot{B}} + t_{A\dot{B}}\varepsilon_{\dot{A}\dot{C}})\varepsilon_{BC}] \\ &= \tfrac{1}{2}[(t_{C\dot{A}}\varepsilon_{AB} - t_{B\dot{A}}\varepsilon_{AC} + 2t_{B\dot{A}}\varepsilon_{AC})\varepsilon_{\dot{B}\dot{C}} \\ &\quad - (t_{A\dot{C}}\varepsilon_{\dot{A}\dot{B}} - t_{A\dot{B}}\varepsilon_{\dot{A}\dot{C}} + 2t_{A\dot{B}}\varepsilon_{\dot{A}\dot{C}})\varepsilon_{BC}] \\ &= \tfrac{1}{2}[(t^{R}{}_{\dot{A}}\varepsilon_{AR}\varepsilon_{CB} + 2t_{B\dot{A}}\varepsilon_{AC})\varepsilon_{\dot{B}\dot{C}} \\ &\quad - (t_{A}{}^{\dot{R}}\varepsilon_{\dot{A}\dot{R}}\varepsilon_{\dot{C}\dot{B}} + 2t_{A\dot{B}}\varepsilon_{\dot{A}\dot{C}})\varepsilon_{BC}] \\ &= t_{B\dot{A}}\varepsilon_{AC}\varepsilon_{\dot{B}\dot{C}} - t_{A\dot{B}}\varepsilon_{BC}\varepsilon_{\dot{A}\dot{C}}, \end{aligned} \tag{1.58}$$

for some $t_{B\dot{A}}$.

If now t_{abcd} is a totally antisymmetric four-index tensor, then as a consequence of the antisymmetry on the first three indices, according to (1.58) we can write

$$t_{A\dot{A}B\dot{B}C\dot{C}D\dot{D}} = \phi_{B\dot{A}D\dot{D}}\varepsilon_{AC}\varepsilon_{\dot{B}\dot{C}} - \phi_{A\dot{B}D\dot{D}}\varepsilon_{\dot{A}\dot{C}}\varepsilon_{BC}, \tag{1.59}$$

and the antisymmetry of t_{abcd} in the last two indices is then equivalent to

$$\phi_{B\dot{A}D\dot{D}}\varepsilon_{AC}\varepsilon_{\dot{B}\dot{C}} - \phi_{A\dot{B}D\dot{D}}\varepsilon_{\dot{A}\dot{C}}\varepsilon_{BC} = -\phi_{B\dot{A}C\dot{C}}\varepsilon_{AD}\varepsilon_{\dot{B}\dot{D}} + \phi_{A\dot{B}C\dot{C}}\varepsilon_{\dot{A}\dot{D}}\varepsilon_{BD};$$

hence, contracting with ε^{AC}, we have

$$2\phi_{B\dot{A}D\dot{D}}\varepsilon_{\dot{B}\dot{C}} - \phi_{B\dot{B}D\dot{D}}\varepsilon_{\dot{A}\dot{C}} = -\phi_{B\dot{A}D\dot{C}}\varepsilon_{\dot{B}\dot{D}} + \phi^{R}{}_{\dot{B}R\dot{C}}\varepsilon_{\dot{A}\dot{D}}\varepsilon_{BD}, \tag{1.60}$$

and contracting this last equation with $\varepsilon^{\dot{A}\dot{C}}$, we obtain

$$0 = -\phi_{B}{}^{\dot{R}}{}_{D\dot{R}}\varepsilon_{\dot{B}\dot{D}} + \phi^{R}{}_{\dot{B}R\dot{D}}\varepsilon_{BD},$$

which implies that

$$\phi_{B}{}^{\dot{R}}{}_{D\dot{R}} = 2a\,\varepsilon_{BD}, \qquad \phi^{R}{}_{\dot{B}R\dot{D}} = 2a\,\varepsilon_{\dot{B}\dot{D}}, \tag{1.61}$$

for some scalar a.

On the other hand, contracting (1.60) with $\varepsilon^{\dot{B}\dot{D}}$, rearranging, and using (1.61), one finds that

$$\begin{aligned} 4\phi_{B\dot{A}D\dot{C}} &= \phi_{B}{}^{\dot{B}}{}_{D\dot{B}}\varepsilon_{\dot{A}\dot{C}} + \phi^{R\dot{B}}{}_{R\dot{C}}\varepsilon_{\dot{A}\dot{B}}\varepsilon_{BD} \\ &= 2a(\varepsilon_{BD}\varepsilon_{\dot{A}\dot{C}} + \varepsilon_{\dot{A}\dot{C}}\varepsilon_{BD}) \\ &= 4a\varepsilon_{BD}\varepsilon_{\dot{A}\dot{C}}. \end{aligned} \tag{1.62}$$

Substituting (1.62) into (1.59), one finds that the spinor equivalent of an antisymmetric four-index tensor is of the form

$$t_{A\dot{A}B\dot{B}C\dot{C}D\dot{D}} = a(\varepsilon_{BD}\varepsilon_{\dot{A}\dot{D}}\varepsilon_{AC}\varepsilon_{\dot{B}\dot{C}} - \varepsilon_{AD}\varepsilon_{\dot{A}\dot{C}}\varepsilon_{BC}\varepsilon_{\dot{B}\dot{D}}), \tag{1.63}$$

for some scalar a.

The spinor equivalent $\varepsilon_{A\dot{A}B\dot{B}C\dot{C}D\dot{D}}$ of the Levi-Civita symbol ε_{abcd}, which is antisymmetric with $\varepsilon_{1234} = 1$, must be proportional to $\varepsilon_{AC}\varepsilon_{BD}\varepsilon_{\dot{A}\dot{D}}\varepsilon_{\dot{B}\dot{C}} - \varepsilon_{AD}\varepsilon_{BC}\varepsilon_{\dot{A}\dot{C}}\varepsilon_{\dot{B}\dot{D}}$. Since $\varepsilon^{abcd}\varepsilon_{abcd} = (-1)^q 24$, where q is the number of negative entries in the diagonal matrix (g_{ab}), the proportionality factor is real or pure imaginary depending on the signature of the metric tensor, and for a given signature, this factor is defined up to sign. From $\varepsilon_{1234} = 1$ one finds that, with the connection symbols given by (1.33)–(1.35) and (1.37),

$$\varepsilon_{A\dot{A}B\dot{B}C\dot{C}D\dot{D}} = \mathrm{i}^q(\varepsilon_{AC}\varepsilon_{BD}\varepsilon_{\dot{A}\dot{D}}\varepsilon_{\dot{B}\dot{C}} - \varepsilon_{AD}\varepsilon_{BC}\varepsilon_{\dot{A}\dot{C}}\varepsilon_{\dot{B}\dot{D}}). \tag{1.64}$$

Therefore, the spinor equivalent of the *dual* of an antisymmetric two-index tensor t_{ab}, defined by ${}^*t_{ab} \equiv \frac{1}{2}\varepsilon_{abcd}t^{cd}$, is [see (1.50)]

$${}^*t_{A\dot{A}B\dot{B}} = \mathrm{i}^q(-\tau_{AB}\varepsilon_{\dot{A}\dot{B}} + \tau_{\dot{A}\dot{B}}\varepsilon_{AB}), \tag{1.65}$$

with τ_{AB} and $\tau_{\dot{A}\dot{B}}$ defined by (1.51), and from this formula we readily obtain the well-known fact that

$${}^*({}^*t_{ab}) = (-1)^q t_{ab}. \tag{1.66}$$

(Note that the factor i^q appears in (1.64) and (1.65) as a consequence of the specific choice for the connection symbols (1.33)–(1.35) and (1.37); by contrast, (1.66) follows from the definition of the dual of t_{ab}.) The terms $\tau_{\dot{A}\dot{B}}\varepsilon_{AB}$ and $\tau_{AB}\varepsilon_{\dot{A}\dot{B}}$ will be referred to as the *self-dual* and the *anti-self-dual* parts of $t_{AB\dot{A}\dot{B}}$, respectively, though in the case of ultrahyperbolic signature, $\tau_{AB}\varepsilon_{\dot{A}\dot{B}}$ is the spinor equivalent of an antisymmetric tensor that coincides with its dual.

From (1.50) and (1.65) we see that for an arbitrary antisymmetric two-index tensor (or *bivector*) t_{ab}, we have

$$\begin{aligned} {}^*t_{ab}t^{ab} &= -\mathrm{i}^q(\tau_{AB}\varepsilon_{\dot{A}\dot{B}} - \tau_{\dot{A}\dot{B}}\varepsilon_{AB})(\tau^{AB}\varepsilon^{\dot{A}\dot{B}} + \tau^{\dot{A}\dot{B}}\varepsilon^{AB}) \\ &= -2\mathrm{i}^q(\tau_{AB}\tau^{AB} - \tau_{\dot{A}\dot{B}}\tau^{\dot{A}\dot{B}}). \end{aligned}$$

The bivector t_{ab} is *simple* if there exist vectors v_a, w_a such that $t_{ab} = v_a w_b - v_b w_a$. Hence, for a simple bivector, ${}^*t_{ab}t^{ab} = \frac{1}{2}\varepsilon_{abcd}t^{cd}t^{ab} = 2\varepsilon_{abcd}v^c w^d v^a w^b = 0$. Thus, we have proved the following result.

Proposition 1.2. *If the bivector t_{ab} is simple, then*

$$\tau_{AB}\tau^{AB} = \tau_{\dot{A}\dot{B}}\tau^{\dot{A}\dot{B}}. \tag{1.67}$$

The converse of this proposition is also true, and we shall give a constructive proof in Section 1.7. Combining (1.65) and (1.67), one concludes that a bivector t_{ab} is simple if and only if its dual is simple. (See also Exercise 1.6.)

Spin-Tensors and Other Derived Objects

Substituting (1.50) into (1.44), one finds that an antisymmetric two-index tensor can be expressed in the form

$$\begin{aligned} t_{ab} &= \tfrac{1}{2}\sigma_a{}^{A\dot{A}}\sigma_b{}^{B\dot{B}}(\tau_{AB}\varepsilon_{\dot{A}\dot{B}} + \tau_{\dot{A}\dot{B}}\varepsilon_{AB}) \\ &= \tfrac{1}{2}\tau_{AB}\,\sigma_a{}^{(A|\dot{R}|}\sigma_b{}^{B)}{}_{\dot{R}} + \tfrac{1}{2}\tau_{\dot{A}\dot{B}}\,\sigma_a{}^{R(\dot{A}}\sigma_{bR}{}^{\dot{B})} \\ &= \tfrac{1}{2}\tau_{AB}S_{ab}{}^{AB} + \tfrac{1}{2}\tau_{\dot{A}\dot{B}}S_{ab}{}^{\dot{A}\dot{B}}, \end{aligned} \tag{1.68}$$

where

$$S_{ab}{}^{AB} \equiv \sigma_a{}^{(A|\dot{R}|}\sigma_b{}^{B)}{}_{\dot{R}}, \qquad S_{ab}{}^{\dot{A}\dot{B}} \equiv \sigma_a{}^{R(\dot{A}}\sigma_{bR}{}^{\dot{B})}. \tag{1.69}$$

Making use of (1.23), we obtain the alternative expression

$$\begin{aligned} S_{ab}{}^{AB} &= \tfrac{1}{2}(\sigma_a{}^{A\dot{R}}\sigma_b{}^{B}{}_{\dot{R}} + \sigma_a{}^{B\dot{R}}\sigma_b{}^{A}{}_{\dot{R}}) \\ &= \tfrac{1}{2}(\sigma_a{}^{A\dot{R}}\sigma_b{}^{B}{}_{\dot{R}} - \sigma_b{}^{A\dot{R}}\sigma_a{}^{B}{}_{\dot{R}}) \\ &= \sigma_{[a}{}^{A\dot{R}}\sigma_{b]}{}^{B}{}_{\dot{R}}, \end{aligned} \tag{1.70}$$

where the brackets denote antisymmetrization on the indices enclosed. In a similar way, we have

$$S_{ab}{}^{\dot{A}\dot{B}} = \sigma_{[a}{}^{R\dot{A}}\sigma_{b]R}{}^{\dot{B}}.$$

There are many relations satisfied by the spin-tensors $S_{ab}{}^{AB}$ and $S_{ab}{}^{\dot{A}\dot{B}}$ that follow directly from (1.21). For example,

$$\begin{aligned} S_{ab}{}^{AB}S^{abCD} &= 4(\varepsilon^{AC}\varepsilon^{BD} + \varepsilon^{AD}\varepsilon^{BC}), \\ S_{ab}{}^{AB}S^{ab\dot{C}\dot{D}} &= 0, \\ S_{ab}{}^{\dot{A}\dot{B}}S^{ab\dot{C}\dot{D}} &= 4(\varepsilon^{\dot{A}\dot{C}}\varepsilon^{\dot{B}\dot{D}} + \varepsilon^{\dot{A}\dot{D}}\varepsilon^{\dot{B}\dot{C}}). \end{aligned} \tag{1.71}$$

Hence, from (1.68) we see that

$$\tau^{AB} = \tfrac{1}{4}t_{ab}S^{abAB}, \qquad \tau^{\dot{A}\dot{B}} = \tfrac{1}{4}t_{ab}S^{ab\dot{A}\dot{B}}. \tag{1.72}$$

On the other hand, from (1.65) we obtain

$${}^*t_{ab} = -\mathrm{i}^q(\tfrac{1}{2}\tau_{AB}S_{ab}{}^{AB} - \tfrac{1}{2}\tau_{\dot{A}\dot{B}}S_{ab}{}^{\dot{A}\dot{B}}), \tag{1.73}$$

and therefore, combining (1.68) and (1.73), we obtain, for example,

$$t_{ab} - (-\mathrm{i})^q\,{}^*t_{ab} = \tau_{AB}S_{ab}{}^{AB} = \tfrac{1}{2}\big(g_{ac}g_{bd} - g_{ad}g_{bc} - (-\mathrm{i})^q\varepsilon_{abcd}\big)t^{cd},$$

and making use of (1.72), we have $\tau_{AB}S_{ab}{}^{AB} = \frac{1}{4}t^{cd}S_{cdAB}S_{ab}{}^{AB}$. In this manner we conclude that

$$S_{ab}{}^{AB}S_{cdAB} = 2\big(g_{ac}g_{bd} - g_{ad}g_{bc} - (-\mathrm{i})^q\varepsilon_{abcd}\big) \tag{1.74}$$

and, in a similar way,

$$S_{ab}{}^{\dot{A}\dot{B}} S_{cd\dot{A}\dot{B}} = 2\big(g_{ac}g_{bd} - g_{ad}g_{bc} + (-\mathrm{i})^q \varepsilon_{abcd}\big). \tag{1.75}$$

Substituting (1.68) and (1.73) into the definition ${}^*t_{ab} = \frac{1}{2}\varepsilon_{abcd}t^{cd}$, it follows that

$$\tfrac{1}{2}\varepsilon_{abcd}S^{cdAB} = -\mathrm{i}^q S_{ab}{}^{AB}, \qquad \tfrac{1}{2}\varepsilon_{abcd}S^{cd\dot{A}\dot{B}} = \mathrm{i}^q S_{ab}{}^{\dot{A}\dot{B}}. \tag{1.76}$$

The second equation (1.71) implies that the contraction $m^{ab}t_{ab}$ of a self-dual bivector with an anti-self-dual one vanishes. Furthermore, (1.71) implies that if $m^{ab}t_{ab} = 0$ for all self-dual [resp. anti-self-dual] bivectors m^{ab}, then t_{ab} is anti-self-dual [resp. self-dual].

The complex scalars $S_{ab}{}^A{}_B$ and $S_{ab}{}^{\dot{A}}{}_{\dot{B}}$ can be arranged in 2×2 matrices, employing the spinor indices to label the entries of these matrices. A straightforward computation, using the explicit expressions (1.33)–(1.35) and (1.37), yields

$$\begin{aligned}
&\big((S_{23}{}^A{}_B),(S_{31}{}^A{}_B),(S_{12}{}^A{}_B),(S_{14}{}^A{}_B),(S_{24}{}^A{}_B),(S_{34}{}^A{}_B)\big)\\
&\qquad = \begin{cases}
(\mathrm{i}\sigma_1,-\mathrm{i}\sigma_2,\mathrm{i}\sigma_3,-\mathrm{i}\sigma_1,\mathrm{i}\sigma_2,-\mathrm{i}\sigma_3) & \text{if the } \sigma^a{}_{A\dot{B}} \text{ are given by (1.33)},\\
(\mathrm{i}\sigma_1,-\mathrm{i}\sigma_2,\mathrm{i}\sigma_3,-\sigma_1,\sigma_2,-\sigma_3) & \text{if the } \sigma^a{}_{A\dot{B}} \text{ are given by (1.34)},\\
(\sigma_1,-\sigma_3,-\mathrm{i}\sigma_2,-\sigma_1,\sigma_3,-\mathrm{i}\sigma_2) & \text{if the } \sigma^a{}_{A\dot{B}} \text{ are given by (1.35)},\\
(-\sigma_2,\sigma_1,-\mathrm{i}\sigma_3,\sigma_2,-\sigma_1,-\mathrm{i}\sigma_3) & \text{if the } \sigma^a{}_{A\dot{B}} \text{ are given by (1.37)},
\end{cases}
\end{aligned} \tag{1.77}$$

where the σ_i are the standard Pauli matrices,

$$\sigma_1 = \begin{pmatrix} 0 & 1\\ 1 & 0\end{pmatrix}, \qquad \sigma_2 = \begin{pmatrix} 0 & -\mathrm{i}\\ \mathrm{i} & 0\end{pmatrix}, \qquad \sigma_3 = \begin{pmatrix} 1 & 0\\ 0 & -1\end{pmatrix}, \tag{1.78}$$

and similarly,

$$\begin{aligned}
&\big((S_{23}{}^{\dot{A}}{}_{\dot{B}}),(S_{31}{}^{\dot{A}}{}_{\dot{B}}),(S_{12}{}^{\dot{A}}{}_{\dot{B}}),(S_{14}{}^{\dot{A}}{}_{\dot{B}}),(S_{24}{}^{\dot{A}}{}_{\dot{B}}),(S_{34}{}^{\dot{A}}{}_{\dot{B}})\big)\\
&\qquad = \begin{cases}
(-\mathrm{i}\sigma_1,-\mathrm{i}\sigma_2,-\mathrm{i}\sigma_3,-\mathrm{i}\sigma_1,-\mathrm{i}\sigma_2,-\mathrm{i}\sigma_3) & \text{if the } \sigma^a{}_{A\dot{B}} \text{ are given by (1.33)},\\
(-\mathrm{i}\sigma_1,-\mathrm{i}\sigma_2,-\mathrm{i}\sigma_3,-\sigma_1,-\sigma_2,-\sigma_3) & \text{if the } \sigma^a{}_{A\dot{B}} \text{ are given by (1.34)},\\
(-\sigma_1,\sigma_3,-\mathrm{i}\sigma_2,-\sigma_1,\sigma_3,\mathrm{i}\sigma_2) & \text{if the } \sigma^a{}_{A\dot{B}} \text{ are given by (1.35)},\\
(-\sigma_2,-\sigma_1,\mathrm{i}\sigma_3,-\sigma_2,-\sigma_1,-\mathrm{i}\sigma_3) & \text{if the } \sigma^a{}_{A\dot{B}} \text{ are given by (1.37)}.
\end{cases}
\end{aligned} \tag{1.79}$$

As we shall see in Section 1.3, the matrices $(S_{ab}{}^A{}_B)$ and $(S_{ab}{}^{\dot{A}}{}_{\dot{B}})$ span two irreducible representations of the Lie algebra of $\mathrm{SO}(p,q)$ [see, e.g., (1.124) and (1.158)]. Note, however, that only when the metric has Lorentzian signature are the sets of 2×2 matrices $\{(S_{ab}{}^A{}_B)\}$ and $\{(S_{ab}{}^{\dot{A}}{}_{\dot{B}})\}$, with $1 \leqslant a < b \leqslant 4$, linearly independent over $\mathbb{R}$ [see (1.76)].

In the case of a totally antisymmetric three-index tensor t_{abc}, (1.44) and (1.58) give

$$t_{abc} = -\frac{1}{2\sqrt{2}}\sigma_a{}^{A\dot{A}}\sigma_b{}^{B\dot{B}}\sigma_c{}^{C\dot{C}}(t_{B\dot{A}}\varepsilon_{AC}\varepsilon_{\dot{B}\dot{C}} - t_{A\dot{B}}\varepsilon_{BC}\varepsilon_{\dot{A}\dot{C}}) = -\frac{1}{\sqrt{2}}\check{\sigma}_{abc}{}^{A\dot{B}}t_{A\dot{B}},$$

where

$$\check{\sigma}_{abc}{}^{A\dot{B}} \equiv \sigma_{[a}{}^{R\dot{B}}\sigma_{b}{}^{A\dot{S}}\sigma_{c]R\dot{S}}. \tag{1.80}$$

A convenient alternative expression of the symbols $\check{\sigma}_{abc}{}^{A\dot{B}}$ is obtained as follows. Making use of (1.40) and (1.64), one obtains

$$\begin{aligned}
\frac{1}{6}\varepsilon^{abcd}\sigma_{a}{}^{R\dot{B}}\sigma_{b}{}^{A\dot{S}}\sigma_{cR\dot{S}} &= -\frac{1}{12}\varepsilon^{abce}\sigma_{a}{}^{R\dot{B}}\sigma_{b}{}^{A\dot{S}}\sigma_{cR\dot{S}}\sigma_{e}{}^{P\dot{Q}}\sigma^{d}{}_{P\dot{Q}} \\
&= -\frac{1}{3}\varepsilon^{R\dot{B}A\dot{S}}{}_{R\dot{S}}{}^{P\dot{Q}}\sigma^{d}{}_{P\dot{Q}} \\
&= -\frac{1}{3}\mathrm{i}^{q}(\varepsilon^{R}{}_{R}\varepsilon^{AP}\varepsilon^{\dot{B}\dot{Q}}\varepsilon^{\dot{S}}{}_{\dot{S}} - \varepsilon^{RP}\varepsilon^{A}{}_{R}\varepsilon^{\dot{B}}{}_{\dot{S}}\varepsilon^{\dot{S}\dot{Q}})\sigma^{d}{}_{P\dot{Q}} \\
&= -\mathrm{i}^{q}\sigma^{dA\dot{B}},
\end{aligned} \tag{1.81}$$

i.e.,

$$\frac{1}{6}\varepsilon^{abcd}\check{\sigma}_{abc}{}^{A\dot{B}} = -\mathrm{i}^{q}\sigma^{dA\dot{B}}. \tag{1.82}$$

Furthermore, from (1.58) we see that $t^{A}{}_{\dot{A}B}{}^{\dot{B}}{}_{A\dot{B}} = 3t_{B\dot{A}}$. Hence, using (1.42),

$$t_{B\dot{A}} = \frac{1}{6\sqrt{2}}\sigma^{aA}{}_{\dot{A}}\sigma^{b}{}_{B}{}^{\dot{B}}\sigma^{c}{}_{A\dot{B}}t_{abc} = \frac{1}{6\sqrt{2}}\check{\sigma}^{abc}{}_{B\dot{A}}t_{abc}. \tag{1.83}$$

By combining the preceding formulas, one finds that

$$\check{\sigma}^{abc}{}_{B\dot{A}}\check{\sigma}_{abc}{}^{Q\dot{P}} = -12\delta^{Q}_{B}\delta^{\dot{P}}_{\dot{A}},$$

which, owing to (1.82), is equivalent to (1.21).

1.3 Spinorial Representation of the Orthogonal Transformations

As we shall show in this section, with the aid of the spinor algebra, one can find simple explicit expressions for all the orthogonal transformations in terms of 2×2 matrices that act on the one-index spinors. We shall also show that any orthogonal transformation with negative determinant can be expressed in terms of two vectors.

According to the definition (1.42) and (1.21), the spinor equivalent of the metric tensor g_{ab} is

$$\frac{1}{2}\sigma^{a}{}_{A\dot{A}}\sigma^{b}{}_{B\dot{B}}g_{ab} = -\varepsilon_{AB}\varepsilon_{\dot{A}\dot{B}};$$

therefore, (1.4) is equivalent to

$$\varepsilon_{AB}\varepsilon_{\dot{A}\dot{B}} = L^{C\dot{C}}{}_{A\dot{A}}L^{D\dot{D}}{}_{B\dot{B}}\varepsilon_{CD}\varepsilon_{\dot{C}\dot{D}}, \tag{1.84}$$

where $L^{A\dot{A}}{}_{B\dot{B}}$ is the spinor equivalent of $L^a{}_b$. Condition (1.84) implies, in particular, that

$$L^{C\dot{C}}{}_{1\dot{1}} L^{D\dot{D}}{}_{1\dot{1}} \varepsilon_{CD} \varepsilon_{\dot{C}\dot{D}} = 0,$$

or equivalently,

$$L^{C\dot{C}}{}_{1\dot{1}} L_{C\dot{C}1\dot{1}} = 0. \tag{1.85}$$

In order to determine the consequences of this relation we shall establish the following useful result.

Proposition 1.3. *If $\psi_{A\dot{B}}$ satisfies $\psi_{A\dot{B}}\psi^{A\dot{B}} = 0$, then there exist α_A and $\beta_{\dot{A}}$ such that*

$$\psi_{A\dot{B}} = \alpha_A \beta_{\dot{B}}. \tag{1.86}$$

Proof. Making use of the rules (1.22), one finds that $\psi_{A\dot{B}}\psi^{A\dot{B}} = 2\psi_{1\dot{1}}\psi_{2\dot{2}} - 2\psi_{1\dot{2}}\psi_{2\dot{1}}$ $= 2\det(\psi_{A\dot{B}})$; therefore $\psi_{A\dot{B}}\psi^{A\dot{B}} = 0$ if and only if the 2×2 matrix $(\psi_{A\dot{B}})$ has vanishing determinant, which means that its rows are linearly dependent and that its columns are linearly dependent, and hence $\psi_{A\dot{B}}$ must be of the form (1.86) (see also Exercise 1.5). □

Thus, from (1.85) we conclude that

$$L^{C\dot{C}}{}_{1\dot{1}} = \alpha^C \beta^{\dot{C}}, \tag{1.87}$$

for some α^A, $\beta^{\dot{A}}$. In an entirely analogous manner one finds that

$$L^{C\dot{C}}{}_{2\dot{2}} = \gamma^C \delta^{\dot{C}}, \tag{1.88}$$

for some γ^A, $\delta^{\dot{A}}$. Substituting (1.87) and (1.88) into $L^{C\dot{C}}{}_{1\dot{1}} L^{D\dot{D}}{}_{2\dot{2}} \varepsilon_{CD} \varepsilon_{\dot{C}\dot{D}} = 1$, which follows from (1.84), one obtains the condition

$$(\alpha^A \gamma_A)(\beta^{\dot{B}} \delta_{\dot{B}}) = 1. \tag{1.89}$$

This condition implies that the two sets $\{\alpha^A, \gamma^A\}$ and $\{\beta^{\dot{A}}, \delta^{\dot{A}}\}$ are linearly independent. (Indeed, if one assumes, for instance, that $\gamma^A = \lambda\alpha^A$, then $\alpha^A\gamma_A = \lambda\alpha^A\alpha_A = 0$.) Furthermore, making use of (1.27), we see that

$$(\alpha_A\gamma_B - \alpha_B\gamma_A)(\beta_{\dot{A}}\delta_{\dot{B}} - \beta_{\dot{B}}\delta_{\dot{A}}) = (\alpha^R\gamma_R)\varepsilon_{AB}(\beta^{\dot{S}}\delta_{\dot{S}})\varepsilon_{\dot{A}\dot{B}} = \varepsilon_{AB}\varepsilon_{\dot{A}\dot{B}}. \tag{1.90}$$

On the other hand, the spinor equivalent of (1.7) is given by

$$\varepsilon_{AB}\varepsilon_{\dot{A}\dot{B}} = L_{A\dot{A}}{}^{C\dot{C}} L_{B\dot{B}}{}^{D\dot{D}} \varepsilon_{CD}\varepsilon_{\dot{C}\dot{D}}. \tag{1.91}$$

Then substituting (1.90) into the left-hand side of (1.91), and (1.87) and (1.88) into the right-hand side, after some simplifications one obtains

$$\alpha_A\delta_{\dot{A}}\gamma_B\beta_{\dot{B}} + \gamma_A\beta_{\dot{A}}\alpha_B\delta_{\dot{B}} = L_{A\dot{A}2\dot{1}}L_{B\dot{B}1\dot{2}} + L_{A\dot{A}1\dot{2}}L_{B\dot{B}2\dot{1}}. \tag{1.92}$$

Using the fact that the sets $\{\alpha^A, \gamma^A\}$ and $\{\beta^{\dot{A}}, \delta^{\dot{A}}\}$ are linearly independent we can expand $L_{A\dot{A}1\dot{2}}$ and $L_{A\dot{A}2\dot{1}}$ in the form

$$\begin{aligned} L_{A\dot{A}1\dot{2}} &= a_1\alpha_A\beta_{\dot{A}} + a_2\alpha_A\delta_{\dot{A}} + a_3\gamma_A\beta_{\dot{A}} + a_4\gamma_A\delta_{\dot{A}}, \\ L_{A\dot{A}2\dot{1}} &= b_1\alpha_A\beta_{\dot{A}} + b_2\alpha_A\delta_{\dot{A}} + b_3\gamma_A\beta_{\dot{A}} + b_4\gamma_A\delta_{\dot{A}}, \end{aligned} \tag{1.93}$$

where $a_1,\ldots,a_4,b_1,\ldots,b_4$ are some (possibly complex) scalars. Then, substituting (1.93) into (1.92), one finds that $a_1 = a_4 = b_1 = b_4 = 0$ and that only the following two cases are possible:

$$\text{(i)} \quad L_{A\dot{A}1\dot{2}} = a_2\alpha_A\delta_{\dot{A}}, \qquad L_{A\dot{A}2\dot{1}} = \frac{1}{a_2}\gamma_A\beta_{\dot{A}}, \tag{1.94}$$

$$\text{(ii)} \quad L_{A\dot{A}1\dot{2}} = a_3\gamma_A\beta_{\dot{A}}, \qquad L_{A\dot{A}2\dot{1}} = \frac{1}{a_3}\alpha_A\delta_{\dot{A}}. \tag{1.95}$$

In the case (i), by defining

$$\begin{aligned} K^A{}_1 &\equiv \left(\frac{a_2}{\alpha^R\gamma_R}\right)^{1/2}\alpha^A, & K^A{}_2 &\equiv \left(\frac{1}{a_2\alpha^R\gamma_R}\right)^{1/2}\gamma^A, \\ M^{\dot{A}}{}_{\dot{1}} &\equiv \left(\frac{\alpha^R\gamma_R}{a_2}\right)^{1/2}\beta^{\dot{A}}, & M^{\dot{A}}{}_{\dot{2}} &\equiv \left(a_2\alpha^R\gamma_R\right)^{1/2}\delta^{\dot{A}}, \end{aligned}$$

from (1.87), (1.88), and (1.94) we obtain

$$\text{(i)} \qquad L^{A\dot{A}}{}_{B\dot{B}} = K^A{}_B M^{\dot{A}}{}_{\dot{B}}, \tag{1.96}$$

and

$$\det(K^A{}_B) = 1, \qquad \det(M^{\dot{A}}{}_{\dot{B}}) = 1 \tag{1.97}$$

[see (1.89)]. In the case (ii), making

$$\begin{aligned} K^A{}_{\dot{1}} &\equiv \left(\frac{1}{a_3\alpha^R\gamma_R}\right)^{1/2}\alpha^A, & K^A{}_{\dot{2}} &\equiv \left(\frac{a_3}{\alpha^R\gamma_R}\right)^{1/2}\gamma^A, \\ M^{\dot{A}}{}_1 &\equiv \left(a_3\alpha^R\gamma_R\right)^{1/2}\beta^{\dot{A}}, & M^{\dot{A}}{}_2 &\equiv \left(\frac{\alpha^R\gamma_R}{a_3}\right)^{1/2}\delta^{\dot{A}}, \end{aligned}$$

we have

$$\text{(ii)} \qquad L^{A\dot{A}}{}_{B\dot{B}} = K^A{}_{\dot{B}} M^{\dot{A}}{}_B, \tag{1.98}$$

with

$$\det(K^A{}_{\dot{B}}) = 1, \qquad \det(M^{\dot{A}}{}_B) = 1. \tag{1.99}$$

Using (1.96) and (1.30) one finds that (1.84) reduces to the condition

$$\det(K^A{}_B)\det(M^{\dot{A}}{}_{\dot{B}}) = 1,$$

which is satisfied as a consequence of (1.97), and similarly, substituting (1.98) into (1.84), it follows that

$$\det(K^A{}_{\dot{B}})\det(M^{\dot{A}}{}_B) = 1.$$

The determinant of the matrix $(L^a{}_b)$ can be computed by means of the general expression

$$\varepsilon_{abcd} L^a{}_e L^b{}_f L^c{}_g L^d{}_h = \det(L^r{}_s)\,\varepsilon_{efgh} \tag{1.100}$$

[cf. (1.30)]. Hence, by virtue of (1.64),

$$\begin{aligned}(\varepsilon_{AC}\varepsilon_{BD}\varepsilon_{\dot{A}\dot{D}}\varepsilon_{\dot{B}\dot{C}} - \varepsilon_{AD}\varepsilon_{BC}\varepsilon_{\dot{A}\dot{C}}\varepsilon_{\dot{B}\dot{D}})L^{A\dot{A}}{}_{E\dot{E}}L^{B\dot{B}}{}_{F\dot{F}}L^{C\dot{C}}{}_{G\dot{G}}L^{D\dot{D}}{}_{H\dot{H}}& \\ = \det(L^r{}_s)\,(\varepsilon_{EG}\varepsilon_{FH}\varepsilon_{\dot{E}\dot{H}}\varepsilon_{\dot{F}\dot{G}} - \varepsilon_{EH}\varepsilon_{FG}\varepsilon_{\dot{E}\dot{G}}\varepsilon_{\dot{F}\dot{H}}),&\end{aligned}$$

and making use of (1.30) one finds that (1.96) is the spinor equivalent of an orthogonal transformation $(L^a{}_b)$ with determinant equal to $+1$, while (1.98) is the spinor equivalent of a matrix $(L^a{}_b)$ with determinant equal to -1. The orthogonal transformations with positive [resp. negative] determinant are called proper [resp. improper] transformations.

If $\sigma^a{}_{A\dot{B}}$ and $\widetilde{\sigma}^a{}_{A\dot{B}}$ are two sets of Infeld–van der Waerden symbols (satisfying (1.21) with the same metric tensor g_{ab}) and $\{\mathbf{e}_1, \mathbf{e}_2, \mathbf{e}_3, \mathbf{e}_4\}$ is an orthonormal basis of V as above, then

$$\mathbf{e}'_a = -\tfrac{1}{2}\widetilde{\sigma}_a{}^{A\dot{B}}\sigma^b{}_{A\dot{B}}\mathbf{e}_b$$

is also an orthonormal basis of V (or of its complexification); in fact, using (1.21) and (1.40),

$$\begin{aligned} g(\mathbf{e}'_a, \mathbf{e}'_b) &= \tfrac{1}{4}\widetilde{\sigma}_a{}^{A\dot{B}}\sigma^c{}_{A\dot{B}}\widetilde{\sigma}_b{}^{C\dot{D}}\sigma^d{}_{C\dot{D}}g_{cd} \\ &= -\tfrac{1}{2}\widetilde{\sigma}_a{}^{A\dot{B}}\widetilde{\sigma}_b{}^{C\dot{D}}\varepsilon_{AC}\varepsilon_{\dot{B}\dot{D}} \\ &= -\tfrac{1}{2}\widetilde{\sigma}_{aA\dot{B}}\widetilde{\sigma}_b{}^{A\dot{B}} \\ &= g_{ab}. \end{aligned}$$

Hence, $\mathbf{e}_a = L^b{}_a\mathbf{e}'_b$ for some 4×4 matrix $(L^a{}_b)$ satisfying (1.4), and by virtue of the definition of the $\mathbf{e}'_a$, we have $\mathbf{e}_a = -\frac{1}{2}L^b{}_a\widetilde{\sigma}_b{}^{A\dot{B}}\sigma^c{}_{A\dot{B}}\mathbf{e}_c$, that is, $-\frac{1}{2}\widetilde{\sigma}_b{}^{A\dot{B}}\sigma^c{}_{A\dot{B}} = (L^{-1})^c{}_b = L_b{}^c$, or

$$\begin{aligned} \widetilde{\sigma}_b{}^{A\dot{B}} &= L_b{}^c\sigma_c{}^{A\dot{B}} \\ &= -\tfrac{1}{2}L_a{}^c\sigma^a{}_{C\dot{D}}\sigma_b{}^{C\dot{D}}\sigma_c{}^{A\dot{B}} \\ &= -L_{C\dot{D}}{}^{A\dot{B}}\sigma_b{}^{C\dot{D}}. \end{aligned}$$

Then, making use of (1.96) or (1.98), one proves that $\sigma^a{}_{A\dot{B}}$ and $\widetilde{\sigma}^a{}_{A\dot{B}}$ must be related as in (1.31) or (1.32).

According to the preceding results, *any* orthogonal transformation with positive determinant, that is, any element of $\mathrm{SO}(p,q)$, is of the form

$$L^a{}_b = \frac{1}{2}\sigma^a{}_{A\dot{A}}\sigma_b{}^{B\dot{B}}K^A{}_B M^{\dot{A}}{}_{\dot{B}}, \tag{1.101}$$

where $(K^A{}_B)$ and $(M^{\dot{A}}{}_{\dot{B}})$ belong to SL(2,$\mathbb{C}$), the group formed by the unimodular 2×2 complex matrices with the usual matrix multiplication. Since for a 4×4 matrix $(L^a{}_b)$, $\det(L^a{}_b) = \det(-L^a{}_b)$, any element of SO$(p,q)$ can also be expressed as

$$L^a{}_b = -\frac{1}{2}\sigma^a{}_{A\dot{A}}\sigma_b{}^{B\dot{B}}K^A{}_B M^{\dot{A}}{}_{\dot{B}}. \tag{1.102}$$

This last expression is more convenient than (1.101) because when $(K^A{}_B)$ and $(M^{\dot{A}}{}_{\dot{B}})$ are both the identity 2×2 matrix, $(L^a{}_b)$ given by (1.102) is the identity 4×4 matrix [see (1.40)]. However, in order for $(L^a{}_b)$ to be real, the matrices $(K^A{}_B)$ and $(M^{\dot{A}}{}_{\dot{B}})$ must be suitably restricted. Since the Infeld–van der Waerden symbols are related to their complex conjugates in a way that depends on the signature of the metric tensor, we need to deal with each case separately. Note in passing that according to (1.21), from (1.102) we have

$$\operatorname{tr}(L^a{}_b) = \operatorname{tr}(K^A{}_B)\operatorname{tr}(M^{\dot{A}}{}_{\dot{B}}). \tag{1.103}$$

1.3.1 Euclidean Signature

When the metric of V has Euclidean signature, the connection symbols have been chosen in such a way that $\overline{\sigma^a{}_{A\dot{B}}} = -\sigma^{aA\dot{B}}$ [see (1.38)]; therefore, the entries (1.102) are real if and only if

$$\sigma^a{}_{A\dot{A}}\sigma_b{}^{B\dot{B}}K^A{}_B M^{\dot{A}}{}_{\dot{B}} = \sigma^{aA\dot{A}}\sigma_{bB\dot{B}}\overline{K^A{}_B}\,\overline{M^{\dot{A}}{}_{\dot{B}}}.$$

Writing the left-hand side of this equation in the form $\sigma^{aA\dot{A}}\sigma_{bB\dot{B}}K_A{}^B M_{\dot{A}}{}^{\dot{B}}$, one obtains the equivalent condition

$$K_A{}^B M_{\dot{A}}{}^{\dot{B}} = \overline{K^A{}_B}\,\overline{M^{\dot{A}}{}_{\dot{B}}}.$$

Since the indices A, B, $\dot{A}$, and $\dot{B}$ are independent of each other, from this last relation it follows that $K_A{}^B = \lambda\,\overline{K^A{}_B}$ and $M_{\dot{A}}{}^{\dot{B}} = \lambda^{-1}\overline{M^{\dot{A}}{}_{\dot{B}}}$, for some scalar λ. Taking into account the fact that the determinant of $(K^A{}_B)$ and that of its complex conjugate are equal to 1, it follows that $\lambda^2 = 1$. On the other hand, by virtue of (1.22), $\det(K^A{}_B) = K^1{}_1K^2{}_2 - K^1{}_2K^2{}_1 = -K^1{}_1K_1{}^1 - K^1{}_2K_1{}^2 = -K^1{}_1\lambda\,\overline{K^1{}_1} - K^1{}_2\lambda\,\overline{K^1{}_2} = -\lambda\,(|K^1{}_1|^2 + |K^1{}_2|^2)$, which implies that λ must be equal to -1. Thus, we are left with

$$K_A{}^B = -\overline{K^A{}_B} \tag{1.104}$$

and, similarly, $M_{\dot{A}}{}^{\dot{B}} = -\overline{M^{\dot{A}}{}_{\dot{B}}}$. These conditions mean that $(K^A{}_B)$ and $(M^{\dot{A}}{}_{\dot{B}})$ are unitary. Indeed, from (1.30) and (1.22) we have, for an arbitrary 2×2 matrix $(M^A{}_B)$,

$$M^A{}_C M_{AD} = \det(M^R{}_S)\,\varepsilon_{CD}, \tag{1.105}$$

and hence, raising the index D, $M_A{}^D M^A{}_C = -\det(M^R{}_S)\,\delta^D_C$, which means that the inverse of $(M^A{}_B)$ has entries

$$(M^{-1})^D{}_A = -\frac{1}{\det(M^R{}_S)} M_A{}^D \tag{1.106}$$

(assuming, of course, that $\det(M^R{}_S) \neq 0$). In particular, for a unimodular 2×2 matrix,

$$(M^{-1})^D{}_A = -M_A{}^D \tag{1.107}$$

[cf. (1.6)]. Hence, the condition (1.104) amounts to $(K^{-1})^B{}_A = \overline{K^A{}_B}$, thus showing that $(K^A{}_B)$ is unitary.

In summary, any SO(4) transformation is of the form (1.102), where $(K^A{}_B)$ and $(M^{\dot{A}}{}_{\dot{B}})$ belong to SU(2), the group of unimodular unitary 2×2 matrices with matrix multiplication. However, the two pairs of SU(2) matrices $\big((K^A{}_B),(M^{\dot{A}}{}_{\dot{B}})\big)$ and $\big(-(K^A{}_B),-(M^{\dot{A}}{}_{\dot{B}})\big)$ correspond to the same SO(4) transformation. The mapping from SU(2) $\times$ SU(2) onto SO(4) given by (1.102) is a group homomorphism whose kernel is $\{(I,I),(-I,-I)\}$, where I denotes the 2×2 unit matrix; hence, we can write

$$\mathrm{SO}(4) \simeq \big(\mathrm{SU}(2) \times \mathrm{SU}(2)\big)/\mathbb{Z}_2, \tag{1.108}$$

where $\mathbb{Z}_2$ denotes the group with two elements, which is identified here with $\{(I,I),(-I,-I)\}$. In effect, if $(L^a{}_b)$ is given by (1.102) and

$$\widetilde{L}^a{}_b = -\frac{1}{2}\sigma^a{}_{A\dot{A}}\sigma_b{}^{B\dot{B}}\widetilde{K}^A{}_B\widetilde{M}^{\dot{A}}{}_{\dot{B}},$$

then the product of the matrices $(\widetilde{L}^a{}_b)$ and $(L^a{}_b)$ is given by

$$\begin{aligned}\widetilde{L}^a{}_c L^c{}_b &= \tfrac{1}{4}\sigma^a{}_{A\dot{A}}\sigma_c{}^{C\dot{C}}\widetilde{K}^A{}_C\widetilde{M}^{\dot{A}}{}_{\dot{C}}\sigma^c{}_{D\dot{D}}\sigma_b{}^{B\dot{B}}K^D{}_B M^{\dot{D}}{}_{\dot{B}} \\ &= -\tfrac{1}{2}\sigma^a{}_{A\dot{A}}\sigma_b{}^{B\dot{B}}\widetilde{K}^A{}_C K^C{}_B\widetilde{M}^{\dot{A}}{}_{\dot{C}}M^{\dot{C}}{}_{\dot{B}},\end{aligned}$$

which is the SO(4) transformation corresponding to the pair of SU(2) matrices $(\widetilde{K}K,\widetilde{M}M)$, where $K = (K^A{}_B)$, $\widetilde{K} = (\widetilde{K}^A{}_B)$, $M = (M^A{}_B)$, $\widetilde{M} = (\widetilde{M}^A{}_B)$ [recall that the group operation in SU(2) $\times$ SU(2) is defined by $(\widetilde{K},\widetilde{M})(K,M) = (\widetilde{K}K,\widetilde{M}M)$].

The group SU(2), whose underlying manifold can be identified with the unit sphere S^3, is connected and so is $\mathrm{SU}(2) \times \mathrm{SU}(2)$; hence, $\mathrm{SU}(2) \times \mathrm{SU}(2)$ is a *double covering group* of SO(4).

The existence of a homomorphism between SO(4) and $\mathrm{SU}(2) \times \mathrm{SU}(2)$ is a well-known result, which is often obtained by considering the corresponding Lie algebra homomorphism. If the 4×4 matrices $(L^a{}_b) = \exp t(\omega^a{}_b)$, for $t \in \mathbb{R}$, are orthogonal, then the entries of the matrix $(\omega^a{}_b)$ must satisfy

$$\omega_{ab} = -\omega_{ba}.$$

(This condition can be obtained by substituting $(L^a{}_b) = \exp t(\omega^a{}_b)$ into (1.4) and taking the derivative of both sides of the equation with respect to t at $t = 0$.) Therefore, the 4×4 *real* matrices J_{ab} with entries

$$(J_{ab})^c{}_d \equiv \delta^c_a g_{bd} - \delta^c_b g_{ad} \tag{1.109}$$

generate $\mathfrak{so}(4)$, the Lie algebra of SO(4) (note that $J_{ab} = -J_{ba}$). (Actually, expressions (1.109) generate the Lie algebra of SO(p,q) if the entries g_{ab} appearing on the right-hand side are those of the metric with signature (p,q).) The commutators between these matrices are readily found to be

$$[J_{ab}, J_{cd}] = g_{ac}J_{db} - g_{bc}J_{da} + g_{ad}J_{bc} - g_{bd}J_{ac}, \tag{1.110}$$

or, making use of the definitions

$$L_1 \equiv -J_{23}, \qquad L_2 \equiv -J_{31}, \qquad L_3 \equiv -J_{12},$$

and

$$M_1 \equiv J_{14}, \qquad M_2 \equiv J_{24}, \qquad M_3 \equiv J_{34},$$

one finds that the commutation relations (1.110) are equivalent to

$$[L_i, L_j] = \varepsilon_{ijk}L_k, \qquad [L_i, M_j] = \varepsilon_{ijk}M_k, \qquad [M_i, M_j] = \varepsilon_{ijk}L_k,$$

where the lowercase Latin indices i, j, k range from 1 to 3 and there is summation over repeated indices. Finally, letting

$$A_i \equiv \tfrac{1}{2}(L_i + M_i), \qquad B_i \equiv \tfrac{1}{2}(L_i - M_i),$$

one finds that

$$[A_i, A_j] = \varepsilon_{ijk}A_k, \qquad [A_i, B_j] = 0, \qquad [B_i, B_j] = \varepsilon_{ijk}B_k.$$

Thus, the sets of 4×4 matrices $\{A_1, A_2, A_3\}$ and $\{B_1, B_2, B_3\}$, which together form a basis of $\mathfrak{so}(4)$, are bases of Lie algebras isomorphic to that of SU(2). Making use of the definitions above, (1.69) and (1.76), one finds that the spinor equivalents of the matrices A_i and B_i are

$$(A_i)^{A\dot{B}}{}_{C\dot{D}} = -\frac{1}{4}\varepsilon_{ijk}\delta^{\dot{B}}_{\dot{D}}S_{jk}{}^A{}_C, \qquad (B_i)^{A\dot{B}}{}_{C\dot{D}} = -\frac{1}{4}\varepsilon_{ijk}\delta^A_C S_{jk}{}^{\dot{B}}{}_{\dot{D}}.$$

That is, each matrix A_i or B_i is the identity transformation on one of the spin spaces. We shall not follow this approach, based on the Lie algebras, centering our attention on the groups directly. However, the connection with the Lie algebras will arise naturally in what follows [see, e.g., (1.124)].

The SU(2) matrices can be parameterized in several convenient ways (see also (1.156) below). For instance, given $(K^A{}_B) \in \mathrm{SU}(2)$, we can look for its eigenspinors. The unitarity of $(K^A{}_B)$ implies that the eigenvalues of $(K^A{}_B)$ are of the form $e^{i\alpha}$, for some $\alpha \in \mathbb{R}$, and the fact that $\det(K^A{}_B) = 1$ implies that the product of the two eigenvalues of $(K^A{}_B)$ must be equal to 1; therefore, there exist α_A and β_A such that

$$K^A{}_B \alpha^B = e^{i\theta/2} \alpha^A, \qquad K^A{}_B \beta^B = e^{-i\theta/2} \beta^A, \tag{1.111}$$

where the factor 1/2 has been introduced for convenience. Hence [see (1.24) and (1.28)],

$$\begin{aligned} K^A{}_B &= \frac{1}{\alpha^R \beta_R} (e^{i\theta/2} \alpha^A \beta_B - e^{-i\theta/2} \beta^A \alpha_B) \\ &= \frac{1}{\alpha^R \beta_R} \left[\cos\tfrac{1}{2}\theta \, (\alpha^A \beta_B - \beta^A \alpha_B) + i \sin\tfrac{1}{2}\theta \, (\alpha^A \beta_B + \beta^A \alpha_B) \right] \\ &= \cos\tfrac{1}{2}\theta \, \delta^A_B + i \sin\tfrac{1}{2}\theta \, \frac{\alpha^A \beta_B + \beta^A \alpha_B}{\alpha^R \beta_R}. \end{aligned} \tag{1.112}$$

As a consequence of the relation (1.104), the one-index spinors α_A and β_A are related to each other. Indeed, taking the complex conjugate of the first equation in (1.111), we have $\overline{K^A{}_B \alpha^B} = e^{-i\theta/2} \overline{\alpha^A}$, hence $-K_A{}^B \overline{\alpha^B} = e^{-i\theta/2} \overline{\alpha^A}$, and, introducing the definition

$$\widehat{\alpha}_C \equiv \overline{\alpha^C} \tag{1.113}$$

(i.e., $\widehat{\alpha}_1 = -\overline{\alpha_2}$, $\widehat{\alpha}_2 = \overline{\alpha_1}$) we obtain $-K_A{}^B \widehat{\alpha}_B = e^{-i\theta/2} \widehat{\alpha}_A$, or equivalently, $K^A{}_B \widehat{\alpha}^B = e^{-i\theta/2} \widehat{\alpha}^A$. By comparing with the second equation in (1.111) one concludes that $\beta^A = \lambda \widehat{\alpha}^A$, for some $\lambda \in \mathbb{C}$. Then, (1.112) can be written in the form

$$K^A{}_B = \cos\tfrac{1}{2}\theta \, \delta^A_B + i \sin\tfrac{1}{2}\theta \, \frac{\alpha^A \widehat{\alpha}_B + \widehat{\alpha}^A \alpha_B}{\alpha^R \widehat{\alpha}_R}. \tag{1.114}$$

Note that $\alpha^R \widehat{\alpha}_R = |\alpha_1|^2 + |\alpha_2|^2 \geqslant 0$, and therefore, if $\alpha_A \neq 0$, the set $\{\alpha_A, \widehat{\alpha}_A\}$ is linearly independent. Equation (1.114) yields

$$K^A{}_A = 2 \cos\tfrac{1}{2}\theta. \tag{1.115}$$

Thus, any SO(4) transformation is given by (1.102) with $(K^A{}_B)$ and $(M^{\dot{A}}{}_{\dot{B}})$ being two SU(2) matrices. The matrix $(K^A{}_B)$ can be expressed in the form (1.114), and in a similar manner, there exists $\gamma^{\dot{A}}$ such that

$$\begin{aligned} M^{\dot{A}}{}_{\dot{B}} &= \frac{1}{\gamma^{\dot{R}} \widehat{\gamma}_{\dot{R}}} (e^{i\varphi/2} \gamma^{\dot{A}} \widehat{\gamma}_{\dot{B}} - e^{-i\varphi/2} \widehat{\gamma}^{\dot{A}} \gamma_{\dot{B}}) \\ &= \cos\tfrac{1}{2}\varphi \, \delta^{\dot{A}}_{\dot{B}} + i \sin\tfrac{1}{2}\varphi \, \frac{\gamma^{\dot{A}} \widehat{\gamma}_{\dot{B}} + \widehat{\gamma}^{\dot{A}} \gamma_{\dot{B}}}{\gamma^{\dot{R}} \widehat{\gamma}_{\dot{R}}}, \end{aligned} \tag{1.116}$$

for some $\varphi \in \mathbb{R}$, where

$$\widehat{\gamma}_{\dot{C}} \equiv \overline{\gamma^{\dot{C}}}. \tag{1.117}$$

Then, $M^{\dot{A}}{}_{\dot{B}}\gamma^{\dot{B}} = e^{i\varphi/2}\gamma^{\dot{A}}$, $M^{\dot{A}}{}_{\dot{B}}\widehat{\gamma}^{\dot{B}} = e^{-i\varphi/2}\widehat{\gamma}^{\dot{A}}$, and

$$M^{\dot{A}}{}_{\dot{A}} = 2\cos\tfrac{1}{2}\varphi. \tag{1.118}$$

The vector equivalents of $\alpha^A\gamma^{\dot{A}}$, $\widehat{\alpha}^A\widehat{\gamma}^{\dot{A}}$, $\alpha^A\widehat{\gamma}^{\dot{A}}$, and $\widehat{\alpha}^A\gamma^{\dot{A}}$ are (complex) eigenvectors of $(L^a{}_b)$ with eigenvalues $e^{i(\theta+\varphi)/2}$, $e^{-i(\theta+\varphi)/2}$, $e^{i(\theta-\varphi)/2}$, and $e^{-i(\theta-\varphi)/2}$, respectively.

Classification of the SO(4) Transformations

The SO(4) transformations can be classified according to whether the traces $K^A{}_A$ and $M^{\dot{A}}{}_{\dot{A}}$ coincide or not. In view of (1.115) and (1.118), $K^A{}_A = M^{\dot{A}}{}_{\dot{A}}$ if and only if $e^{i\theta/2} = e^{i\varphi/2}$ or $e^{i\theta/2} = e^{-i\varphi/2}$. Hence, the traces $K^A{}_A$ and $M^{\dot{A}}{}_{\dot{A}}$ coincide if and only if two eigenvalues of $(L^a{}_b)$ are equal to 1. As is well known, any SO(3) transformation has at least one eigenvalue equal to 1, which means that there exists a one-dimensional subspace (the axis of the rotation) pointwise invariant under the transformation. The SO(4) transformations $(L^a{}_b)$ such that $K^A{}_A = M^{\dot{A}}{}_{\dot{A}}$ (which will be called *simple*) are analogous to the SO(3) transformations in the sense that there exists a proper subspace pointwise invariant under $(L^a{}_b)$.

Assuming $\alpha^R\widehat{\alpha}_R = 1 = \gamma^{\dot{R}}\widehat{\gamma}_{\dot{R}}$, we define

$$v^{A\dot{A}} = \frac{1}{\sqrt{2}}(\alpha^A\gamma^{\dot{A}} - \widehat{\alpha}^A\widehat{\gamma}^{\dot{A}}), \qquad w^{A\dot{A}} = \frac{i}{\sqrt{2}}(\alpha^A\gamma^{\dot{A}} + \widehat{\alpha}^A\widehat{\gamma}^{\dot{A}}). \tag{1.119}$$

Then, one can readily verify that $v^{A\dot{A}}$ and $w^{A\dot{A}}$ are the spinor equivalents of two mutually orthogonal real unit vectors v^a, w^a. When $e^{i\theta/2} = e^{i\varphi/2}$ we obtain

$$\begin{aligned} K^A{}_B M^{\dot{A}}{}_{\dot{B}} v^{B\dot{B}} &= \cos\theta\, v^{A\dot{A}} + \sin\theta\, w^{A\dot{A}}, \\ K^A{}_B M^{\dot{A}}{}_{\dot{B}} w^{B\dot{B}} &= -\sin\theta\, v^{A\dot{A}} + \cos\theta\, w^{A\dot{A}}, \end{aligned}$$

which means that the corresponding orthogonal transformation is a rotation through the angle θ in the plane spanned by $\{v^a, w^a\}$. In a similar manner, when $e^{i\theta/2} = e^{-i\varphi/2}$, the SO(4) transformation defined by the pair of SU(2) matrices (1.114) and (1.116) is a rotation through the angle θ in the two-dimensional plane generated by the vector equivalents of

$$\frac{1}{\sqrt{2}}(\alpha^A\widehat{\gamma}^{\dot{A}} + \widehat{\alpha}^A\gamma^{\dot{A}}), \qquad \frac{i}{\sqrt{2}}(\alpha^A\widehat{\gamma}^{\dot{A}} - \widehat{\alpha}^A\gamma^{\dot{A}})$$

(which are real and orthogonal to v^a and w^a).

Since $\theta = \frac{1}{2}(\theta+\varphi) + \frac{1}{2}(\theta-\varphi)$ and $\varphi = \frac{1}{2}(\theta+\varphi) - \frac{1}{2}(\theta-\varphi)$, any SO(4) transformation, which is defined by a pair of matrices of the form (1.114) and (1.116), is the composition of two simple SO(4) transformations, one of them representing a rotation through $\frac{1}{2}(\theta+\varphi)$ in a two-dimensional plane and the other representing a rotation through $\frac{1}{2}(\theta - \varphi)$ in the orthogonal complement of the first two-dimensional plane.

The SU(2) matrices $(K^A{}_B)$ and $(M^{\dot{A}}{}_{\dot{B}})$ can be expressed in terms of the unit vectors v_a, w_a, introduced above. Making use of the definitions (1.119) and (1.69), we have

$$\begin{aligned} v^a w^b S_{abAB} &= v^a w^b \sigma_{a(A}{}^{\dot{R}} \sigma_{|b|B)\dot{R}} \\ &= 2v_{(A}{}^{\dot{R}} w_{B)\dot{R}} \\ &= \mathrm{i}(\alpha_A \widehat{\alpha}_B + \widehat{\alpha}_A \alpha_B), \end{aligned} \tag{1.120}$$

and therefore (1.114) can be written as

$$K^A{}_B = \cos\tfrac{1}{2}\theta \; \delta^A_B + \sin\tfrac{1}{2}\theta \; v^a w^b S_{ab}{}^A{}_B. \tag{1.121}$$

In a similar way, we have

$$\begin{aligned} v^a w^b S_{ab\dot{A}\dot{B}} &= v^a w^b \sigma_a{}^R{}_{(\dot{A}} \sigma_{|bR|\dot{B})} \\ &= 2v^R{}_{(\dot{A}} w_{|R|\dot{B})} \\ &= \mathrm{i}(\gamma_{\dot{A}} \widehat{\gamma}_{\dot{B}} + \widehat{\gamma}_{\dot{A}} \gamma_{\dot{B}}); \end{aligned}$$

hence, (1.116) is equivalent to

$$M^{\dot{A}}{}_{\dot{B}} = \cos\tfrac{1}{2}\varphi \; \delta^{\dot{A}}_{\dot{B}} + \sin\tfrac{1}{2}\varphi \; v^a w^b S_{ab}{}^{\dot{A}}{}_{\dot{B}}.$$

With $\alpha^R \widehat{\alpha}_R = 1$, the square of the matrix $(\alpha^A \widehat{\alpha}_B + \widehat{\alpha}^A \alpha_B)$ is the 2×2 identity matrix,

$$(\alpha^A \widehat{\alpha}_C + \widehat{\alpha}^A \alpha_C)(\alpha^C \widehat{\alpha}_B + \widehat{\alpha}^C \alpha_B) = \alpha^A \widehat{\alpha}_B - \widehat{\alpha}^A \alpha_B = \delta^A_B,$$

and therefore, from (1.120) it follows that the powers of the matrix $\left(v^a w^b S_{ab}{}^A{}_B\right)$ are given by

$$\left(v^a w^b S_{ab}{}^A{}_B\right)^{2k} = (-1)^k (\delta^A_B), \qquad \left(v^a w^b S_{ab}{}^A{}_B\right)^{2k+1} = (-1)^k \left(v^a w^b S_{ab}{}^A{}_B\right)$$

for $k = 0, 1, 2, \ldots$, and making use of the usual series expansion for the exponential, we obtain the matrix equality

$$\exp\left(\tfrac{1}{2}\theta \; v^a w^b S_{ab}{}^A{}_B\right) = \cos\tfrac{1}{2}\theta \; \left(\delta^A_B\right) + \sin\tfrac{1}{2}\theta \; \left(v^a w^b S_{ab}{}^A{}_B\right).$$

Thus, comparing with (1.121), we conclude that

$$(K^A{}_B) = \exp\left(\tfrac{1}{2}\theta \; v^a w^b S_{ab}{}^A{}_B\right),$$

or equivalently,

$$(K^A{}_B) = \exp\left(\mathrm{i}\theta \; \tfrac{1}{2}(\alpha^A \widehat{\alpha}_B + \widehat{\alpha}^A \alpha_B)\right). \tag{1.122}$$

In an analogous manner, we find that

$$(M^{\dot{A}}{}_{\dot{B}}) = \exp\left(\tfrac{1}{2}\varphi\, v^a w^b S_{ab}{}^{\dot{A}}{}_{\dot{B}}\right),$$

or

$$(M^{\dot{A}}{}_{\dot{B}}) = \exp\left(\mathrm{i}\varphi\, \tfrac{1}{2}(\gamma^{\dot{A}}\widehat{\gamma}_{\dot{B}} + \widehat{\gamma}^{\dot{A}}\gamma_{\dot{B}})\right). \tag{1.123}$$

These expressions show that any SU(2) matrix is the exponential of some (traceless) matrix.

We can get expressions for $(K^A{}_B)$ and $(M^{\dot{A}}{}_{\dot{B}})$ that show more similarities. Indeed, making use of (1.76), we have

$$\theta\, v^a w^b S_{ab}{}^A{}_B = \left[\tfrac{1}{2}(\theta+\varphi)\, v^a w^b - \tfrac{1}{2}(\theta-\varphi)\, \tfrac{1}{2} v_c w_d \varepsilon^{cdab}\right] S_{ab}{}^A{}_B$$

and

$$\varphi\, v^a w^b S_{ab}{}^{\dot{A}}{}_{\dot{B}} = \left[\tfrac{1}{2}(\theta+\varphi)\, v^a w^b - \tfrac{1}{2}(\theta-\varphi)\, \tfrac{1}{2} v_c w_d \varepsilon^{cdab}\right] S_{ab}{}^{\dot{A}}{}_{\dot{B}},$$

and therefore, we can also write

$$\begin{aligned}(K^A{}_B) &= \exp\left(\tfrac{1}{4} t^{ab} S_{ab}{}^A{}_B\right),\\ (M^{\dot{A}}{}_{\dot{B}}) &= \exp\left(\tfrac{1}{4} t^{ab} S_{ab}{}^{\dot{A}}{}_{\dot{B}}\right),\end{aligned} \tag{1.124}$$

where we have introduced the bivector

$$t^{ab} \equiv (\theta+\varphi)\, v^{[a} w^{b]} - (\theta-\varphi)\, \tfrac{1}{2}\varepsilon^{abcd} v_c w_d. \tag{1.125}$$

Making use of (1.52) and (1.65), one finds that the spinor equivalent of the bivector t_{ab} is

$$t_{A\dot{A}B\dot{B}} = \mathrm{i}\theta\, \alpha_{(A}\widehat{\alpha}_{B)}\varepsilon_{\dot{A}\dot{B}} + \mathrm{i}\varphi\, \gamma_{(\dot{A}}\widehat{\gamma}_{\dot{B})}\varepsilon_{AB}. \tag{1.126}$$

As shown below, the spinor equivalent of any real bivector is of the form (1.126) [see (1.232)]; therefore, any real bivector is of the form (1.125), and for any real bivector t_{ab}, equations (1.124) define an SO(4) transformation.

Any matrix whose spinor equivalent is of the form $\tau^A{}_B\varepsilon^{\dot{A}}{}_{\dot{B}}$ commutes with any matrix whose spinor equivalent is of the form $\varepsilon^A{}_B\tau^{\dot{A}}{}_{\dot{B}}$; therefore

$$\exp\left((\tau^A{}_B\varepsilon^{\dot{A}}{}_{\dot{B}}) + (\varepsilon^A{}_B\tau^{\dot{A}}{}_{\dot{B}})\right) = \exp\left(\tau^A{}_B\varepsilon^{\dot{A}}{}_{\dot{B}}\right)\exp\left(\varepsilon^A{}_B\tau^{\dot{A}}{}_{\dot{B}}\right),$$

and taking $\tau_{AB} = \mathrm{i}\theta\, \alpha_{(A}\widehat{\alpha}_{B)}$, $\tau_{\dot{A}\dot{B}} = \mathrm{i}\varphi\, \gamma_{(\dot{A}}\widehat{\gamma}_{\dot{B})}$, from (1.122) and (1.123) we have

$$\exp\left((\tau^A{}_B\varepsilon^{\dot{A}}{}_{\dot{B}}) + (\varepsilon^A{}_B\tau^{\dot{A}}{}_{\dot{B}})\right) = \left(M^A{}_B K^{\dot{A}}{}_{\dot{B}}\right),$$

which implies that the SO(4) transformation corresponding to the pair of SU(2) matrices (1.124) can be also expressed as

$$(L^a{}_b) = \exp(-t^a{}_b). \tag{1.127}$$

Thus, any SO(4) transformation can be written as the exponential of some matrix. Furthermore, making use of Proposition 1.2 and the preceding results one can verify that the bivector t^{ab} is simple if and only if the SO(4) transformation $(L^a{}_b) = \exp(-t^a{}_b)$ is simple.

Improper O(4) Transformations

Following a procedure similar to that leading to (1.104), one concludes that any O(4) transformation with negative determinant can be expressed in the form

$$L^a{}_b = -\frac{1}{2}\sigma^a{}_{A\dot{A}}\sigma_b{}^{B\dot{B}}K^A{}_{\dot{B}}M^{\dot{A}}{}_B, \tag{1.128}$$

where $(K^A{}_{\dot{B}})$ and $(M^{\dot{A}}{}_B)$ are SU(2) matrices; given $(L^a{}_b)$, the matrices $(K^A{}_{\dot{B}})$ and $(M^{\dot{A}}{}_B)$ are defined up to a common sign.

The matrices $(K^A{}_{\dot{B}})$ and $(M^{\dot{A}}{}_B)$ appearing in (1.128) can be regarded as the spinor equivalents of two vectors. Owing to (1.104) and (1.47), these two vectors are pure imaginary. Hence, letting

$$K_{A\dot{B}} = \mathrm{i}\sqrt{2}\,v_{A\dot{B}}, \qquad M_{\dot{A}B} = \mathrm{i}\sqrt{2}\,w_{B\dot{A}}, \tag{1.129}$$

one finds that $\det(K^A{}_{\dot{B}}) = \frac{1}{2}K^{A\dot{B}}K_{A\dot{B}} = -v^{A\dot{B}}v_{A\dot{B}} = 1$ and, similarly, $-w^{A\dot{B}}w_{A\dot{B}} = 1$; therefore, the vector equivalents v_a and w_a of $v_{A\dot{B}}$ and $w_{A\dot{B}}$, respectively, are real unit vectors [see (1.46)]. Then, making use of (1.27) we obtain the identity

$$\begin{aligned}
v_{A\dot{B}}w_{B\dot{A}} &= \tfrac{1}{2}(v_{A\dot{B}}w_{B\dot{A}} - v_{A\dot{A}}w_{B\dot{B}} + v_{A\dot{A}}w_{B\dot{B}} \\
&\qquad + v_{A\dot{B}}w_{B\dot{A}} - v_{B\dot{B}}w_{A\dot{A}} + v_{B\dot{B}}w_{A\dot{A}}) \\
&= \tfrac{1}{2}(-\varepsilon_{\dot{A}\dot{B}}v_A{}^{\dot{R}}w_{B\dot{R}} + v_{A\dot{A}}w_{B\dot{B}} + \varepsilon_{AB}v^R{}_{\dot{B}}w_{R\dot{A}} + v_{B\dot{B}}w_{A\dot{A}}) \\
&= \tfrac{1}{2}\big[-\varepsilon_{\dot{A}\dot{B}}(v_{(A}{}^{\dot{R}}w_{B)\dot{R}} + \tfrac{1}{2}\varepsilon_{AB}v^{S\dot{R}}v_{S\dot{R}}) + v_{A\dot{A}}w_{B\dot{B}} \\
&\qquad + \varepsilon_{AB}(v^R{}_{(\dot{B}}w_{|R|\dot{A})} + \tfrac{1}{2}\varepsilon_{\dot{B}\dot{A}}v^{R\dot{S}}w_{R\dot{S}}) + v_{B\dot{B}}w_{A\dot{A}}\big] \\
&= -\tfrac{1}{2}(v^{R\dot{S}}w_{R\dot{S}})\varepsilon_{AB}\varepsilon_{\dot{A}\dot{B}} + \tfrac{1}{2}(v_{A\dot{A}}w_{B\dot{B}} + v_{B\dot{B}}w_{A\dot{A}}) \\
&\qquad - \tfrac{1}{2}(v_{(A}{}^{\dot{R}}w_{B)\dot{R}}\varepsilon_{\dot{A}\dot{B}} - v^R{}_{(\dot{A}}w_{|R|\dot{B})}\varepsilon_{AB}).
\end{aligned} \tag{1.130}$$

According to (1.52) and (1.65), $-(v_{(A}{}^{\dot{R}}w_{B)\dot{R}}\varepsilon_{\dot{A}\dot{B}} - v^R{}_{(\dot{A}}w_{|R|\dot{B})}\varepsilon_{AB})$, which appears in the last line of this identity, is the spinor equivalent of the dual of the antisymmetrized product $v_aw_b - v_bw_a$. Hence, substituting (1.129) and (1.130) into (1.128), we obtain

$$L^a{}_b = -v^cw_c\delta^a_b + v^aw_b + w^av_b + g^{ae}\varepsilon_{ebcd}v^cw^d, \tag{1.131}$$

which is, therefore, the tensor equivalent of $L^{A\dot{A}}{}_{B\dot{B}} = 2v^A{}_{\dot{B}}w_B{}^{\dot{A}}$.

Thus, for each improper orthogonal transformation there exist two (not necessarily distinct) unit vectors v_a, w_a, defined up to a common sign, such that

(1.131) holds. Expression (1.131) shows that if $\{v_a, w_a\}$ is linearly independent, there are two two-dimensional subspaces of V invariant under the orthogonal transformation $(L^a{}_b)$; one of them is the subspace spanned by $\{v_a, w_a\}$ (note that $L^a{}_b v^b = w^a$ and $L^a{}_b w^b = v^a$), and the other is formed by the vectors simultaneously orthogonal to v_a and w_a. When $\{v_a, w_a\}$ is linearly dependent (i.e., $w_a = \mp v_a$), the last term in (1.131) vanishes, and $(L^a{}_b)$ corresponds to the reflection on the hyperplane orthogonal to v_a or to the negative of this reflection (see Section 1.4).

Conversely, given two real unit vectors v_a, w_a, equation (1.131) yields an improper orthogonal transformation.

1.3.2 Lorentzian Signature

In the case that the metric has Lorentzian signature, the connection symbols have been chosen in such a way that $\overline{\sigma^a{}_{A\dot{B}}} = \sigma^a{}_{B\dot{A}}$; therefore, the unimodular matrix $(L^a{}_b)$ given by (1.102) is real if and only if

$$\sigma^a{}_{A\dot{B}}\sigma_b{}^{C\dot{D}}K^A{}_C M^{\dot{B}}{}_{\dot{D}} = \sigma^a{}_{B\dot{A}}\sigma_b{}^{D\dot{C}}\overline{K^A{}_C}\,\overline{M^{\dot{B}}{}_{\dot{D}}},$$

which is equivalent to $K^A{}_C M^{\dot{B}}{}_{\dot{D}} = \overline{K^B{}_D}\,\overline{M^{\dot{A}}{}_{\dot{C}}}$. Hence, $\overline{M^{\dot{A}}{}_{\dot{C}}} = \lambda K^A{}_C$ and $\overline{K^B{}_D} = \lambda^{-1} M^{\dot{B}}{}_{\dot{D}}$, for some scalar λ. Since both matrices $(K^A{}_B)$ and $(M^{\dot{A}}{}_{\dot{B}})$ have unit determinant, we have $\lambda^2 = 1$, so that finally we obtain the condition

$$M^{\dot{A}}{}_{\dot{B}} = \pm\overline{K^A{}_B}. \tag{1.132}$$

Thus, any SO(3,1) transformation can be expressed in the form

$$L^a{}_b = \mp\tfrac{1}{2}\sigma^a{}_{A\dot{B}}\sigma_b{}^{C\dot{D}}K^A{}_C\overline{K^B{}_D}, \tag{1.133}$$

with $(K^A{}_B) \in \mathrm{SL}(2,\mathbb{C})$. The SO(3,1) transformations of the form (1.133) with the upper sign form a subgroup of SO(3,1) denoted by $\mathrm{SO}^{\uparrow}(3,1)$ or $\mathrm{SO}(3,1)_0$ (being the connected component of SO(3,1) containing the identity). Furthermore, the mapping

$$(K^A{}_B) \mapsto (L^a{}_b), \qquad \text{with} \qquad L^a{}_b = -\tfrac{1}{2}\sigma^a{}_{A\dot{B}}\sigma_b{}^{C\dot{D}}K^A{}_C\overline{K^B{}_D}, \tag{1.134}$$

is a two-to-one homomorphism of $\mathrm{SL}(2,\mathbb{C})$ onto $\mathrm{SO}^{\uparrow}(3,1)$. In fact, making use of (1.21), one finds that the product of the $\mathrm{SO}^{\uparrow}(3,1)$ matrices

$$L^a{}_b = -\tfrac{1}{2}\sigma^a{}_{A\dot{B}}\sigma_b{}^{C\dot{D}}K^A{}_C\overline{K^B{}_D}, \qquad M^a{}_b = -\tfrac{1}{2}\sigma^a{}_{A\dot{B}}\sigma_b{}^{C\dot{D}}Q^A{}_C\overline{Q^B{}_D}$$

is given by

$$L^a{}_b M^b{}_c = -\tfrac{1}{2}\sigma^a{}_{A\dot{B}}\sigma_c{}^{G\dot{H}}K^A{}_C Q^C{}_G\overline{K^B{}_D Q^D{}_H},$$

which is the $SO^{\uparrow}(3,1)$ matrix corresponding to the product of the matrices $(K^A{}_B)$ and $(Q^A{}_B)$ under the mapping (1.134).

Thus, $SO^{\uparrow}(3,1) \simeq SL(2,\mathbb{C})/\mathbb{Z}_2$, and since $SL(2,\mathbb{C})$ is connected, $SL(2,\mathbb{C})$ is a double covering group of $SO^{\uparrow}(3,1)$.

Making use of the explicit expressions for the Infeld–van der Waerden symbols (1.33), from (1.133) one finds that

$$L^4{}_4 = \pm\tfrac{1}{2}\left(|K^1{}_1|^2 + |K^1{}_2|^2 + |K^2{}_1|^2 + |K^2{}_2|^2\right).$$

Therefore, $SO^{\uparrow}(3,1)$ is formed by the orthogonal transformations with unit determinant such that $L^4{}_4 > 0$.

In the context of special relativity, the $O(3,1)$ transformations are identified with the Lorentz transformations that can be defined as those relating the Cartesian spacetime coordinates measured by two inertial frames (see, e.g., Jackson 1975, Rindler 1977). The elements of $SO^{\uparrow}(3,1)$ are called orthochronous proper Lorentz transformations; $SO^{\uparrow}(3,1)$ is called the restricted Lorentz group. The $SO(3,1)$ transformations (1.133) with the lower sign produce a time inversion ($L^4{}_4 < 0$) and a space inversion.

Lorentz Boosts

The elementary Lorentz transformation

$$\begin{aligned} x' &= \gamma(x - vt), \\ y' &= y, \\ z' &= z, \\ ct' &= \gamma(ct - vx/c), \end{aligned} \tag{1.135}$$

where c denotes the speed of light in vacuum and $\gamma \equiv (1 - v^2/c^2)^{-1/2}$, gives the relationship between the Cartesian coordinates of an event measured by two inertial reference frames, assuming that the primed reference frame S$'$ moves with respect to the unprimed one, S, along the x-axis with velocity v. The Lorentz transformation (1.135) is equivalent to $x'^a = L^a{}_b x^b$ if we identify $(x,y,z,ct) = (x^1,x^2,x^3,x^4)$, and similarly for the primed coordinates, with

$$L^a{}_b = \delta^a_b + (\gamma - 1)\left(\frac{v^a v_b}{v^c v_c} - T^a T_b\right) + \frac{\gamma}{c}(v^a T_b - T^a v_b). \tag{1.136}$$

Here the T^a are the components of a (timelike) vector such that $T^a T_a = -1$, v^a is a (spacelike) vector orthogonal to T^a, $v^a T_a = 0$, which gives the relative velocity of S$'$ with respect to S, and $\gamma = (1 - v^a v_a/c^2)^{-1/2}$. In fact, it can readily be seen that (1.136) reduces to (1.135) when $(T^a) = (0,0,0,1)$ and $(v^a) = (v,0,0,0)$. The matrix $(L^a{}_b)$ given by (1.136) represents an orthogonal transformation and therefore must be of the form (1.133). (It is customary to define $x^0 = ct$, with the tensor indices $a,b,\ldots$ running from 0 to 3; we do not follow here that convention in order to employ a uniform convention for all signatures of the metric.)

As is well known, it is convenient to write $(v^a v_a)^{1/2} = c \tanh \zeta$, for some $\zeta \in \mathbb{R}$ (called *rapidity*); then $\gamma = \cosh \zeta$, and (1.136) amounts to

$$\begin{aligned} L^a{}_b &= \delta^a_b + (\cosh \zeta - 1) \left(\frac{v^a}{(v^c v_c)^{1/2}} - T^a \right) \left(\frac{v_b}{(v^c v_c)^{1/2}} + T_b \right) \\ &\quad - (\cosh \zeta - 1 - \sinh \zeta) \left(\frac{v^a T_b - T^a v_b}{(v^c v_c)^{1/2}} \right) \\ &= \delta^a_b - 2(\cosh \zeta - 1)\, n^a l_b - (\cosh \zeta - 1 - \sinh \zeta)(l^a n_b - n^a l_b) \\ &= \delta^a_b - (\mathrm{e}^{-\zeta} - 1)\, l^a n_b - (\mathrm{e}^{\zeta} - 1)\, n^a l_b, \end{aligned} \tag{1.137}$$

where we have defined

$$l^a \equiv \frac{1}{\sqrt{2}} \left(T^a + \frac{v^a}{(v^c v_c)^{1/2}} \right), \qquad n^a \equiv \frac{1}{\sqrt{2}} \left(T^a - \frac{v^a}{(v^c v_c)^{1/2}} \right).$$

By virtue of the definitions of T^a and v^a, the vectors l^a and n^a satisfy

$$l^a l_a = 0 = n^a n_a, \qquad l^a n_a = -1. \tag{1.138}$$

As can readily be seen, the null vectors l^a and n^a are eigenvectors of $(L^a{}_b)$ with eigenvalues $\mathrm{e}^{-\zeta}$ and e^{ζ}, respectively.

Then, for instance, $l^{A\dot{B}}$, the spinor equivalent of the null vector l^a, satisfies $l^{A\dot{B}} l_{A\dot{B}} = 0$, and according to (1.86), there exist α_A and $\gamma_{\dot{A}}$ such that $l_{A\dot{B}} = \alpha_A \gamma_{\dot{B}}$. Since the components l^a are real, $\overline{l_{A\dot{B}}} = l_{B\dot{A}}$, which implies that $\gamma_{\dot{A}} = \lambda \overline{\alpha_A}$, for some $\lambda \in \mathbb{R}$. Hence, absorbing the absolute value of λ into α_A, it follows that

$$l_{A\dot{B}} = \pm \alpha_A \overline{\alpha}_{\dot{B}}, \tag{1.139}$$

where

$$\overline{\alpha}_{\dot{B}} \equiv \overline{\alpha_B}. \tag{1.140}$$

(An explicit proof is given below; see (1.161).)

According to the conventions (1.34) and (1.44), the sign in (1.139) is positive [resp. negative] if the component l^4 is positive [resp. negative]. (From (1.34), (1.44), (1.139), and (1.140) we have $l^4 = 2^{-1/2}(l^{1\dot{1}} + l^{2\dot{2}}) = \pm 2^{-1/2}(|\alpha^1|^2 + |\alpha^2|^2)$.) In a similar manner, there exists β_A such that $n_{A\dot{B}} = \pm \beta_A \overline{\beta}_{\dot{B}}$. Taking

$$l_{A\dot{B}} = \alpha_A \overline{\alpha}_{\dot{B}}, \qquad n_{A\dot{B}} = \beta_A \overline{\beta}_{\dot{B}}, \tag{1.141}$$

from the last condition in (1.138) we obtain $1 = l^{A\dot{B}} n_{A\dot{B}} = \alpha^A \overline{\alpha}^{\dot{B}} \beta_A \overline{\beta}_{\dot{B}} = |\alpha^A \beta_A|^2$. Since expressions (1.141) are left unchanged if we multiply α_A or β_A by any phase factor, we can assume that

$$\alpha^A \beta_A = 1. \tag{1.142}$$

Making use of (1.141), (1.28), and (1.142), one finds that the spinor equivalent of (1.137) is given by

$$\begin{aligned} L^{A\dot{A}}{}_{B\dot{B}} &= -\delta^A_B \delta^{\dot{A}}_{\dot{B}} - (\mathrm{e}^{-\zeta} - 1)\alpha^A \overline{\alpha}^{\dot{A}} \beta_B \overline{\beta}_{\dot{B}} - (\mathrm{e}^{\zeta} - 1)\beta^A \overline{\beta}^{\dot{A}} \alpha_B \overline{\alpha}_{\dot{B}} \\ &= -(\alpha^A \beta_B - \beta^A \alpha_B)(\overline{\alpha}^{\dot{A}} \overline{\beta}_{\dot{B}} - \overline{\beta}^{\dot{A}} \overline{\alpha}_{\dot{B}}) - (\mathrm{e}^{-\zeta} - 1)\alpha^A \overline{\alpha}^{\dot{A}} \beta_B \overline{\beta}_{\dot{B}} \\ &\quad - (\mathrm{e}^{\zeta} - 1)\beta^A \overline{\beta}^{\dot{A}} \alpha_B \overline{\alpha}_{\dot{B}} \\ &= -(\mathrm{e}^{-\zeta/2}\alpha^A \beta_B - \mathrm{e}^{\zeta/2}\beta^A \alpha_B)(\mathrm{e}^{-\zeta/2}\overline{\alpha}^{\dot{A}} \overline{\beta}_{\dot{B}} - \mathrm{e}^{\zeta/2}\overline{\beta}^{\dot{A}} \overline{\alpha}_{\dot{B}}) \\ &= -K^A{}_B \overline{K^A{}_B}, \end{aligned} \tag{1.143}$$

where

$$\begin{aligned} K^A{}_B &= \pm\,(\mathrm{e}^{-\zeta/2}\alpha^A \beta_B - \mathrm{e}^{\zeta/2}\beta^A \alpha_B) \\ &= \pm\left[\cosh\tfrac{1}{2}\zeta\,\delta^A_B - \sinh\tfrac{1}{2}\zeta\,(\alpha^A \beta_B + \beta^A \alpha_B)\right] \end{aligned} \tag{1.144}$$

[cf. (1.112)]. One can verify directly that $\det(K^A{}_B) = 1$. The rapidity is determined by the absolute value of the trace of $(K^A{}_B)$, $|K^A{}_A| = 2\cosh\frac{1}{2}\zeta$.

The direction of the velocity of S′ with respect to S, represented by v^a, is hidden in the definitions of l^a, n^a, and of the one-index spinors α_A, β_A appearing in (1.141) and (1.144); however, it may be noticed that by virtue of the antisymmetry of S_{abAB} on the tensor indices a, b, we have [see (1.69), (1.141), and (1.142)]

$$\begin{aligned} (v^c v_c)^{-1/2} v^a T^b S_{abAB} = l^a n^b S_{abAB} &= l^a n^b \sigma_{a(A}{}^{\dot{R}} \sigma_{|b|B)\dot{R}} = 2 l_{(A}{}^{\dot{R}} n_{B)\dot{R}} \\ &= 2\alpha_{(A} \overline{\alpha}^{\dot{R}} \beta_{B)} \overline{\beta}_{\dot{R}} = 2\alpha_{(A}\beta_{B)}; \end{aligned} \tag{1.145}$$

thus, expression (1.144) is equivalent to

$$K^A{}_B = \pm\left(\cosh\tfrac{1}{2}\zeta\,\delta^A_B - \sinh\tfrac{1}{2}\zeta\,u^a T^b S_{ab}{}^A{}_B\right), \tag{1.146}$$

where $u^a \equiv v^a/(v^c v_c)^{1/2}$. Hence, with $T^a = \delta^a_4$ as above, making use of (1.77),

$$(K^A{}_B) = \pm\begin{pmatrix} \cosh\frac{1}{2}\zeta + u^3 \sinh\frac{1}{2}\zeta & (u^1 + \mathrm{i}u^2)\sinh\frac{1}{2}\zeta \\ (u^1 - \mathrm{i}u^2)\sinh\frac{1}{2}\zeta & \cosh\frac{1}{2}\zeta - u^3 \sinh\frac{1}{2}\zeta \end{pmatrix}. \tag{1.147}$$

The complex conjugate of the matrix (1.147) can be expressed as

$$\overline{K} = (\overline{K^A{}_B}) = \pm\left[(\cosh\tfrac{1}{2}\zeta)I + (\sinh\tfrac{1}{2}\zeta)u^j \sigma_j\right],$$

where I is the 2×2 identity matrix, and the σ_i are the Pauli matrices $(i, j, \ldots = 1,2,3)$. Then, the components of the unit vector u^i are given in terms of $\overline{K}$ by

$$\mathrm{tr}\,(\overline{K}\sigma_i) = \pm 2(\sinh\tfrac{1}{2}\zeta)u^i.$$

As a consequence of (1.142), the square of the matrix $(\alpha^A \beta_B + \beta^A \alpha_B) = (u^a T^b S_{ab}{}^A{}_B)$ is the identity; therefore the matrix (1.146) can be expressed as

$$(K^A{}_B) = \pm\exp\left(-\tfrac{1}{2}\zeta\,u^a T^b S_{ab}{}^A{}_B\right) = \pm\exp\left(-\tfrac{1}{2}\zeta\,l^a n^b S_{ab}{}^A{}_B\right). \tag{1.148}$$

As we shall see below, this last equation implies that the boost (1.136), (1.137) can be written in the form

$$(L^a{}_b) = \exp\left(\zeta(l^a n_b - n^a l_b)\right)$$

[see (1.159)], which, among other things, shows the additive character of the rapidity ζ for boosts in the same direction.

Classification of the Orthochronous Proper Lorentz Transformations

The matrix (1.144) does not represent the most general SL(2, $\mathbb{C}$) matrix. If $(K^A{}_B)$ is an arbitrary SL(2, $\mathbb{C}$) matrix, we can write $K_{AB} = \frac{1}{2}(K_{AB} - K_{BA}) + \frac{1}{2}(K_{AB} + K_{BA}) = \frac{1}{2}K^C{}_C\varepsilon_{AB} + K_{(AB)}$, i.e., $K_{AB} = \nu\varepsilon_{AB} + \mu_{AB}$, where ν is some complex scalar and $\mu_{AB} = \mu_{BA}$. As shown below [see (1.227)], the fundamental theorem of algebra implies the existence of α_A, β_A, such that $\mu_{AB} = \alpha_{(A}\beta_{B)}$. Then α^A and β^A are eigenspinors of $(K^A{}_B)$; in fact,

$$K^A{}_B\alpha^B = \left[\nu\delta^A_B + \tfrac{1}{2}(\alpha^A\beta_B + \beta^A\alpha_B)\right]\alpha^B = (\nu + \tfrac{1}{2}\alpha^B\beta_B)\alpha^A,$$

and $K^A{}_B\beta^B = (\nu - \frac{1}{2}\alpha^B\beta_B)\beta^A$; thus, if $\{\alpha^A, \beta^A\}$ is linearly independent, then

$$K^A{}_B\alpha^B = \lambda\alpha^A, \qquad K^A{}_B\beta^B = \lambda^{-1}\beta^A, \tag{1.149}$$

for some $\lambda \in \mathbb{C}$, since the determinant of $(K^A{}_B)$ is equal to 1. The contraction $\alpha^A\beta_A$ is different from zero; hence, by rescaling α^A or β^A, we can impose the condition $\alpha^A\beta_A = 1$; then (1.149) are equivalent to

$$K^A{}_B = \lambda\alpha^A\beta_B - \lambda^{-1}\beta^A\alpha_B \tag{1.150}$$

[cf. (1.144)].

From (1.134) and (1.149) it follows that the products $\alpha^A\overline{\alpha}^{\dot{A}}$, $\beta^A\overline{\beta}^{\dot{A}}$, $\alpha^A\overline{\beta}^{\dot{A}}$, and $\beta^A\overline{\alpha}^{\dot{A}}$ are the spinor equivalents of eigenvectors of $(L^a{}_b)$ with eigenvalues

$$|\lambda|^2, \qquad |\lambda|^{-2}, \qquad \lambda/\overline{\lambda}, \qquad \overline{\lambda}/\lambda, \tag{1.151}$$

respectively; $\alpha^A\overline{\alpha}^{\dot{A}}$, $\beta^A\overline{\beta}^{\dot{A}}$ are the spinor equivalents of two real null vectors [see (1.139)], while $\alpha^A\overline{\beta}^{\dot{A}}$ and $\beta^A\overline{\alpha}^{\dot{A}}$ are the spinor equivalents of two complex-conjugate null vectors. Thus, if $(K^A{}_B)$ is of the form (1.150), then $(L^a{}_b)$ possesses two real null eigenvectors, and the vector equivalents of $\alpha^A\overline{\beta}^{\dot{A}} + \beta^A\overline{\alpha}^{\dot{A}}$ and $\mathrm{i}(\alpha^A\overline{\beta}^{\dot{A}} - \beta^A\overline{\alpha}^{\dot{A}})$, which are real, span a two-dimensional invariant subspace on which $(L^a{}_b)$ is a rotation through the angle $\arg(\lambda/\overline{\lambda})$.

Two eigenvalues of $(L^a{}_b)$ are equal to 1 if and only if $|\lambda| = 1$ or $\lambda \in \mathbb{R}$ [see (1.151)]. The orthochronous proper Lorentz transformations of this type, which leave pointwise invariant a two-dimensional subspace, are called *simple* (Lounesto 1997, Sec. 9.6). When $|\lambda| = 1$, $(L^a{}_b)$ represents a rotation; the vector equivalents of $\alpha^A\overline{\alpha}^{\dot{A}}$ and $\beta^A\overline{\beta}^{\dot{A}}$ span a real two-dimensional subspace pointwise invariant under

$(L^a{}_b)$, the trace $K^A{}_A$ is real, and $|K^A{}_A| < 2$. When λ is real, $(L^a{}_b)$ represents a boost, the vector equivalents of $\alpha^A\overline{\beta}^{\dot{A}} + \beta^A\overline{\alpha}^{\dot{A}}$ and $\mathrm{i}(\alpha^A\overline{\beta}^{\dot{A}} - \beta^A\overline{\alpha}^{\dot{A}})$ span a real two-dimensional subspace pointwise invariant under $(L^a{}_b)$, the trace $K^A{}_A$ is real, and $|K^A{}_A| > 2$.

On the other hand, it can be readily verified that $K^A{}_A = \lambda + \lambda^{-1}$ is real if and only if $|\lambda| = 1$ or $\lambda \in \mathbb{R}$. Therefore, the SL$(2,\mathbb{C})$ matrix (1.150) corresponds to a simple orthochronous proper Lorentz transformation if and only if the trace $K^A{}_A$ is real.

In the remaining case, in which $\{\alpha^A, \beta^A\}$ is linearly dependent, we have

$$K^A{}_B = \nu\delta^A_B + \mu\alpha^A\alpha_B,$$

where ν and μ are two complex numbers. Then, the condition $\det(K^A{}_B) = 1$ gives $\nu = \pm 1$, and by rescaling α_A we can write

$$K^A{}_B = \pm(\delta^A_B - \tfrac{1}{2}\zeta\,\alpha^A\alpha_B), \tag{1.152}$$

where ζ is a *real* number. Now α^A is the only eigenspinor of $(K^A{}_B)$, and $\alpha^A\overline{\alpha}^{\dot{A}}$ is the spinor equivalent of a real null eigenvector of the corresponding orthogonal transformation $(L^a{}_b)$, with eigenvalue 1 (these Lorentz transformations are called *null rotations*).

If γ_A is a one-index spinor such that $\alpha^A\gamma_A = 1$, then the vector equivalent of $\mathrm{i}(\alpha^A\overline{\gamma}^{\dot{A}} - \gamma^A\overline{\alpha}^{\dot{A}})$ is a (spacelike) real eigenvector of $(L^a{}_b)$ with eigenvalue 1; hence, for the SO$^\uparrow(3,1)$ transformation corresponding to (1.152) there is always a real two-dimensional subspace pointwise invariant under $(L^a{}_b)$ (that is, $(L^a{}_b)$ is simple). (Note that γ_A is defined up to the transformation $\gamma_A \mapsto \gamma_A + \kappa\alpha_A$, with $\kappa \in \mathbb{C}$ arbitrary, but the two-dimensional subspace pointwise invariant under $(L^a{}_b)$ is not affected by this ambiguity.) From (1.152) we see that $K^A{}_A = \pm 2$.

Substituting (1.152) into (1.102), with the aid of (1.28), we obtain

$$\begin{aligned}
L^a{}_b &= -\tfrac{1}{2}\sigma^a{}_{A\dot{A}}\sigma_b{}^{B\dot{B}}(\delta^A_B - \tfrac{1}{2}\zeta\,\alpha^A\alpha_B)(\delta^{\dot{A}}_{\dot{B}} - \tfrac{1}{2}\zeta\,\overline{\alpha}^{\dot{A}}\overline{\alpha}_{\dot{B}}) \\
&= \delta^a_b - \tfrac{1}{2}\sigma^a{}_{A\dot{A}}\sigma_b{}^{B\dot{B}}\big[-\tfrac{1}{2}\zeta\,\alpha^A\overline{\alpha}^{\dot{A}}(\alpha_B\overline{\gamma}_{\dot{B}} + \gamma_B\overline{\alpha}_{\dot{B}}) \\
&\qquad + \tfrac{1}{2}\zeta\,(\alpha^A\overline{\gamma}^{\dot{A}} + \gamma^A\overline{\alpha}^{\dot{A}})\alpha_B\overline{\alpha}_{\dot{B}} + \tfrac{1}{4}\zeta^2\alpha^A\overline{\alpha}^{\dot{A}}\alpha_B\overline{\alpha}_{\dot{B}}\big] \\
&= \delta^a_b + \frac{1}{\sqrt{2}}\zeta(l^a s_b - s^a l_b) - \frac{1}{4}\zeta^2 l^a l_b,
\end{aligned} \tag{1.153}$$

where l_a and s_a are the vector equivalents of $\alpha_A\overline{\alpha}_{\dot{A}}$ and $(1/\sqrt{2})(\alpha_A\overline{\gamma}_{\dot{A}} + \gamma_A\overline{\alpha}_{\dot{A}})$, respectively. The vectors l_a and s_a are real and orthogonal to each other; l_a is null and $s^a s_a = 1$. (A similar classification of the proper orthochronous Lorentz transformations, making use of quaternions, is obtained in Lanczos (1970); see also Penrose and Rindler 1984, Hall 2004.)

The foregoing discussion proves the following result.

Proposition 1.4. *The trace of the* SL$(2,\mathbb{C})$ *matrix* $(K^A{}_B)$ *is real if and only if there exists a real two-dimensional subspace pointwise invariant under the corresponding*

$SO^{\uparrow}(3,1)$ *transformation* $(L^a{}_b)$. *Furthermore, in that case,* $(L^a{}_b)$ *is a boost or a rotation if* $|K^A{}_A| > 2$ *or* $|K^A{}_A| < 2$, *respectively. When* $|K^A{}_A| = 2$, *with* $(K^A{}_B) \neq \pm(\delta^A_B)$, $(L^a{}_b)$ *has only one real null eigenvector, and the corresponding eigenvalue is equal to* 1.

Since any nonzero complex number λ can be expressed as the product of a positive real number and a complex number of unit modulus, $\lambda = |\lambda| e^{i\theta/2}$, for some $\theta \in \mathbb{R}$, from (1.149) it follows that the $SO^{\uparrow}(3,1)$ transformation defined by (1.150) is the composition of two simple orthochronous Lorentz transformations. Furthermore, making use of (1.145), one finds that (1.150) can also be expressed in the form

$$K^A{}_B = \tfrac{1}{2}(\lambda + \lambda^{-1})\, \delta^A_B + \tfrac{1}{2}(\lambda - \lambda^{-1})\, l^a n^b S_{ab}{}^A{}_B, \tag{1.154}$$

with l_a and n_a defined by (1.141). For instance, taking $\lambda = e^{i\theta/2}$ and $T^a = \delta^a_4$, we have (see Exercise 1.9)

$$(K^A{}_B) = \begin{pmatrix} \cos\frac{1}{2}\theta - iu^3 \sin\frac{1}{2}\theta & -i(u^1 + iu^2)\sin\frac{1}{2}\theta \\ -i(u^1 - iu^2)\sin\frac{1}{2}\theta & \cos\frac{1}{2}\theta + iu^3 \sin\frac{1}{2}\theta \end{pmatrix}, \tag{1.155}$$

where now u^a is a unit vector defining the axis of the rotation [cf. (1.147)]. Alternatively, denoting by m^a the vector equivalent of $\alpha^A \overline{\beta}^{\dot{A}}$, we have

$$m^a \overline{m^b} S_{abAB} = m^a \overline{m^b} \sigma_{a(A}{}^{\dot{R}} \sigma_{|b|B)\dot{R}} = 2\alpha_{(A} \overline{\beta}^{\dot{R}} \beta_{B)} \overline{\alpha}_{\dot{R}} = -2\alpha_{(A}\beta_{B)}$$

[cf. (1.145)]; hence, (1.150) is also equivalent to

$$K^A{}_B = \tfrac{1}{2}(\lambda + \lambda^{-1})\, \delta^A_B - \tfrac{1}{2}(\lambda - \lambda^{-1})\, m^a \overline{m^b}\, S_{ab}{}^A{}_B.$$

With the conventions adopted here, the composition of a rotation about the z-axis through an angle φ followed by a rotation about the y-axis through an angle θ, followed finally by a rotation about the z-axis through an angle χ, corresponds to the matrix

$$\begin{aligned} (K^A{}_B) &= \begin{pmatrix} e^{-i\chi/2} & 0 \\ 0 & e^{i\chi/2} \end{pmatrix} \begin{pmatrix} \cos\frac{1}{2}\theta & \sin\frac{1}{2}\theta \\ -\sin\frac{1}{2}\theta & \cos\frac{1}{2}\theta \end{pmatrix} \begin{pmatrix} e^{-i\varphi/2} & 0 \\ 0 & e^{i\varphi/2} \end{pmatrix} \\ &= \begin{pmatrix} e^{-i(\varphi+\chi)/2}\cos\frac{1}{2}\theta & e^{i(\varphi-\chi)/2}\sin\frac{1}{2}\theta \\ -e^{i(\chi-\varphi)/2}\sin\frac{1}{2}\theta & e^{i(\varphi+\chi)/2}\cos\frac{1}{2}\theta \end{pmatrix} \end{aligned} \tag{1.156}$$

and its negative. (Note the order of the factors.) As one can readily verify, this matrix also belongs to SU(2), and this last expression gives a parameterization of the SU(2) matrices by Euler angles (φ, θ, χ).

It is convenient to express the eigenvalue λ of $(K^A{}_B)$ in the form $\lambda = e^{-z/2}$; then $(K^A{}_B)$ corresponds to a simple $SO^{\uparrow}(3,1)$ transformation when z is real or pure imaginary, and from (1.154) we also have

$$K^A{}_B = \cosh\tfrac{1}{2}z\, \delta^A_B - \sinh\tfrac{1}{2}z\, l^a n^b S_{ab}{}^A{}_B.$$

On the other hand, using the fact that the square of the matrix $(l^a n^b S_{ab}{}^A{}_B)$ is the identity [see (1.145)], we find that

$$(K^A{}_B) = \exp\left(-\tfrac{1}{2}z\, l^a n^b S_{ab}{}^A{}_B\right)$$

and, by virtue of (1.76), letting $z = x + \mathrm{i}y$,

$$\begin{aligned} z\, l^a n^b S_{ab}{}^A{}_B &= (x + \mathrm{i}y) l^a n^b S_{ab}{}^A{}_B \\ &= \left(x l_c n_d - y \tfrac{1}{2} l^a n^b \varepsilon_{abcd}\right) S^{cdA}{}_B \\ &= -\tfrac{1}{2} t_{ab} S^{abA}{}_B, \end{aligned}$$

where t_{ab} is the *real* bivector

$$t_{ab} \equiv -2x l_{[a} n_{b]} + y\, \varepsilon_{abcd} l^c n^d. \tag{1.157}$$

Thus,

$$(K^A{}_B) = \exp\left(\tfrac{1}{4} t^{ab} S_{ab}{}^A{}_B\right) \tag{1.158}$$

[cf. (1.124)]. Making use of the criterion given in Proposition 1.2, one can verify that t_{ab} is simple if and only if x or y vanishes; therefore, the bivector (1.157) is simple if and only if the $\mathrm{SL}(2,\mathbb{C})$ matrix (1.158) corresponds to a simple $\mathrm{SO}^{\uparrow}(3,1)$ transformation. Since for the Lorentzian signature we have $\overline{S_{abAB}} = S_{ab\dot{A}\dot{B}}$, it follows that $(M^{\dot{A}}{}_{\dot{B}}) = \exp\left(\tfrac{1}{4} t^{ab} S_{ab}{}^{\dot{A}}{}_{\dot{B}}\right)$ [cf. (1.124)]; hence, as in the foregoing subsection, it follows that

$$(L^a{}_b) = \exp(-t^a{}_b). \tag{1.159}$$

Finally, in the case of the null rotations,

$$\frac{1}{\sqrt{2}} l^a s^b S_{abAB} = \frac{1}{\sqrt{2}} l^a s^b \sigma_{a(A}{}^{\dot{R}} \sigma_{|b|B)\dot{R}} = \alpha_A \alpha_B,$$

where l_a and s_a are the vector equivalents of $\alpha_A \overline{\alpha}_{\dot{A}}$ and $(1/\sqrt{2})(\alpha_A \overline{\gamma}_{\dot{A}} + \gamma_A \overline{\alpha}_{\dot{A}})$, respectively. Hence (1.152) can also be expressed in the form

$$K^A{}_B = \pm\left(\delta^A_B - \frac{1}{2\sqrt{2}} \zeta\, l^a s^b S_{ab}{}^A{}_B\right),$$

or equivalently,

$$(K^A{}_B) = \pm \exp\left(-\frac{1}{2\sqrt{2}} \zeta\, l^a s^b S_{ab}{}^A{}_B\right), \tag{1.160}$$

since the square of the matrix $(\alpha^A \alpha_B)$ is equal to zero. This expression is of the form (1.158), where t_{ab} is the simple bivector $-\sqrt{2}\zeta\, l_{[a} s_{b]}$. As we have seen, (1.152) corresponds to a simple Lorentz transformation, and therefore, in all cases, the $\mathrm{SL}(2,\mathbb{C})$ matrix $(K^A{}_B) = \exp\left(\tfrac{1}{4} t^{ab} S_{ab}{}^A{}_B\right)$ corresponds to a simple $\mathrm{SO}^{\uparrow}(3,1)$ transformation if and only if the bivector t_{ab} is simple. (One can directly verify that in this case, (1.159) reduces to (1.153).)

Note that in order to obtain (1.124), (1.158), and (1.160), the explicit form of the matrices $(S_{ab}{}^{A}{}_{B})$ and $(S_{ab}{}^{\dot{A}}{}_{\dot{B}})$, given by (1.77) and (1.79), has not been required.

Parameterization of Null Vectors, Aberration of Light

By definition, a (real) null vector l^a satisfies the condition $l^a l_a = 0$ and therefore

$$l^4 = \pm\left[(l^1)^2 + (l^2)^2 + (l^3)^2\right]^{1/2}.$$

This means that apart from the ambiguity in the sign of l^4, the components of the null vector l^a are determined by the point $(l^1, l^2, l^3) \in \mathbb{R}^3$, which, in turn, can be parameterized by the usual spherical coordinates (r, θ, φ), according to

$$(l^1, l^2, l^3) = (r\sin\theta\cos\varphi, r\sin\theta\sin\varphi, r\cos\theta).$$

Assuming $l^4 > 0$, the components of the spinor equivalent of l^a are then given by [see (1.34)]

$$\begin{aligned}
\begin{pmatrix} l^{1\dot{1}} & l^{1\dot{2}} \\ l^{2\dot{1}} & l^{2\dot{2}} \end{pmatrix} &= \frac{r}{\sqrt{2}} \begin{pmatrix} 1-\cos\theta & -\mathrm{e}^{\mathrm{i}\varphi}\sin\theta \\ -\mathrm{e}^{-\mathrm{i}\varphi}\sin\theta & 1+\cos\theta \end{pmatrix} \\
&= \sqrt{2}r \begin{pmatrix} \sin^2\frac{1}{2}\theta & -\mathrm{e}^{\mathrm{i}\varphi}\sin\frac{1}{2}\theta\cos\frac{1}{2}\theta \\ -\mathrm{e}^{-\mathrm{i}\varphi}\sin\frac{1}{2}\theta\cos\frac{1}{2}\theta & \cos^2\frac{1}{2}\theta \end{pmatrix} \\
&= \sqrt{2}r \begin{pmatrix} -\mathrm{e}^{\mathrm{i}\varphi/2}\sin\frac{1}{2}\theta \\ \mathrm{e}^{-\mathrm{i}\varphi/2}\cos\frac{1}{2}\theta \end{pmatrix} \begin{pmatrix} -\mathrm{e}^{-\mathrm{i}\varphi/2}\sin\frac{1}{2}\theta & \mathrm{e}^{\mathrm{i}\varphi/2}\cos\frac{1}{2}\theta \end{pmatrix} \\
&= \begin{pmatrix} \alpha^1 \\ \alpha^2 \end{pmatrix} \begin{pmatrix} \overline{\alpha}^{\dot{1}} & \overline{\alpha}^{\dot{2}} \end{pmatrix},
\end{aligned} \tag{1.161}$$

which shows explicitly that the spinor equivalent of a real null vector with $l^4 > 0$ is of the form $l^{A\dot{A}} = \alpha^A \overline{\alpha}^{\dot{A}}$ [see (1.139)]. From (1.161) we see that

$$\begin{pmatrix} \alpha^1 \\ \alpha^2 \end{pmatrix} = 2^{1/4} r^{1/2} \mathrm{e}^{-\mathrm{i}\chi/2} \begin{pmatrix} -\mathrm{e}^{\mathrm{i}\varphi/2}\sin\frac{1}{2}\theta \\ \mathrm{e}^{-\mathrm{i}\varphi/2}\cos\frac{1}{2}\theta \end{pmatrix}, \tag{1.162}$$

where χ is an arbitrary real number.

The wave four-vector of an electromagnetic plane wave k^a is a future-pointing null vector, that is, $k^4 > 0$ (see, e.g., Jackson 1975), and therefore, its spinor equivalent is of the form $\alpha^A \overline{\alpha}^{\dot{A}}$, with α^A given by

$$\begin{pmatrix} \alpha^1 \\ \alpha^2 \end{pmatrix} = 2^{1/4} k^{1/2} \begin{pmatrix} -\mathrm{e}^{\mathrm{i}\varphi/2}\sin\frac{1}{2}\theta \\ \mathrm{e}^{-\mathrm{i}\varphi/2}\cos\frac{1}{2}\theta \end{pmatrix}, \tag{1.163}$$

where $k = [(k^1)^2 + (k^2)^2 + (k^3)^2]^{1/2}$ is the wave number, and θ and φ are the standard polar and azimuth angles of the direction of propagation of the wave with respect to some inertial reference frame S. The spinor equivalent of k'^a the wave four-vector of the wave with respect to a second inertial reference frame S′, which moves with velocity v with respect to S along the z-axis, is also of the form $\alpha'^A \overline{\alpha'^{\dot{A}}}$, with $\alpha'^A = K^A{}_B \alpha^B$, and $(K^A{}_B)$ given by [see (1.147)]

$$(K^A{}_B) = \begin{pmatrix} \cosh\frac{1}{2}\zeta + \sinh\frac{1}{2}\zeta & 0 \\ 0 & \cosh\frac{1}{2}\zeta - \sinh\frac{1}{2}\zeta \end{pmatrix} = \begin{pmatrix} e^{\zeta/2} & 0 \\ 0 & e^{-\zeta/2} \end{pmatrix},$$

with $v = c\tanh\zeta$; hence, expressing α'^A in the form (1.163), we immediately obtain

$$k'^{1/2} \sin\tfrac{1}{2}\theta' \, e^{i\varphi'/2} = e^{\zeta/2} k^{1/2} \sin\tfrac{1}{2}\theta \, e^{i\varphi/2},$$

$$k'^{1/2} \cos\tfrac{1}{2}\theta' \, e^{-i\varphi'/2} = e^{-\zeta/2} k^{1/2} \cos\tfrac{1}{2}\theta \, e^{-i\varphi/2},$$

which imply $\varphi' = \varphi$ and

$$\tan\tfrac{1}{2}\theta' = e^{\zeta} \tan\tfrac{1}{2}\theta = \sqrt{\frac{c+v}{c-v}} \tan\tfrac{1}{2}\theta$$

(see also Misner, Thorne, and Wheeler 1973, Penrose and Rindler 1984). (Note that θ is the angle between the direction of propagation of the wave and the direction of motion of S′ with respect to S.) Furthermore, recalling that the angular frequency ω of the wave is related to k by $\omega = ck$, we have

$$\omega' = c\left[\left(e^{\zeta/2} k^{1/2} \sin\tfrac{1}{2}\theta\right)^2 + \left(e^{-\zeta/2} k^{1/2} \cos\tfrac{1}{2}\theta\right)^2\right] = \gamma\omega\left(1 - \frac{v}{c}\cos\theta\right),$$

which gives the relativistic Doppler shift.

Improper Lorentz Transformations. Spatial Inversion

Proceeding as at the beginning of this subsection, one finds that the O(3,1) transformations with determinant -1 are of the form

$$L^a{}_b = \mp\tfrac{1}{2} \sigma^a{}_{A\dot{B}} \sigma_b{}^{C\dot{D}} K^A{}_{\dot{D}} \overline{K^B{}_{\dot{C}}}, \tag{1.164}$$

where $(K^A{}_{\dot{B}}) \in \mathrm{SL}(2,\mathbb{C})$. Making use of the explicit expression (1.33), from (1.164) one finds that

$$L^4{}_4 = \pm\tfrac{1}{2}\left(|K^1{}_{\dot{1}}|^2 + |K^1{}_{\dot{2}}|^2 + |K^2{}_{\dot{1}}|^2 + |K^2{}_{\dot{2}}|^2\right),$$

which shows that $(L^a{}_b)$ preserves [resp. reverses] the time orientation if we take the upper [resp. lower] sign in (1.164).

As in the case of Euclidean signature, the entries of the matrix $(K^A{}_{\dot{B}})$ appearing in (1.164) can be regarded as the spinor equivalent of a vector, but in the present case that vector is complex (according to (1.47), $K_{A\dot{B}}$ is the spinor equivalent of a real vector if and only if $\overline{K_{A\dot{B}}} = K_{B\dot{A}}$). Since $\det(K^A{}_{\dot{B}}) = \frac{1}{2}K^{A\dot{B}}K_{A\dot{B}} = 1$, denoting by K_a the vector equivalent of $K_{A\dot{B}}$, we have $K^aK_a = -2$, which means that the real and the imaginary parts of K_a are orthogonal to each other. Making use of (1.130), (1.52), and (1.65), from (1.164) we obtain

$$\begin{aligned} L^a{}_b &= \mp\tfrac{1}{2}\big(-K^c\overline{K_c}\,\delta^a_b + K^a\overline{K_b} + \overline{K^a}K_b - \mathrm{i}g^{ae}\varepsilon_{ebcd}K^c\overline{K^d}\big) \\ &= \mp\big[-\tfrac{1}{2}(v^cv_c + w^cw_c)\delta^a_b + v^av_b + w^aw_b - g^{ae}\varepsilon_{ebcd}v^cw^d\big], \end{aligned} \tag{1.165}$$

where v_a and w_a are the real and the imaginary parts of K_a, respectively [cf. (1.131)] (see also Penrose and Rindler 1984).

From $K^aK_a = -2$, it follows that $v^av_a - w^aw_a = -2$, and therefore, from (1.165) we see that $L^a{}_bv^b = \pm v^a$ and $L^a{}_bw^b = \mp w^a$. If $\{v^a, w^a\}$ is linearly independent, this set spans a two-dimensional subspace invariant under $(L^a{}_b)$.

When $\{v^a, w^a\}$ is linearly dependent, the orthogonality of v^a and w^a implies that v^a or w^a is equal to zero; then $(L^a{}_b)$ corresponds to the reflection in a three-dimensional hyperplane passing through the origin or to the negative of such a reflection (see also Section 1.4, below). If v^a is equal to zero, w^a is spacelike, and if w^a is equal to zero, v^a is timelike. Thus, in the latter case, $(L^a{}_b)$ corresponds to the reflection in the three-dimensional hyperplane orthogonal to the unit timelike vector $T^a = v^a/\sqrt{2}$, or to the negative of this reflection. Hence, if $T_{A\dot{B}}$ is the spinor equivalent of T_a, from (1.164) it follows that

$$L^a{}_b = -\sigma^a{}_{A\dot{B}}\sigma_b{}^{C\dot{D}}T^A{}_{\dot{D}}T_C{}^{\dot{B}} \tag{1.166}$$

corresponds to the improper Lorentz transformation that leaves invariant the timelike vector T^a and reverses the sign of the vectors orthogonal to T^a. This transformation is the spatial inversion for an observer for which T^a (or $-T^a$) represents its time axis [cf. also (1.190)].

1.3.3 Ultrahyperbolic Signature

When $(g_{ab}) = \mathrm{diag}(1,1,-1,-1)$, the Infeld–van der Waerden symbols can all be real [see (1.35)]. If these symbols are real, the matrix (1.102) is real if and only if the unimodular matrices $(K^A{}_B)$ and $(M^{\dot{A}}{}_{\dot{B}})$ are both real or both pure imaginary. The 4×4 matrices of the form

$$L^a{}_b = -\tfrac{1}{2}\sigma^a{}_{A\dot{A}}\sigma_b{}^{B\dot{B}}K^A{}_BM^{\dot{A}}{}_{\dot{B}}, \tag{1.167}$$

with $\sigma^a{}_{A\dot{B}}$ given by (1.35) and $(K^A{}_B), (M^{\dot{A}}{}_{\dot{B}}) \in \mathrm{SL}(2,\mathbb{R})$, the group of unimodular real 2×2 matrices with matrix multiplication, form a connected subgroup of SO(2,2) that contains the identity of this group and will be denoted by $\mathrm{SO}(2,2)_0$.

Furthermore, the mapping $\big((K^A{}_B),(M^{\dot{A}}{}_{\dot{B}})\big) \mapsto (L^a{}_b)$, with $L^a{}_b$ given by (1.167), is a two-to-one homomorphism of $\mathrm{SL}(2,\mathbb{R}) \times \mathrm{SL}(2,\mathbb{R})$ onto $\mathrm{SO}(2,2)_0$, i.e.,

$$\mathrm{SO}(2,2)_0 \simeq \big(\mathrm{SL}(2,\mathbb{R}) \times \mathrm{SL}(2,\mathbb{R})\big)/\mathbb{Z}_2. \tag{1.168}$$

Following a procedure analogous to those employed in the preceding two subsections, we can find the possible forms of the $\mathrm{SL}(2,\mathbb{R})$ matrices. As shown in Section 1.7, below, for a given 2×2 matrix $(K^A{}_B)$ there exist α_A and β_B such that $K^A{}_B = \nu\delta^A_B + \frac{1}{2}(\alpha^A\beta_B + \beta^A\alpha_B)$, for some scalar ν. Then α_A and β_A are eigenspinors of $(K^A{}_B)$, and, assuming that $\{\alpha_A, \beta_A\}$ is linearly independent, the condition $\det(K^A{}_B) = 1$ implies that

$$K^A{}_B\alpha^B = \lambda\alpha^A, \qquad K^A{}_B\beta^B = \lambda^{-1}\beta^A, \tag{1.169}$$

for some $\lambda \in \mathbb{C}$ [cf. (1.149)]. If we now assume that $(K^A{}_B)$ belongs to $\mathrm{SL}(2,\mathbb{R})$, the complex conjugate of the first equation in (1.169) gives

$$K^A{}_B\widehat{\alpha}^B = \overline{\lambda}\widehat{\alpha}^A,$$

where, in the present case, the mate (conjugate or adjoint) of α_A is defined by

$$\widehat{\alpha}_A \equiv \overline{\alpha_A} \tag{1.170}$$

[cf. (1.113)]. This means that if α_A is an eigenspinor of an $\mathrm{SL}(2,\mathbb{R})$ matrix, so is its mate $\widehat{\alpha}_A$, and now we have to consider two possible subcases according to whether $\alpha^A\widehat{\alpha}_A$ is different from zero or not. (Note that $\alpha^A\widehat{\alpha}_A = \alpha_1\overline{\alpha_2} - \alpha_2\overline{\alpha_1}$ is pure imaginary and that $\mathrm{i}\alpha^A\widehat{\alpha}_A = 2\,\mathrm{Re}\,(\mathrm{i}\alpha_1\overline{\alpha_2})$ can take any real value.)

When $\alpha^A\widehat{\alpha}_A \neq 0$, $\{\alpha_A, \widehat{\alpha}_A\}$ is linearly independent, the equations above imply that β_A must be proportional to $\widehat{\alpha}_A$ and $\overline{\lambda} = \lambda^{-1}$, which means that $\lambda = \mathrm{e}^{\mathrm{i}\theta/2}$, for some $\theta \in \mathbb{R}$. Hence, from (1.169) we have

$$\begin{aligned} K^A{}_B &= \frac{1}{\alpha^R\widehat{\alpha}_R}\left(\mathrm{e}^{\mathrm{i}\theta/2}\alpha^A\widehat{\alpha}_B - \mathrm{e}^{-\mathrm{i}\theta/2}\widehat{\alpha}^A\alpha_B\right) \\ &= \cos\tfrac{1}{2}\theta\;\delta^A_B + \mathrm{i}\sin\tfrac{1}{2}\theta\;\frac{\alpha^A\widehat{\alpha}_B + \widehat{\alpha}^A\alpha_B}{\alpha^R\widehat{\alpha}_R}. \end{aligned} \tag{1.171}$$

When $\alpha^A\widehat{\alpha}_A = 0$, $\widehat{\alpha}_A$ is proportional to α_A, and hence, $\overline{\lambda} = \lambda$. Writing $\widehat{\alpha}_A = \mu\alpha_A$, with $\mu \in \mathbb{C}$, then $\widehat{\widehat{\alpha}}_A = (\mu\alpha_A)\widehat{\ } = \overline{\mu}\widehat{\alpha}_A = \overline{\mu}\mu\alpha_A$, and using the fact that $\widehat{\widehat{\alpha}}_A = \alpha_A$, it follows that $\mu = \mathrm{e}^{\mathrm{i}\chi}$, for some $\chi \in \mathbb{R}$; thus, $(\mathrm{e}^{\mathrm{i}\chi/2}\alpha_A)\widehat{\ } = \mathrm{e}^{\mathrm{i}\chi/2}\alpha_A$, and absorbing the factor $\mathrm{e}^{\mathrm{i}\chi/2}$ into α_A, we have $\widehat{\alpha}_A = \alpha_A$. Similarly, $\widehat{\beta}_A$ must be also proportional to β_A and we can assume that $\widehat{\beta}_A = \beta_A$. Then, expressing λ in the form $\lambda = \pm\mathrm{e}^{\theta/2}$, from (1.169) it follows that

$$\begin{aligned} K^A{}_B &= \pm\frac{1}{\alpha^R\beta_R}\left(\mathrm{e}^{\theta/2}\alpha^A\beta_B - \mathrm{e}^{-\theta/2}\beta^A\alpha_B\right) \\ &= \pm\left(\cosh\tfrac{1}{2}\theta\;\delta^A_B + \sinh\tfrac{1}{2}\theta\;\frac{\alpha^A\beta_B + \beta^A\alpha_B}{\alpha^R\beta_R}\right). \end{aligned} \tag{1.172}$$

Finally, when $\{\alpha_A, \beta_A\}$ is linearly dependent we have $K^A{}_B = \nu\delta^A_B + \mu\alpha^A\alpha_B$, and the condition $\det(K^A{}_B) = 1$ requires $\nu = \pm 1$. Since $(K^A{}_B)$ must be real, $\widehat{\alpha}_A$ has to be proportional to α_A and we can assume that $\widehat{\alpha}_A = \alpha_A$; hence, in this case,

$$K^A{}_B = \pm(\delta^A_B + \tfrac{1}{2}\theta\,\alpha^A\alpha_B),$$

with $\theta \in \mathbb{R}$.

Summarizing, any $SL(2,\mathbb{R})$ matrix different from the unit matrix has one of the forms

$$\begin{cases} \cos\tfrac{1}{2}\theta\;\delta^A_B + \mathrm{i}\sin\tfrac{1}{2}\theta\;\dfrac{\alpha^A\widehat{\alpha}_B + \widehat{\alpha}^A\alpha_B}{\alpha^R\widehat{\alpha}_R}, & \text{with } \alpha^A\widehat{\alpha}_A \neq 0, \\ \pm\left(\cosh\tfrac{1}{2}\theta\;\delta^A_B + \sinh\tfrac{1}{2}\theta\;\dfrac{\alpha^A\beta_B + \beta^A\alpha_B}{\alpha^R\beta_R}\right), & \text{with } \widehat{\alpha}_A = \alpha_A, \widehat{\beta}_A = \beta_A, \\ \pm(\delta^A_B + \tfrac{1}{2}\theta\,\alpha^A\alpha_B), & \text{with } \widehat{\alpha}_A = \alpha_A. \end{cases} \tag{1.173}$$

These three cases correspond to $|K^A{}_A|$ less than 2, greater than 2, and equal to 2, respectively.

Similarly, one concludes that any $SL(2,\mathbb{R})$ matrix $(M^{\dot{A}}{}_{\dot{B}})$ different from the unit matrix can have one of the forms

$$\begin{cases} \cos\tfrac{1}{2}\varphi\;\delta^{\dot{A}}_{\dot{B}} + \mathrm{i}\sin\tfrac{1}{2}\varphi\;\dfrac{\alpha^{\dot{A}}\widehat{\alpha}_{\dot{B}} + \widehat{\alpha}^{\dot{A}}\alpha_{\dot{B}}}{\alpha^{\dot{R}}\widehat{\alpha}_{\dot{R}}}, & \text{with } \alpha^{\dot{A}}\widehat{\alpha}_{\dot{A}} \neq 0, \\ \pm\left(\cosh\tfrac{1}{2}\varphi\;\delta^{\dot{A}}_{\dot{B}} + \sinh\tfrac{1}{2}\varphi\;\dfrac{\alpha^{\dot{A}}\beta_{\dot{B}} + \beta^{\dot{A}}\alpha_{\dot{B}}}{\alpha^{\dot{R}}\beta_{\dot{R}}}\right), & \text{with } \widehat{\alpha}_{\dot{A}} = \alpha_{\dot{A}}, \widehat{\beta}_{\dot{A}} = \beta_{\dot{A}}, \\ \pm(\delta^{\dot{A}}_{\dot{B}} + \tfrac{1}{2}\varphi\,\alpha^{\dot{A}}\alpha_{\dot{B}}), & \text{with } \widehat{\alpha}_{\dot{A}} = \alpha_{\dot{A}}, \end{cases} \tag{1.174}$$

where

$$\widehat{\alpha}_{\dot{A}} \equiv \overline{\alpha_{\dot{A}}}. \tag{1.175}$$

Since the two $SL(2,\mathbb{R})$ matrices $(K^A{}_B)$, $(M^{\dot{A}}{}_{\dot{B}})$ appearing in (1.167) are independent of each other and each of these matrices can have one of the forms (1.173) and (1.174), there is a huge variety of $SO(2,2)_0$ transformations.

Improper O(2, 2) Transformations

With the Infeld–van der Waerden symbols given by (1.35), all the orthogonal transformations with determinant equal to -1 are given by

$$L^a{}_b = -\frac{1}{2}\sigma^a{}_{A\dot{A}}\sigma_b{}^{B\dot{B}}K^A{}_{\dot{B}}M^{\dot{A}}{}_B, \tag{1.176}$$

where the unimodular matrices $(K^A{}_{\dot{B}})$ and $(M^{\dot{A}}{}_B)$ are both real or both pure imaginary. $K_{A\dot{B}}$ and $M_{\dot{A}B}$ can be regarded as the spinor equivalents of two real vectors or two pure imaginary vectors, respectively; in the first case, writing

$$K_{A\dot{B}} = \sqrt{2}\,v_{A\dot{B}}, \qquad M_{\dot{A}B} = \sqrt{2}\,w_{B\dot{A}},$$

one finds that $v_{A\dot{B}}$ and $w_{A\dot{B}}$ are the spinor equivalents of two real vectors v_a and w_a such that $v^a v_a = -1 = w^a w_a$. Making use of (1.52), (1.65), and (1.130), one finds that (1.176) is equivalent to

$$L^a{}_b = v^c w_c \delta^a_b - v^a w_b - w^a v_b + g^{ae} \varepsilon_{ebcd} v^c w^d.$$

From this last expression one finds that $L^a{}_b v^b = w^a$ and $L^a{}_b w^b = v^a$.

In the second case, $K_{A\dot{B}}$ and $M_{\dot{A}B}$ can be regarded as the spinor equivalents of two pure imaginary vectors; letting

$$K_{A\dot{B}} = \mathrm{i}\sqrt{2}\,v_{A\dot{B}}, \qquad M_{\dot{A}B} = \mathrm{i}\sqrt{2}\,w_{B\dot{A}},$$

one finds that $v_{A\dot{B}}$ and $w_{A\dot{B}}$ are the spinor equivalents of two real vectors v_a and w_a such that $v^a v_a = 1 = w^a w_a$, and (1.176) amounts to

$$L^a{}_b = -v^c w_c \delta^a_b + v^a w_b + w^a v_b - g^{ae} \varepsilon_{ebcd} v^c w^d.$$

Here we also have $L^a{}_b v^b = w^a$ and $L^a{}_b w^b = v^a$.

As in the cases of the improper O(4) and O(3,1) transformations, when $\{v^a, w^a\}$ is linearly independent, the two-dimensional subspace spanned by $\{v^a, w^a\}$ is invariant under $(L^a{}_b)$ as well as its orthogonal complement, and when $\{v^a, w^a\}$ is linearly dependent, $(L^a{}_b)$ or $-(L^a{}_b)$ is a reflection in a three-dimensional hyperplane passing through the origin.

Alternative Basis

Apart from the real Infeld–van der Waerden symbols (1.35), another advantageous choice is given by the complex scalars (1.37) because they satisfy $\overline{\sigma^a{}_{A\dot{A}}} = \eta_{AB}\eta_{\dot{A}\dot{B}}\sigma^{aB\dot{B}}$, with η_{AB} and $\eta_{\dot{A}\dot{B}}$ defined as in (1.39),

$$(\eta_{AB}) \equiv \begin{pmatrix} 1 & 0 \\ 0 & -1 \end{pmatrix} \equiv (\eta_{\dot{A}\dot{B}}).$$

Then, the entries of the orthogonal transformation $(L^a{}_b)$ given by (1.102) are real if and only if

$$\sigma^a{}_{A\dot{A}} \sigma_b{}^{B\dot{B}} K^A{}_B M^{\dot{A}}{}_{\dot{B}} = \eta_{AC}\eta_{\dot{A}\dot{C}}\sigma^{aC\dot{C}}\eta^{BD}\eta^{\dot{B}\dot{D}}\sigma_{bD\dot{D}}\overline{K^A{}_B M^{\dot{A}}{}_{\dot{B}}},$$

where, following the general rules (1.22),

$$(\eta^{AB}) = \begin{pmatrix} -1 & 0 \\ 0 & 1 \end{pmatrix} = (\eta^{\dot{A}\dot{B}}).$$

Thus

$$K_C{}^D M_{\dot{C}}{}^{\dot{D}} = \eta_{AC}\eta_{\dot{A}\dot{C}}\eta^{BD}\eta^{\dot{B}\dot{D}}\overline{K^A{}_B M^{\dot{A}}{}_{\dot{B}}},$$

i.e., $K_C{}^D = \lambda\eta_{AC}\eta^{BD}\overline{K^A{}_B}$ and $M_{\dot{C}}{}^{\dot{D}} = \lambda^{-1}\eta_{\dot{A}\dot{C}}\eta^{\dot{B}\dot{D}}\overline{M^{\dot{A}}{}_{\dot{B}}}$, for some $\lambda \in \mathbb{C}$. Again, since $\det(K^A{}_B)$ and $\det(M^{\dot{A}}{}_{\dot{B}})$ are equal to 1, $\lambda = \pm 1$. Hence, making use of (1.107), one obtains the conditions

$$\eta_{CA}K^C{}_S\overline{K^A{}_R} = \pm\eta_{SR}, \qquad \eta_{\dot{C}\dot{A}}M^{\dot{C}}{}_{\dot{S}}\overline{M^{\dot{A}}{}_{\dot{R}}} = \pm\eta_{\dot{S}\dot{R}}. \tag{1.177}$$

The unimodular 2×2 complex matrices $(K^A{}_B)$ or $(M^{\dot{A}}{}_{\dot{B}})$ satisfying (1.177) with the positive sign form a group with the usual matrix multiplication that is denoted by SU(1,1). Since the pairs $\big((K^A{}_B),(M^{\dot{A}}{}_{\dot{B}})\big)$ and $\big(-(K^A{}_B),-(M^{\dot{A}}{}_{\dot{B}})\big)$ give the same $\mathrm{SO}(2,2)_0$ transformation, $\mathrm{SO}(2,2)_0$ can be identified with $\big(\mathrm{SU}(1,1)\times\mathrm{SU}(1,1)\big)/\mathbb{Z}_2$.

The components of the spinor equivalent $v_{A\dot{B}}$ of a vector v_a computed according to (1.42) with the aid of the real Infeld–van der Waerden symbols (1.35) are related to the components of the spinor equivalent $\widetilde{v}_{A\dot{B}}$ of the same vector computed using the complex Infeld–van der Waerden symbols (1.37), which will be denoted by $\widetilde{\sigma}^a{}_{A\dot{B}}$, by means of [see (1.31)]

$$\widetilde{v}_{A\dot{B}} = \frac{1}{\sqrt{2}}\widetilde{\sigma}^a{}_{A\dot{B}}v_a = \frac{1}{\sqrt{2}}\overset{0}{K}{}^C{}_A\overset{0}{M}{}^{\dot{D}}{}_{\dot{B}}\sigma^a{}_{C\dot{D}}v_a = \overset{0}{K}{}^C{}_A\overset{0}{M}{}^{\dot{D}}{}_{\dot{B}}v_{C\dot{D}}, \tag{1.178}$$

with $(\overset{0}{K}{}^A{}_B)$ and $(\overset{0}{M}{}^{\dot{A}}{}_{\dot{B}})$ given by (1.36). Consistently with this relationship, we shall assume that the components of one-index spinors are related by

$$\widetilde{\phi}_A = \overset{0}{K}{}^B{}_A\phi_B, \qquad \widetilde{\psi}_{\dot{A}} = \overset{0}{M}{}^{\dot{B}}{}_{\dot{A}}\psi_{\dot{B}}, \tag{1.179}$$

where $\widetilde{\phi}_A$, $\widetilde{\psi}_{\dot{A}}$ are the components with respect to the spinor basis associated with the complex Infeld–van der Waerden symbols (1.37), and ϕ_A, $\psi_{\dot{A}}$ are the components with respect to the basis associated with the real Infeld–van der Waerden symbols (1.35).

Given the components of a one-index spinor ϕ_A with respect to the basis associated with the real Infeld–van der Waerden symbols (1.35), the operation $\phi_A \mapsto \overline{\phi_A}$ produces another one-index spinor, which has been denoted by $\widehat{\phi}_A$ [see (1.170)]. According to (1.179), the components of $\widehat{\phi}_A$ with respect to the spinor basis associated with the complex Infeld–van der Waerden symbols (1.37) are given by $\widetilde{\widehat{\phi}}_A = \overset{0}{K}{}^B{}_A\widehat{\phi}_B = \overset{0}{K}{}^B{}_A\overline{\phi_B}$, but $\phi_B = (\overset{0}{K}{}^{-1})^C{}_B\widetilde{\phi}_C = -\overset{0}{K}{}_B{}^C\widetilde{\phi}_C$, and hence

$$\widetilde{\widehat{\phi}}_A = \overset{0}{K}{}^B{}_A\overline{(-\overset{0}{K}{}_B{}^C\widetilde{\phi}_C)} = (\overset{0}{K}{}^{-1})_A{}^B\overline{\overset{0}{K}{}_B{}^C\widetilde{\phi}_C}.$$

On the other hand, an explicit computation using (1.36) shows that

$$\overline{\overset{0}{K}{}^{A}{}_{B}} = -\mathrm{i}\varepsilon_{BC}\eta^{CD}\,\overset{0}{K}{}^{A}{}_{D}, \tag{1.180}$$

and therefore

$$\widetilde{\widehat{\phi}}_A = (\overset{0}{K}{}^{-1})_A{}^B(-\mathrm{i}\varepsilon^{CD}\eta_{DE}\,\overset{0}{K}{}_B{}^E)\overline{\widetilde{\phi}_C} = -\mathrm{i}\eta_{AD}\overline{\widetilde{\phi}^D}.$$

Hence, dropping the tildes, with respect to the basis associated with the complex Infeld–van der Waerden symbols (1.37), the mate (conjugate or adjoint) of α_A has to be defined by

$$\widehat{\alpha}_A \equiv -\mathrm{i}\,\eta_{AB}\overline{\alpha^B} \tag{1.181}$$

(i.e., $\widehat{\alpha}_1 = \mathrm{i}\overline{\alpha_2}$, $\widehat{\alpha}_2 = \mathrm{i}\overline{\alpha_1}$) [cf. (1.113)]. One can verify that as a consequence of (1.177), if α_A is an eigenspinor of an SU(1,1) matrix $(K^A{}_B)$, then so is its mate $\widehat{\alpha}_A$.

Depending on the value of the trace $K^A{}_A$, an SU(1,1) matrix $(K^A{}_B)$ has one of the forms (1.173). (Note that the trace $K^A{}_A$ is real.)

Since $(\overset{0}{M}{}^{\dot{A}}{}_{\dot{B}})$ is the complex conjugate of $(\overset{0}{K}{}^{A}{}_{B})$, from (1.179) and (1.180) one finds that with respect to the basis associated with the complex Infeld–van der Waerden symbols (1.37), the definition of the mate (conjugate or adjoint) of a spinor with one dotted index (1.175) is equivalent to

$$\widehat{\alpha}_{\dot{A}} \equiv \mathrm{i}\,\eta_{\dot{A}\dot{B}}\overline{\alpha^{\dot{B}}}. \tag{1.182}$$

Spin Transformations

Given two orthonormal bases $\{\mathbf{e}_1,\mathbf{e}_2,\mathbf{e}_3,\mathbf{e}_4\}$ and $\{\mathbf{e}'_1,\mathbf{e}'_2,\mathbf{e}'_3,\mathbf{e}'_4\}$ satisfying (1.2), any vector $v \in V$ can be expressed in the form $v = v^a\mathbf{e}_a$ and $v = v'^a\mathbf{e}'_a$, where the v^a and v'^a are real numbers related by the "transformation law"

$$v'^a = L^a{}_b v^b, \tag{1.183}$$

which follows from (1.3), with $(L^a{}_b) \in \mathrm{O}(p,q)$. Hence, if $(L^a{}_b) \in \mathrm{SO}(p,q)_0$ [the connected component of the identity in SO(p,q)], the spinor equivalents of v^a and v'^a, denoted by $v^{A\dot{B}}$ and $v'^{A\dot{B}}$, respectively, are related through [see (1.102)]

$$v'^{A\dot{B}} = K^A{}_C M^{\dot{B}}{}_{\dot{D}} v^{C\dot{D}}, \tag{1.184}$$

where

$$\begin{cases} (K^A{}_B), (M^{\dot{A}}{}_{\dot{B}}) \in \mathrm{SU}(2) & \text{if the } \sigma^a{}_{A\dot{B}} \text{ are given by (1.33)},\\ (K^A{}_B) \in \mathrm{SL}(2,\mathbb{C}), (M^{\dot{A}}{}_{\dot{B}}) = \overline{(K^A{}_B)} & \text{if the } \sigma^a{}_{A\dot{B}} \text{ are given by (1.34)},\\ (K^A{}_B), (M^{\dot{A}}{}_{\dot{B}}) \in \mathrm{SL}(2,\mathbb{R}) & \text{if the } \sigma^a{}_{A\dot{B}} \text{ are given by (1.35)},\\ (K^A{}_B), (M^{\dot{A}}{}_{\dot{B}}) \in \mathrm{SU}(1,1) & \text{if the } \sigma^a{}_{A\dot{B}} \text{ are given by (1.37)}. \end{cases} \tag{1.185}$$

Similarly, since the components of a tensor with respect to the bases $\{\mathbf{e}_1, \mathbf{e}_2, \mathbf{e}_3, \mathbf{e}_4\}$ and $\{\mathbf{e}'_1, \mathbf{e}'_2, \mathbf{e}'_3, \mathbf{e}'_4\}$ are related by $t'^{ab\ldots} = L^a{}_c L^b{}_d \cdots t^{cd\ldots}$, their spinor equivalents are related by means of

$$t'^{AB\ldots\dot{A}\dot{B}\ldots} = K^A{}_C M^{\dot{A}}{}_{\dot{C}} K^B{}_D M^{\dot{B}}{}_{\dot{D}} \cdots t^{CD\ldots\dot{C}\dot{D}\ldots}. \tag{1.186}$$

The transformations (1.184) and (1.186) can be seen as consequences of the *spin transformations*

$$\psi'^A = K^A{}_C \psi^C, \qquad \phi'^{\dot{A}} = M^{\dot{A}}{}_{\dot{C}} \phi^{\dot{C}} \tag{1.187}$$

on the one-index spinors. Whereas in the transformations (1.184) and (1.186) the two pairs of matrices $\big((K^A{}_B), (M^{\dot{A}}{}_{\dot{B}})\big)$ and $\big(-(K^A{}_B), -(M^{\dot{A}}{}_{\dot{B}})\big)$, corresponding to a given $\mathrm{SO}(p,q)_0$ transformation, yield the same result (as it should), the effect of the spin transformations (1.187) does depend on which pair of matrices restricted by (1.185) we choose.

1.4 Reflections

A special class of orthogonal transformations is that of reflections on hyperplanes passing through the origin. A nonnull vector N_a defines a three-dimensional subspace of V formed by the vectors orthogonal to N_a. The 4×4 real matrix given by

$$L^a{}_b = \delta^a_b - \frac{2N^a N_b}{N^c N_c} \tag{1.188}$$

corresponds to the reflection on the hyperplane passing through the origin orthogonal to N_a and is an orthogonal transformation with negative determinant, as can be verified using (1.4) and (1.100) or, in a simpler manner, by comparing (1.190), below, with (1.98). Making use of (1.42), (1.21), and (1.46), one finds that the spinor equivalent of (1.188) is

$$L^{A\dot{A}}{}_{B\dot{B}} = -\delta^A_B \delta^{\dot{A}}_{\dot{B}} + \frac{2N^{A\dot{A}} N_{B\dot{B}}}{N^{C\dot{C}} N_{C\dot{C}}}, \tag{1.189}$$

and then, by virtue of the identity [see (1.28) and (1.23)]

$$\begin{aligned}
N^{A\dot{A}} N_{B\dot{B}} &= N^{A\dot{A}} N_{B\dot{B}} - N^A{}_{\dot{B}} N_B{}^{\dot{A}} + N^A{}_{\dot{B}} N_B{}^{\dot{A}} \\
&= N^{A\dot{R}} N_{B\dot{R}} \delta^{\dot{A}}_{\dot{B}} + N^A{}_{\dot{B}} N_B{}^{\dot{A}} \\
&= \tfrac{1}{2} (N^{A\dot{R}} N_{B\dot{R}} - N_B{}^{\dot{R}} N^A{}_{\dot{R}}) \delta^{\dot{A}}_{\dot{B}} + N^A{}_{\dot{B}} N_B{}^{\dot{A}} \\
&= \tfrac{1}{2} \delta^A_B \delta^{\dot{A}}_{\dot{B}} N^{S\dot{R}} N_{S\dot{R}} + N^A{}_{\dot{B}} N_B{}^{\dot{A}},
\end{aligned}$$

one obtains the simple expression

$$L^{A\dot{A}}{}_{B\dot{B}} = \frac{2N^{A}{}_{\dot{B}} N_{B}{}^{\dot{A}}}{N^{C\dot{C}} N_{C\dot{C}}}, \tag{1.190}$$

which is of the form (1.98).

The composition of any two reflections on hyperplanes passing through the origin (or of any even number of such reflections) is an orthogonal transformation with unit determinant. In a three-dimensional space, with definite or indefinite metric tensor, any orthogonal transformation with unit determinant is the composition of two reflections (see, e.g., Cartan 1966, Porteous 1995, Lounesto 1997, Torres del Castillo 2003); however, in a four-dimensional space, a similar result does not hold. In fact, the vectors belonging to the two-dimensional subspace formed by the intersection of two different hyperplanes passing through the origin are fixed points of the composition of the reflections on these hyperplanes. Thus, the composition of the reflections in two hyperplanes passing through the origin is a proper orthogonal transformation with a double eigenvalue 1, while, as we have seen, not every proper orthogonal transformation has two eigenvalues equal to 1.

For instance, if the metric tensor of V is positive definite, the spinor equivalent of the composition of the reflections on the hyperplanes orthogonal to the unit vectors N^a and $\widetilde{N}^a$ is [see (1.190); an extra minus sign comes from (1.46)]

$$L^{A\dot{A}}{}_{B\dot{B}} = -\big(-2N^{A}{}_{\dot{C}} N_{C}{}^{\dot{A}}\big)\big(-2\widetilde{N}^{C}{}_{\dot{B}} \widetilde{N}_{B}{}^{\dot{C}}\big) = -4N^{A}{}_{\dot{C}} \widetilde{N}_{B}{}^{\dot{C}} N_{C}{}^{\dot{A}} \widetilde{N}^{C}{}_{\dot{B}},$$

which corresponds to an SO(4) transformation of the form (1.102) with $K^{A}{}_{B} = 2N^{A}{}_{\dot{C}} \widetilde{N}_{B}{}^{\dot{C}}$ and $M^{\dot{A}}{}_{\dot{B}} = 2N_{C}{}^{\dot{A}} \widetilde{N}^{C}{}_{\dot{B}}$. Then, $K^{A}{}_{A} = -2N^{A\dot{C}} \widetilde{N}_{A\dot{C}}$, and $M^{\dot{A}}{}_{\dot{A}} = -2N^{C\dot{A}} \widetilde{N}_{C\dot{A}}$, i.e., $K^{A}{}_{A} = M^{\dot{A}}{}_{\dot{A}}$, which means that the composition of two reflections is a simple SO(4) transformation (see Section 1.3.1). In fact, the converse is also true; we can see explicitly that any simple SO(4) transformation is the composition of two reflections in hyperplanes passing through the origin. Letting

$$\begin{aligned} N_{A\dot{B}} &= \frac{1}{\sqrt{2}}\big(e^{i\theta/4} \alpha_A \gamma_{\dot{B}} - e^{-i\theta/4} \widehat{\alpha}_A \widehat{\gamma}_{\dot{B}}\big), \\ \widetilde{N}_{A\dot{B}} &= \frac{1}{\sqrt{2}}\big(e^{-i\theta/4} \alpha_A \gamma_{\dot{B}} - e^{i\theta/4} \widehat{\alpha}_A \widehat{\gamma}_{\dot{B}}\big), \end{aligned}$$

assuming that $\alpha^R \widehat{\alpha}_R = 1$ and $\gamma^{\dot{R}} \widehat{\gamma}_{\dot{R}} = 1$, one finds that $2N^{A}{}_{\dot{C}} \widetilde{N}_{B}{}^{\dot{C}} = e^{i\theta/2} \alpha^A \widehat{\alpha}_B - e^{-i\theta/2} \widehat{\alpha}^A \alpha_B$ and $2N_{C}{}^{\dot{A}} \widetilde{N}^{C}{}_{\dot{B}} = e^{i\theta/2} \gamma^{\dot{A}} \widehat{\gamma}_{\dot{B}} - e^{-i\theta/2} \widehat{\gamma}^{\dot{A}} \gamma_{\dot{B}}$, which are of the form (1.114) and (1.116) with $\varphi = \theta$, and letting

$$\begin{aligned} N_{A\dot{B}} &= \frac{1}{\sqrt{2}}\big(e^{i\theta/4} \alpha_A \widehat{\gamma}_{\dot{B}} + e^{-i\theta/4} \widehat{\alpha}_A \gamma_{\dot{B}}\big), \\ \widetilde{N}_{A\dot{B}} &= \frac{1}{\sqrt{2}}\big(e^{i\theta/4} \widehat{\alpha}_A \gamma_{\dot{B}} + e^{-i\theta/4} \alpha_A \widehat{\gamma}_{\dot{B}}\big), \end{aligned}$$

one finds that $2N^{A}{}_{\dot{C}} \widetilde{N}_{B}{}^{\dot{C}} = e^{i\theta/2} \alpha^A \widehat{\alpha}_B - e^{-i\theta/2} \widehat{\alpha}^A \alpha_B$ and $2N_{C}{}^{\dot{A}} \widetilde{N}^{C}{}_{\dot{B}} = e^{-i\theta/2} \gamma^{\dot{A}} \widehat{\gamma}_{\dot{B}} - e^{i\theta/2} \widehat{\gamma}^{\dot{A}} \gamma_{\dot{B}}$, which are of the form (1.114) and (1.116) with $\varphi = -\theta$. In both cases, the vector equivalents N^a and $\widetilde{N}^a$ of $N^{A\dot{B}}$ and $\widetilde{N}^{A\dot{B}}$, respectively, are real unit vectors and satisfy $N^a\widetilde{N}_a = \cos\frac{1}{2}\theta$.

As shown in Section 1.3.1, any SO(4) transformation is simple or is the composition of two simple SO(4) transformations, and therefore, any SO(4) transformation can be obtained through the composition of two or four reflections.

On the other hand, the spinor equivalent of the composition of an arbitrary improper O(4) transformation given by (1.128), followed by a reflection given by (1.190), is

$$\frac{2N^A{}_{\dot{B}}N_B{}^{\dot{A}}}{N^{R\dot{R}}N_{R\dot{R}}}K^B{}_{\dot{C}}M^{\dot{B}}{}_C = -\frac{(-2N^A{}_{\dot{B}}w_C{}^{\dot{B}})}{\sqrt{-N^{R\dot{R}}N_{R\dot{R}}}}\frac{(2N_B{}^{\dot{A}}v^B{}_{\dot{C}})}{\sqrt{-N^{R\dot{R}}N_{R\dot{R}}}},$$

where we have made use of the definitions (1.129). This composition corresponds to a simple SO(4) transformation if, for instance, we choose $N_{A\dot{B}}$ orthogonal to $v_{A\dot{B}}$ and $w_{A\dot{B}}$, since the traces

$$\frac{(-2N^A{}_{\dot{B}}w_A{}^{\dot{B}})}{\sqrt{-N^{R\dot{R}}N_{R\dot{R}}}} \qquad \text{and} \qquad \frac{(2N_B{}^{\dot{A}}v^B{}_{\dot{C}})}{\sqrt{-N^{R\dot{R}}N_{R\dot{R}}}}$$

coincide. In effect,

$$-2N^A{}_{\dot{B}}w_A{}^{\dot{B}} = 0 \qquad \text{and} \qquad 2N_B{}^{\dot{A}}v^B{}_{\dot{A}} = 0.$$

(Alternatively, if $v_{A\dot{B}} \neq w_{A\dot{B}}$, one can choose $N_{A\dot{B}} = v_{A\dot{B}} - w_{A\dot{B}}$.) Hence, using the fact that the square of a reflection is equal to the identity, we conclude that any improper O(4) transformation is the composition of a simple SO(4) transformation followed by a reflection, or equivalently, any improper O(4) transformation is a reflection or can be expressed as the composition of three reflections.

Now we shall consider only the $\mathrm{SO}^{\uparrow}(3,1)$ transformations, showing that any simple orthochronous proper Lorentz transformation can be expressed as the composition of two reflections. Hence, any orthochronous proper Lorentz transformation is the composition of two or four reflections.

The Lorentz boosts, given by (1.136) or (1.137), have a double eigenvalue equal to 1, and these transformations can be expressed as the composition of two reflections on hyperplanes passing through the origin. The vector equivalents of

$$N_{A\dot{B}} = \frac{1}{\sqrt{2}}(\mathrm{e}^{-\zeta/4}\alpha_A\overline{\alpha}_{\dot{B}} + \mathrm{e}^{\zeta/4}\beta_A\overline{\beta}_{\dot{B}}),$$

$$\widetilde{N}_{A\dot{B}} = \frac{1}{\sqrt{2}}(\mathrm{e}^{\zeta/4}\alpha_A\overline{\alpha}_{\dot{B}} + \mathrm{e}^{-\zeta/4}\beta_A\overline{\beta}_{\dot{B}}),$$

with α_A and β_A defined by (1.141) and (1.142), are two real vectors N_a and $\widetilde{N}_a$ (i.e., $\overline{N_{A\dot{B}}} = N_{B\dot{A}}$ and $\overline{\widetilde{N}_{A\dot{B}}} = \widetilde{N}_{B\dot{A}}$) such that $N^aN_a = -1$, $\widetilde{N}^a\widetilde{N}_a = -1$, and $\widetilde{N}^aN_a =$

$-\cosh\frac{1}{2}\zeta$. Making use of (1.190), one finds that the composition of the reflections in the hyperplanes orthogonal to N^a and $\widetilde{N}^a$ is determined by

$$\begin{aligned} L^{A\dot{A}}{}_{B\dot{B}} &= -(2N^A{}_{\dot{C}}N_C{}^{\dot{A}})(2\widetilde{N}^C{}_{\dot{B}}\widetilde{N}_B{}^{\dot{C}}) \\ &= -(2N^A{}_{\dot{C}}\widetilde{N}_B{}^{\dot{C}})(2N_C{}^{\dot{A}}\widetilde{N}^C{}_{\dot{B}}) \\ &= -(2N^A{}_{\dot{C}}\widetilde{N}_B{}^{\dot{C}})\overline{(2N^A{}_{\dot{C}}\widetilde{N}_B{}^{\dot{C}})}, \end{aligned} \tag{1.191}$$

but

$$\begin{aligned} 2N^A{}_{\dot{C}}\widetilde{N}_B{}^{\dot{C}} &= (\mathrm{e}^{-\zeta/4}\alpha^A\overline{\alpha}_{\dot{C}} + \mathrm{e}^{\zeta/4}\beta^A\overline{\beta}_{\dot{C}})(\mathrm{e}^{\zeta/4}\alpha_B\overline{\alpha}^{\dot{C}} + \mathrm{e}^{-\zeta/4}\beta_B\overline{\beta}^{\dot{C}}) \\ &= -(\mathrm{e}^{-\zeta/2}\alpha^A\beta_B - \mathrm{e}^{\zeta/2}\beta^A\alpha_B), \end{aligned}$$

and therefore (1.191) coincides with (1.143).

Similarly, any orthochronous proper Lorentz transformation that represents a rotation is the composition of two reflections. For instance, the vector equivalents of

$$\begin{aligned} N_{A\dot{B}} &= \frac{1}{\sqrt{2}}(\mathrm{e}^{\mathrm{i}\theta/4}\alpha_A\overline{\beta}_{\dot{B}} + \mathrm{e}^{-\mathrm{i}\theta/4}\beta_A\overline{\alpha}_{\dot{B}}), \\ \widetilde{N}_{A\dot{B}} &= \frac{1}{\sqrt{2}}(\mathrm{e}^{-\mathrm{i}\theta/4}\alpha_A\overline{\beta}_{\dot{B}} + \mathrm{e}^{\mathrm{i}\theta/4}\beta_A\overline{\alpha}_{\dot{B}}), \end{aligned}$$

with $\alpha^A\beta_A = 1$, are two real vectors N_a and $\widetilde{N}_a$ such that $N^aN_a = 1$, $\widetilde{N}^a\widetilde{N}_a = 1$, and $N^a\widetilde{N}_a = \cos\frac{1}{2}\theta$; furthermore, $2N^A{}_{\dot{C}}\widetilde{N}_B{}^{\dot{C}} = \mathrm{e}^{\mathrm{i}\theta/2}\alpha^A\beta_B - \mathrm{e}^{-\mathrm{i}\theta/2}\beta^A\alpha_B$, which coincides with (1.150) when $\lambda = \mathrm{e}^{\mathrm{i}\theta/2}$.

The orthochronous proper Lorentz transformation corresponding to the $\mathrm{SL}(2,\mathbb{C})$ matrix (1.152) is equal to the composition of the reflections on the hyperplanes orthogonal to the vectors given by

$$\begin{aligned} N_{A\dot{B}} &= \frac{1}{\sqrt{2}}(\alpha_A\overline{\beta}_{\dot{B}} + \beta_A\overline{\alpha}_{\dot{B}}), \\ \widetilde{N}_{A\dot{B}} &= \pm\frac{1}{\sqrt{2}}(-\tfrac{1}{2}\zeta\alpha_A\overline{\alpha}_{\dot{B}} + \alpha_A\overline{\beta}_{\dot{B}} + \beta_A\overline{\alpha}_{\dot{B}}), \end{aligned}$$

where β_A is a one-index spinor such that $\alpha^A\beta_A = 1$. In fact, one finds that $2N^A{}_{\dot{C}}\widetilde{N}_B{}^{\dot{C}} = \pm(\delta^A_B - \frac{1}{2}\zeta\alpha^A\alpha_B)$.

Orthogonal Projection on a Hyperplane

Given a vector v^a, the decomposition into symmetric and antisymmetric parts

$$N_B{}^{\dot{R}}v_{A\dot{R}} = N_{(B}{}^{\dot{R}}v_{A)\dot{R}} + N_{[B}{}^{\dot{R}}v_{A]\dot{R}} = N_{(B}{}^{\dot{R}}v_{A)\dot{R}} + \tfrac{1}{2}\varepsilon_{BA}N^{R\dot{R}}v_{R\dot{R}},$$

which, contracted with $N^B{}_{\dot{A}}$, is equivalent to

$$N^B{}_{\dot{A}}N_B{}^{\dot{R}}v_{A\dot{R}} = N^B{}_{\dot{A}}N_{(B}{}^{\dot{R}}v_{A)\dot{R}} + \tfrac{1}{2}\varepsilon_{BA}N^B{}_{\dot{A}}N^{R\dot{R}}v_{R\dot{R}}$$

or

$$-\frac{1}{2}N^{B\dot{B}}N_{B\dot{B}}\delta^{\dot{R}}_{\dot{A}}v_{A\dot{R}} = \frac{1}{\sqrt{2}}(N^{C\dot{C}}N_{C\dot{C}})^{1/2}N^B{}_{\dot{A}}v^{\perp}{}_{BA} - \frac{1}{2}N^{R\dot{R}}v_{R\dot{R}}N_{A\dot{A}},$$

where

$$v^{\perp}{}_{BA} \equiv \frac{\sqrt{2}N_{(B}{}^{\dot{R}}v_{A)\dot{R}}}{(N^{C\dot{C}}N_{C\dot{C}})^{1/2}}, \tag{1.192}$$

allows us to express (the spinor components of) the vector v^a as the sum of a vector orthogonal to N^a and a vector proportional to N^a,

$$v_{A\dot{A}} = -\frac{\sqrt{2}N^B{}_{\dot{A}}v^{\perp}{}_{AB}}{(N^{C\dot{C}}N_{C\dot{C}})^{1/2}} + \frac{N^{B\dot{B}}v_{B\dot{B}}N_{A\dot{A}}}{N^{C\dot{C}}N_{C\dot{C}}}.$$

(The first term on the right-hand side of the last expression corresponds to a vector orthogonal to N^a, since $N^{A\dot{A}}(N^B{}_{\dot{A}}v^{\perp}{}_{AB}) = \frac{1}{2}N^{R\dot{A}}N_{R\dot{A}}\varepsilon^{AB}v^{\perp}{}_{AB}$, which is equal to zero because of the symmetry of $v^{\perp}{}_{AB}$.) Hence, using the fact that $v^{\perp}{}_{BA}$ is symmetric, the orthogonal projection of v^a on the hyperplane orthogonal to N^a is the vector

$$\frac{\sqrt{2}N^B{}_{\dot{A}}v^{\perp}{}_{AB}\,\mathbf{e}^{A\dot{A}}}{(N^{C\dot{C}}N_{C\dot{C}})^{1/2}} = -v^{\perp AB}\frac{\sqrt{2}N_{(B}{}^{\dot{A}}\,\mathbf{e}_{A)\dot{A}}}{(N^{C\dot{C}}N_{C\dot{C}})^{1/2}},$$

and therefore, a basis of (the complexification of) this hyperplane is formed by the three vectors

$$\mathbf{e}_{AB} \equiv \frac{\sqrt{2}N_{(A}{}^{\dot{A}}\,\mathbf{e}_{B)\dot{A}}}{(N^{C\dot{C}}N_{C\dot{C}})^{1/2}}. \tag{1.193}$$

The vectors $\mathbf{e}_{AB}$ need not be real, and the explicit relationship between the vectors $\mathbf{e}_{AB}$ and their complex conjugates depends on the choice of the Infeld–van der Waerden symbols.

The map that sends an arbitrary vector $v = -v^{A\dot{B}}\mathbf{e}_{A\dot{B}}$ into its orthogonal projection on the hyperplane orthogonal to N^a, $-v^{\perp AB}\mathbf{e}_{AB}$, can be extended to higher-rank tensors assuming that this map is linear and preserves tensor products, keeping in mind that the components of the projected objects are referred to a basis induced by $\{\mathbf{e}_{AB}\}$ (see also Sommers 1980, Sen 1982, Shaw 1983a,b, Ashtekar 1987, 1991, Torres del Castillo 2003). (In a three-dimensional space, with a definite or indefinite metric, only one type of spinor indices is necessary.)

1.5 Clifford Algebra. Dirac Spinors

Spinors are frequently introduced starting from the study of the Clifford algebras (see, e.g., Penrose and Rindler 1986, Lawson and Michelsohn 1989, Chevalley 1996,

Lounesto 1997). In this section we will show how the one-index spinors are related to the Clifford algebras considering only the case of four-dimensional spaces.

As in the preceding sections, g_{ab} will denote the components of the metric tensor of V with respect to an orthonormal basis, and we shall assume that (g_{ab}) is of one of the forms (1.1). The Clifford algebra associated with V can be defined as the associative algebra generated by $1, \gamma_1, \gamma_2, \gamma_3, \gamma_4$, which satisfy the relations

$$\gamma_a \gamma_b + \gamma_b \gamma_a = 2 g_{ab}, \tag{1.194}$$

and 1 is the unit element. The elements of the algebra are the linear combinations with complex coefficients of $1, \gamma_1, \gamma_2, \gamma_3, \gamma_4$, and their products.

The anticommutation relations (1.194) appear in connection with the Dirac equation for the electron. The relation $E^2/c^2 - \mathbf{p}^2 = m_0^2 c^2$ between the relativistic energy E and the relativistic momentum $\mathbf{p}$ of a particle with rest mass m_0, leads to the Klein–Gordon equation

$$\left(g^{ab} \frac{\partial}{\partial x^a} \frac{\partial}{\partial x^b} - \frac{m_0^2 c^2}{\hbar^2} \right) \psi = 0, \tag{1.195}$$

where $(g^{ab}) = \operatorname{diag}(1,1,1,-1)$ and the x^a are Cartesian space-time coordinates associated with an inertial reference frame, when E and $\mathbf{p}$ are replaced by the operators $\mathrm{i}\hbar\partial/\partial t$ and $-\mathrm{i}\hbar\nabla$, respectively, following the standard rules of quantum mechanics. In contrast with the Schrödinger equation, the Klein–Gordon equation involves second derivatives with respect to time and does not lead to a conserved positive definite probability density. The Dirac equation is obtained by looking for a factorization of the operator appearing in the Klein–Gordon equation (1.195) of the form

$$\left(g^{ab} \frac{\partial}{\partial x^a} \frac{\partial}{\partial x^b} - \frac{m_0^2 c^2}{\hbar^2} \right) = \left(\gamma^a \frac{\partial}{\partial x^a} + \frac{m_0 c}{\hbar} \right) \left(\gamma^b \frac{\partial}{\partial x^b} - \frac{m_0 c}{\hbar} \right),$$

which requires the validity of (1.194). (The index of γ_a is raised following the usual rule with the aid of g^{ab}.) The Dirac equation can be expressed as

$$\left(\gamma^a \frac{\partial}{\partial x^a} - \frac{m_0 c}{\hbar} \right) \psi = 0. \tag{1.196}$$

We are interested in finding an irreducible representation of the Clifford algebra, that is, we want to consider the elements of the algebra as linear operators on some complex vector space $\mathfrak{D}$ in such a way that there are no proper subspaces of $\mathfrak{D}$ that are invariant under all these operators. As we shall show, there is an infinite number of equivalent irreducible representations of the algebra in a complex four-dimensional vector space $\mathfrak{D}$, and finding each of these representations amounts to finding a set of Infeld–van der Waerden symbols.

With the components of the metric tensor given by (1.1), the basic relations (1.194) yield

$$(\gamma_a)^2 = \pm I_N, \tag{1.197}$$

where I_N denotes the identity of $\mathfrak{D}$, and

$$\gamma_a \gamma_b = -\gamma_b \gamma_a, \qquad \text{for } a \neq b. \tag{1.198}$$

From (1.197) it follows that each operator γ_a is invertible, ${\gamma_a}^{-1} = \pm\gamma_a$, and from (1.198) one obtains $\det\gamma_a \det\gamma_b = (-1)^N \det\gamma_b \det\gamma_a$, where N denotes the dimension of $\mathfrak{D}$. Hence, N must be even.

Defining

$$\gamma_5 \equiv \mathrm{i}^q \gamma_1 \gamma_2 \gamma_3 \gamma_4, \tag{1.199}$$

where, as above, q is the number of (-1)'s appearing in the diagonal matrix (g_{ab}), with the aid of (1.197) and (1.198) one finds that

$$\gamma_5 \gamma_a = -\gamma_a \gamma_5 \tag{1.200}$$

and $\gamma_5^2 = (-1)^q(\gamma_1)^2(\gamma_2)^2(\gamma_3)^2(\gamma_4)^2 = I_N$. Furthermore, from (1.200) we have $\gamma_5 = -\gamma_a\gamma_5{\gamma_a}^{-1}$ (without summation on a); hence, using the fact that $\operatorname{tr}(AB) = \operatorname{tr}(BA)$, we obtain $\operatorname{tr}\gamma_5 = -\operatorname{tr}(\gamma_a\gamma_5{\gamma_a}^{-1}) = -\operatorname{tr}\gamma_5$, i.e., $\operatorname{tr}\gamma_5 = 0$. The eigenvalues of γ_5 are equal to $+1$ or -1 and from the condition $\operatorname{tr}\gamma_5 = 0$ it follows that there are as many $+1$'s as -1's (which also implies that $\dim\mathfrak{D}$ is even). Hence, there is a basis of $\mathfrak{D}$ formed by eigenvectors of γ_5, with respect to which γ_5 is represented by the matrix

$$\begin{pmatrix} I_{N/2} & 0 \\ 0 & -I_{N/2} \end{pmatrix},$$

where $I_{N/2}$ is the $N/2 \times N/2$ unit matrix. Writing the matrix corresponding to γ_a in block form

$$\begin{pmatrix} A_a & B_a \\ C_a & D_a \end{pmatrix},$$

where A_a, B_a, C_a, and D_a are $N/2 \times N/2$ matrices, from (1.200) one finds that $A_a = D_a = 0$, and (1.194) amounts to

$$B_a C_b + B_b C_a = 2g_{ab} I_{N/2}, \qquad C_a B_b + C_b B_a = 2g_{ab} I_{N/2}. \tag{1.201}$$

These equations imply that $C_a = g_{aa}(B_a)^{-1}$ (without summation on a). Furthermore, for $a \neq b$, $B_a C_b = -B_b C_a$, and therefore $\det B_a \det C_b = (-1)^{N/2} \det B_b \det C_a$. On the other hand, $\det C_a = (g_{aa})^{N/2}(\det B_a)^{-1}$ (without summation on a); hence

$$\frac{(g_{aa})^{N/2}}{(\det B_a)^2} = (-1)^{N/2} \frac{(g_{bb})^{N/2}}{(\det B_b)^2}, \qquad \text{for } a \neq b,$$

which cannot be satisfied if $N/2$ is odd, and we conclude that N, the dimension of $\mathfrak{D}$, must be a multiple of 4.

Assuming $N = 4$, we have

$$B_a = \begin{pmatrix} x_a & y_a \\ z_a & w_a \end{pmatrix}, \tag{1.202}$$

where for $a = 1,2,3,4$, x_a, y_a, z_a, and w_a are scalars, and hence

$$C_a = \lambda_a \begin{pmatrix} -w_a & y_a \\ z_a & -x_a \end{pmatrix} \qquad \text{(without summation on } a\text{)} \tag{1.203}$$

with

$$\lambda_a \equiv -\frac{g_{aa}}{\det B_a} \qquad \text{(without summation on } a\text{)}.$$

Substituting (1.202) and (1.203) into (1.201), one finds that all λ_a must have a common value, λ (say) (otherwise (g_{ab}) would be equal to a matrix of rank 2 at most), and

$$\lambda(-x_a w_b - w_a x_b + y_a z_b + z_a y_b) = 2g_{ab}. \tag{1.204}$$

By means of an appropriate change of basis of $\mathfrak{D}$ we can eliminate the factor λ appearing in (1.203) and (1.204). In effect, applying the similarity transformation

$$\begin{pmatrix} 0 & B_a \\ C_a & 0 \end{pmatrix} \mapsto \begin{pmatrix} \lambda^{1/2} I & 0 \\ 0 & I \end{pmatrix} \begin{pmatrix} 0 & B_a \\ C_a & 0 \end{pmatrix} \begin{pmatrix} \lambda^{-1/2} I & 0 \\ 0 & I \end{pmatrix} = \begin{pmatrix} 0 & \lambda^{1/2} B_a \\ \lambda^{-1/2} C_a & 0 \end{pmatrix},$$

and redefining $\lambda^{1/2} x_a$, $\lambda^{1/2} y_a$, $\lambda^{1/2} z_a$, and $\lambda^{1/2} w_a$ as x_a, y_a, z_a, and w_a, respectively, one concludes that there is a basis of $\mathfrak{D}$, formed by eigenvectors of γ_5, with respect to which the operators γ_a are represented by

$$\gamma_a = \begin{pmatrix} 0 & B_a \\ C_a & 0 \end{pmatrix} \tag{1.205}$$

with

$$B_a = \begin{pmatrix} x_a & y_a \\ z_a & w_a \end{pmatrix}, \qquad C_a = \begin{pmatrix} -w_a & y_a \\ z_a & -x_a \end{pmatrix}, \tag{1.206}$$

and

$$2g_{ab} = -x_a w_b - w_a x_b + y_a z_b + z_a y_b. \tag{1.207}$$

This last relation implies that the four (possibly complex) vectors with components x_a, y_a, z_a, and w_a form essentially a null tetrad as defined in Section 1.2; i.e., they are null vectors, $x^a x_a = y^a y_a = z^a z_a = w^a w_a = 0$, that form a basis of (the complexification of) V, and the only nonvanishing inner products among them are $x^b w_b = -2$ and $y^b z_b = 2$. In fact, from (1.207) it follows that for any vector v_a,

$$v_a = g_{ab} v^b = \tfrac{1}{2}(-v^b w_b x_a - v^b x_b w_a + v^b z_b y_a + v^b y_b z_a),$$

thus showing that x_a, y_a, z_a, and w_a span the complexification of V, and from this last relation one obtains, e.g.,

$$x_a = \tfrac{1}{2}(-x^b w_b x_a - x^b x_b w_a + x^b z_b y_a + x^b y_b z_a),$$

which shows that $x^b x_b = x^b z_b = x^b y_b = 0$ and $x^b w_b = -2$.

Letting

$$\sigma_{a1\dot{1}} \equiv x_a, \qquad \sigma_{a1\dot{2}} \equiv y_a, \qquad \sigma_{a2\dot{1}} \equiv z_a, \qquad \sigma_{a2\dot{2}} \equiv w_a, \tag{1.208}$$

one finds that (1.207) is equivalent to (1.40), and therefore, finding a representation of the Clifford algebra associated with the metric tensor g_{ab} in a four-dimensional complex vector space $\mathfrak{D}$ is equivalent to finding a null tetrad and also to finding Infeld–van der Waerden symbols for the metric tensor g_{ab}. Thus, according to (1.206) and (1.208) we have

$$B_a = \begin{pmatrix} \sigma_{a1\dot{1}} & \sigma_{a1\dot{2}} \\ \sigma_{a2\dot{1}} & \sigma_{a2\dot{2}} \end{pmatrix}, \qquad C_a = -\begin{pmatrix} \sigma_a{}^{1\dot{1}} & \sigma_a{}^{2\dot{1}} \\ \sigma_a{}^{1\dot{2}} & \sigma_a{}^{2\dot{2}} \end{pmatrix}, \tag{1.209}$$

and in order to get agreement with the notation employed in the preceding sections, the components of an element of $\mathfrak{D}$ with respect to a basis such that (1.205)–(1.207) and (1.209) hold will be denoted by

$$\begin{pmatrix} \psi_A \\ \phi^{\dot{A}} \end{pmatrix}. \tag{1.210}$$

The elements of $\mathfrak{D}$ are called *Dirac spinors* or *bispinors*. Owing to (1.205) and (1.209),

$$\gamma_a \begin{pmatrix} \psi_A \\ \phi^{\dot{A}} \end{pmatrix} = \begin{pmatrix} \sigma_{aA\dot{A}}\phi^{\dot{A}} \\ -\sigma_a{}^{A\dot{A}}\psi_A \end{pmatrix}. \tag{1.211}$$

In order to verify that the representation of the Clifford algebra (1.194) given by (1.205)–(1.207) is irreducible, it suffices to show (according to Schur's lemma) that any operator that commutes with all the linear combinations of the operators (1.205) must be a scalar multiple of the identity. Indeed, such an operator, Δ, say, must commute with γ_5 and therefore with respect to the basis of $\mathfrak{D}$ formed by the eigenvectors of γ_5 previously employed is represented by a matrix of the block form

$$\begin{pmatrix} F & 0 \\ 0 & G \end{pmatrix},$$

where $F = (F_A{}^B)$ and $G = (G^{\dot{A}}{}_{\dot{B}})$ are 2×2 matrices. Then, Δ also commutes with γ_a if and only if $FB_a = B_a G$ and $GC_a = C_a F$, which amounts to

$$F_A{}^B \sigma_{aB\dot{C}} = \sigma_{aA\dot{B}} G^{\dot{B}}{}_{\dot{C}}$$

[see (1.209)]. Then, by contracting with $\sigma^{aR\dot{S}}$ one finds that

$$F_A{}^R \delta^{\dot{S}}_{\dot{C}} = G^{\dot{S}}{}_{\dot{C}} \delta^R_A,$$

which implies that F and G are the same scalar multiple of the identity matrix and therefore Δ is a scalar multiple of the identity.

If $(L^a{}_b)$ is an orthogonal transformation and the operators γ_a satisfy (1.194), then the operators $\gamma'_a \equiv L_a{}^b \gamma_b$ also satisfy (1.194) [cf. (1.183)]; in fact, $\gamma'_a\gamma'_b + \gamma'_b\gamma'_a = L_a{}^c\gamma_c L_b{}^d\gamma_d + L_b{}^d\gamma_d L_a{}^c\gamma_c = L_a{}^c L_b{}^d(\gamma_c\gamma_d + \gamma_d\gamma_c) = 2L_a{}^c L_b{}^d g_{cd} I_N = 2g_{ab}I_N$. Thus, the γ_a and the γ'_a form representations of the Clifford algebra, and as a consequence of (1.96), (1.98), and (1.211), these two representations are equivalent to each other in the sense that there exists a unimodular operator U of $\mathfrak{D}$ onto itself such that $\gamma'_a = U^{-1}\gamma_a U$. In effect, if $(L^a{}_b) \in \mathrm{SO}(p,q)$, using (1.102) and (1.211), one finds that

$$\gamma'_a \begin{pmatrix} \psi_A \\ \phi^{\dot{A}} \end{pmatrix} = \begin{pmatrix} L_a{}^b \sigma_{bA\dot{A}} \phi^{\dot{A}} \\ -L_a{}^b \sigma_b{}^{A\dot{A}} \psi_A \end{pmatrix} = \begin{pmatrix} K^C{}_A M^{\dot{C}}{}_{\dot{A}} \sigma_{aC\dot{C}} \phi^{\dot{A}} \\ -K_C{}^A M_{\dot{C}}{}^{\dot{A}} \sigma_a{}^{C\dot{C}} \psi_A \end{pmatrix}$$
$$= U^{-1}\gamma_a U \begin{pmatrix} \psi_A \\ \phi^{\dot{A}} \end{pmatrix},$$

with

$$U \begin{pmatrix} \psi_A \\ \phi^{\dot{A}} \end{pmatrix} = \begin{pmatrix} -K_A{}^C \psi_C \\ M^{\dot{A}}{}_{\dot{C}} \phi^{\dot{C}} \end{pmatrix}, \qquad U^{-1} \begin{pmatrix} \psi_A \\ \phi^{\dot{A}} \end{pmatrix} = \begin{pmatrix} K^C{}_A \psi_C \\ -M_{\dot{C}}{}^{\dot{A}} \phi^{\dot{C}} \end{pmatrix}, \tag{1.212}$$

and one can verify that $U^{-1}\gamma_5 U = \gamma_5$. Recall that the matrices $(K^A{}_B)$ and $(M^{\dot{A}}{}_{\dot{B}})$ are defined by $(L^a{}_b)$ up to a common sign; hence, U is defined by $(L^a{}_b)$ up to sign. Note also that (1.212) coincides with the spin transformations (1.187), and we can identify $\mathfrak{D}$ with the *direct sum* of the spin spaces.

Similarly, in the case of an improper orthogonal transformation given by $L^b{}_a = -\frac{1}{2}\sigma^b{}_{B\dot{B}}\sigma_a{}^{A\dot{A}} K^B{}_{\dot{A}} M^{\dot{B}}{}_A$, we have

$$\gamma'_a \begin{pmatrix} \psi_A \\ \phi^{\dot{A}} \end{pmatrix} = \begin{pmatrix} K^C{}_{\dot{A}} M^{\dot{C}}{}_A \sigma_{aC\dot{C}} \phi^{\dot{A}} \\ -K_C{}^{\dot{A}} M_{\dot{C}}{}^A \sigma_a{}^{C\dot{C}} \psi_A \end{pmatrix} = U^{-1}\gamma_a U \begin{pmatrix} \psi_A \\ \phi^{\dot{A}} \end{pmatrix},$$

with

$$U \begin{pmatrix} \psi_A \\ \phi^{\dot{A}} \end{pmatrix} = \begin{pmatrix} K_{A\dot{C}} \phi^{\dot{C}} \\ -M^{\dot{A}C} \psi_C \end{pmatrix}, \qquad U^{-1} \begin{pmatrix} \psi_A \\ \phi^{\dot{A}} \end{pmatrix} = \begin{pmatrix} -M_{\dot{C}A} \phi^{\dot{C}} \\ K^{C\dot{A}} \psi_C \end{pmatrix}, \tag{1.213}$$

and $U^{-1}\gamma_5 U = -\gamma_5$. Thus, for any $(L^a{}_b) \in \mathrm{O}(p,q)$ there exists a unimodular operator U of $\mathfrak{D}$ onto itself such that

$$U^{-1}\gamma^a U = L^a{}_b \gamma^b. \tag{1.214}$$

The operator U can be chosen in the explicit form (1.212) or (1.213), for a proper or improper orthogonal transformation, respectively. In the first case each spinor space is mapped onto itself, but in the second case one spin space is mapped onto the other. (In fact, given a unimodular operator U satisfying (1.214), the operators $\mathrm{i}U$, $-U$, and $-\mathrm{i}U$ are also unimodular and satisfy (1.214). One can verify directly that the set of unimodular operators U such that (1.214) holds for some $(L^a{}_b) \in \mathrm{O}(p,q)$ form a group under composition and that the map $U \mapsto (L^a{}_b)$ given by (1.214) is a group homomorphism.)

We can prove the covariance of the Dirac equation (1.196) under a Lorentz transformation $x'^a = L^a{}_b x^b$, with $(L^a{}_b) \in \mathrm{O}(3,1)$. Indeed, applying the operator U defined by (1.212) or (1.213) to the Dirac equation (1.196), we obtain

$$\left(U\gamma^a U^{-1}\frac{\partial}{\partial x^a} - \frac{m_0 c}{\hbar}\right) U\psi = 0,$$

but $U\gamma^a U^{-1} = (L^{-1})^a{}_b \gamma^b$ and $\partial/\partial x^a = L^c{}_a(\partial/\partial x'^c)$, and hence the last equation is equivalent to

$$\left(\gamma^a \frac{\partial}{\partial x'^a} - \frac{m_0 c}{\hbar}\right) U\psi = 0,$$

which is of the form (1.196), with the coordinates x'^a in place of x^a and ψ replaced by $U\psi$.

More generally, if $\{\gamma'_1, \gamma'_2, \gamma'_3, \gamma'_4\}$ is any other set of operators on $\mathfrak{D}$ satisfying the anticommutation relations (1.194), then, according to the foregoing results, there exists a basis of $\mathfrak{D}$ formed by eigenvectors of $\gamma'_5 = \mathrm{i}^q \gamma'_1\gamma'_2\gamma'_3\gamma'_4$. Defining an operator S by the condition that S send each eigenvector of γ_5 into an eigenvector of γ'_5 corresponding to the same eigenvalue, we have, $S^{-1}\gamma'_5 S = \gamma_5$. Then, one can readily verify that the operators $S^{-1}\gamma'_a S$ anticommute with γ_5 and satisfy the anticommutation relations (1.194); therefore, with respect to the basis formed by the eigenvectors of γ_5, $S^{-1}\gamma'_a S$ has the form given by (1.205) and (1.209), for some set of Infeld–van der Waerden symbols. As shown in Section 1.3, any two sets of Infeld–van der Waerden symbols are related through (1.31) or (1.32), and one finally concludes that there exists an operator U such that $\gamma'_a = U^{-1}\gamma_a U$. In the context of the Dirac equation, this result is known as Pauli's theorem.

When the metric (g_{ab}) has Lorentzian signature, one can define the Hermitian inner product between bispinors

$$\left\langle \begin{pmatrix} \chi_A \\ \eta^{\dot{A}} \end{pmatrix}, \begin{pmatrix} \psi_A \\ \phi^{\dot{A}} \end{pmatrix} \right\rangle \equiv \mathrm{i}(\overline{\chi}_{\dot{A}}\phi^{\dot{A}} - \overline{\eta}^A \psi_A), \qquad (1.215)$$

which is indefinite and manifestly invariant under $\mathrm{SO}^{\uparrow}(3,1)$. It can readily be shown that this inner product is invariant only under the transformations (1.212) or (1.213) induced by proper or improper Lorentz transformations that preserve the time orientation. The operators γ_a are anti-Hermitian with respect to this inner product. The adjoint of an operator defined by means of this inner product is the so-called *Dirac adjoint*. (Note that γ_5 is also anti-Hermitian but its eigenvalues are real; this is possible owing to the fact that the inner product (1.215) is indefinite.) When the metric has Euclidean or ultrahyperbolic signature, it is also possible to define a Hermitian inner product between bispinors (see Exercise 1.12).

The inner product $\langle\,,\,\rangle_{\rm std}$ between bispinors employed in the standard formulation of the Dirac equation (see, e.g., Messiah 1962, Davydov 1988, Merzbacher 1998) is positive definite, but it is not invariant under $\mathrm{SO}^{\uparrow}(3,1)$ transformations. Up to a constant positive real factor, this inner product is related to the inner product (1.215) by

$$\langle \Phi, \Psi \rangle_{\rm std} = -\mathrm{i}\langle \Phi, \gamma^4 \Psi \rangle.$$

According to (1.211) and (1.69), the action of the commutator $[\gamma_a, \gamma_b]$ on a bispinor (1.210) is given by

$$[\gamma_a, \gamma_b] \begin{pmatrix} \psi_A \\ \phi^{\dot{A}} \end{pmatrix} = \begin{pmatrix} 2S_{abA}{}^{C}\psi_C \\ -2S_{ab}{}^{\dot{A}}{}_{\dot{C}}\phi^{\dot{C}} \end{pmatrix}. \tag{1.216}$$

As we have shown, any SO(4) transformation can be written in the form $(L^a{}_b) = \exp(-t^a{}_b)$, where t_{ab} is an antisymmetric tensor [see (1.127)] and the corresponding SU(2) matrices can be chosen as in (1.124), namely $(K^A{}_B) = \exp\left(\frac{1}{4}t^{ab}S_{ab}{}^{A}{}_{B}\right)$, $(M^{\dot{A}}{}_{\dot{B}}) = \exp\left(\frac{1}{4}t^{ab}S_{ab}{}^{\dot{A}}{}_{\dot{B}}\right)$; hence, from (1.212) and (1.216) we see that the transformation induced on the bispinors is given by

$$U = \exp\left(-\tfrac{1}{8}t^{ab}[\gamma_a, \gamma_b]\right). \tag{1.217}$$

According to the results of Section 1.3.2, any $\mathrm{SO}^{\uparrow}(3,1)$ transformation can also be expressed as $(L^a{}_b) = \exp(-t^a{}_b)$, where t_{ab} is an antisymmetric tensor, and one finds that (1.217) also holds.

As in the case of any vector space, one can make use of other bases of $\mathfrak{D}$ (not formed by eigenvectors of γ_5). When the metric has Lorentzian signature, the bases usually employed in connection with the Dirac equation are formed by eigenvectors of γ_4.

The explicit representation for the γ_a is useful to obtain some identities that do not depend on the representation. For instance, from (1.211) and (1.81) it follows that

$$\frac{1}{6}\varepsilon^{abcd}\gamma_a\gamma_b\gamma_c = \mathrm{i}^q \gamma_5 \gamma^d,$$

or equivalently,

$$\gamma_{[a}\gamma_b\gamma_{c]} = (-\mathrm{i})^q \gamma_5 \varepsilon_{abcd}\gamma^d. \tag{1.218}$$

Making use repeatedly of the basic relations (1.194), one finds that $\gamma_{[a}\gamma_b\gamma_{c]} = \gamma_a\gamma_b\gamma_c - g_{bc}\gamma_a + g_{ac}\gamma_b - g_{ab}\gamma_c$, hence, from (1.218),

$$\gamma_a\gamma_b\gamma_c = g_{bc}\gamma_a - g_{ac}\gamma_b + g_{ab}\gamma_c + (-\mathrm{i})^q\gamma_5\varepsilon_{abcd}\gamma^d$$

(see also Exercise 1.14).

1.6 Inner Products. Mate of a Spinor

As we have seen in Section 1.3, when the signature of the metric is $(++++)$ or $(++--)$, it is useful to define the *mate* (conjugate, or adjoint) of a spinor by

$$\widehat{\psi}_{AB\ldots\dot{C}\dot{D}\ldots} = \begin{cases} \overline{\psi^{AB\ldots\dot{C}\dot{D}\ldots}} & \text{if the } \sigma^a{}_{A\dot{B}} \text{ are given by (1.33)}, \\ \overline{\psi_{AB\ldots\dot{C}\dot{D}\ldots}} & \text{if the } \sigma^a{}_{A\dot{B}} \text{ are given by (1.35)}, \\ (-\mathrm{i})\eta_{AM}(-\mathrm{i})\eta_{BN}\cdots\mathrm{i}\,\eta_{\dot{C}\dot{R}}\,\mathrm{i}\,\eta_{\dot{D}\dot{S}}\cdots\overline{\psi^{MN\ldots\dot{R}\dot{S}\ldots}} & \\ & \text{if the } \sigma^a{}_{A\dot{B}} \text{ are given by (1.37)}, \end{cases} \tag{1.219}$$

or equivalently,

$$\widehat{\psi}^{AB\ldots\dot{C}\dot{D}\ldots} = \begin{cases} (-1)^m\overline{\psi_{AB\ldots\dot{C}\dot{D}\ldots}} & \text{if the } \sigma^a{}_{A\dot{B}} \text{ are given by (1.33)}, \\ \overline{\psi^{AB\ldots\dot{C}\dot{D}\ldots}} & \text{if the } \sigma^a{}_{A\dot{B}} \text{ are given by (1.35)}, \\ (-1)^m(-\mathrm{i})\eta^{AM}(-\mathrm{i})\eta^{BN}\cdots\mathrm{i}\,\eta^{\dot{C}\dot{R}}\,\mathrm{i}\,\eta^{\dot{D}\dot{S}}\cdots\overline{\psi_{MN\ldots\dot{R}\dot{S}\ldots}} & \\ & \text{if the } \sigma^a{}_{A\dot{B}} \text{ are given by (1.37)}, \end{cases} \tag{1.220}$$

where m is the *total* number of indices of $\psi_{AB\ldots\dot{C}\dot{D}\ldots}$. Then one finds that

$$\widehat{\varepsilon}_{AB} = \varepsilon_{AB}, \qquad \widehat{\varepsilon}_{\dot{A}\dot{B}} = \varepsilon_{\dot{A}\dot{B}}, \tag{1.221}$$

and

$$\widehat{\widehat{\psi}}_{AB\ldots\dot{C}\dot{D}\ldots} = \begin{cases} (-1)^m\psi_{AB\ldots\dot{C}\dot{D}\ldots} & \text{if the signature is } (++++), \\ \psi_{AB\ldots\dot{C}\dot{D}\ldots} & \text{if the signature is } (++--). \end{cases} \tag{1.222}$$

From (1.222) it follows that when the signature is Euclidean, a nonzero m-index spinor can be proportional to its mate if and only if m is even; assuming $\widehat{\psi}_{AB\ldots\dot{C}\dot{D}\ldots} = \lambda\psi_{AB\ldots\dot{C}\dot{D}\ldots}$, for some scalar λ, we have $\widehat{\widehat{\psi}}_{AB\ldots\dot{C}\dot{D}\ldots} = \overline{\lambda}\widehat{\psi}_{AB\ldots\dot{C}\dot{D}\ldots} = |\lambda|^2\psi_{AB\ldots\dot{C}\dot{D}\ldots}$. Comparison with (1.222) yields $|\lambda|^2 = (-1)^m$, which implies that m must be even, and if m is even, λ must be of the form $e^{\mathrm{i}\theta}$, for some $\theta \in \mathbb{R}$.

The mapping $\psi_{AB\ldots\dot{C}\dot{D}\ldots} \mapsto \widehat{\psi}_{AB\ldots\dot{C}\dot{D}\ldots}$ is antilinear, and its existence is equivalent to that of a Hermitian inner product given by

$$\langle\phi,\psi\rangle \equiv \begin{cases} \widehat{\phi}_{AB\ldots\dot{R}\dot{S}\ldots}\,\psi^{AB\ldots\dot{R}\dot{S}\ldots} & \text{if the signature is } (++++), \\ \mathrm{i}^m\widehat{\phi}_{AB\ldots\dot{R}\dot{S}\ldots}\,\psi^{AB\ldots\dot{R}\dot{S}\ldots} & \text{if the signature is } (++--), \end{cases} \tag{1.223}$$

for any pair of spinors of the same type, where m is the total number of indices of each spinor. This inner product is invariant under spin transformations. Furthermore, the inner product (1.223) is positive definite if and only if the metric tensor of V is positive definite.

The mate of the sum [resp. tensor product] of two spinors is equal to the sum [resp. tensor product] of their mates, and by virtue of (1.221), we have, for instance, $\overline{\alpha^A\beta_A} = \widehat{\alpha}^A\widehat{\beta}_A$.

It should be remarked that the explicit expression for the components of the mate of a spinor depends on the choice of the Infeld–van der Waerden symbols; each set of Infeld–van der Waerden symbols corresponds to a definite choice for the basis of the spin space.

By comparing (1.47) with (1.219) one finds that $t_{A\dot{A}\ldots D\dot{D}}$ is the spinor equivalent of a *real* n-index tensor $t_{ab\ldots d}$ if and only if

$$\widehat{t}_{A\dot{A}\ldots D\dot{D}} = \begin{cases} (-1)^n t_{A\dot{A}\ldots D\dot{D}} & \text{if the signature is } (++++), \\ t_{A\dot{A}\ldots D\dot{D}} & \text{if the signature is } (++--). \end{cases} \tag{1.224}$$

For instance, the spinor equivalent of a real bivector satisfies $\widehat{t}_{A\dot{A}B\dot{B}} = t_{A\dot{A}B\dot{B}}$, which, by virtue of (1.50) and (1.221), is equivalent to $\widehat{\tau}_{AB} = \tau_{AB}$ and $\widehat{\tau}_{\dot{A}\dot{B}} = \tau_{\dot{A}\dot{B}}$.

The existence of the mate of a spinor in the cases in which the spin group is SU(2) or SU(1,1) (or a group isomorphic to one of them) follows from the fact that for each of these groups, the fundamental representation is equivalent to its conjugate, which amounts to the existence of a nonsingular 2×2 matrix (defined up to a factor) M such that

$$\overline{K} = M^{-1}KM, \tag{1.225}$$

for *all* K in the group. For example, one can readily verify in an explicit manner that any 2×2 complex matrix K belonging to SU(2) is of the form $\left(\begin{smallmatrix}\alpha & \beta \\ -\overline{\beta} & \overline{\alpha}\end{smallmatrix}\right)$, with $|\alpha|^2 + |\beta|^2 = 1$, and that

$$\overline{\begin{pmatrix}\alpha & \beta \\ -\overline{\beta} & \overline{\alpha}\end{pmatrix}} = \begin{pmatrix}0 & -1 \\ 1 & 0\end{pmatrix}^{-1}\begin{pmatrix}\alpha & \beta \\ -\overline{\beta} & \overline{\alpha}\end{pmatrix}\begin{pmatrix}0 & -1 \\ 1 & 0\end{pmatrix}.$$

That is, (1.225) holds with $M = \left(\begin{smallmatrix}0 & -1 \\ 1 & 0\end{smallmatrix}\right)$ or any nonvanishing scalar multiple of this matrix. Similarly, any element of SU(1,1) is of the form $\left(\begin{smallmatrix}\alpha & \beta \\ \overline{\beta} & \overline{\alpha}\end{smallmatrix}\right)$, with $|\alpha|^2 - |\beta|^2 = 1$, and

$$\overline{\begin{pmatrix}\alpha & \beta \\ \overline{\beta} & \overline{\alpha}\end{pmatrix}} = \begin{pmatrix}0 & -\mathrm{i} \\ -\mathrm{i} & 0\end{pmatrix}^{-1}\begin{pmatrix}\alpha & \beta \\ \overline{\beta} & \overline{\alpha}\end{pmatrix}\begin{pmatrix}0 & -\mathrm{i} \\ -\mathrm{i} & 0\end{pmatrix}.$$

The matrix M gives (up to a factor) the relation between the complex conjugates of the components of a spinor and the components of its mate; thus, in the cases of SU(2) and SU(1,1) we have

$$\begin{pmatrix}0 & -1 \\ 1 & 0\end{pmatrix}\begin{pmatrix}\overline{\psi^1} \\ \overline{\psi^2}\end{pmatrix} = \begin{pmatrix}-\overline{\psi^2} \\ \overline{\psi^1}\end{pmatrix} \quad \text{and} \quad \begin{pmatrix}0 & -\mathrm{i} \\ -\mathrm{i} & 0\end{pmatrix}\begin{pmatrix}\overline{\psi^1} \\ \overline{\psi^2}\end{pmatrix} = \begin{pmatrix}-\mathrm{i}\overline{\psi^2} \\ -\mathrm{i}\overline{\psi^1}\end{pmatrix},$$

respectively [cf. (1.219)].

1.7 Principal Spinors. Algebraic Classification

Owing to (1.27) and its dotted version, any spinor with two or more undotted or dotted indices can be expressed as a sum of a totally symmetric spinor in each type of spinor indices plus totally symmetric spinors multiplied by ε's. For example, an object like $\mu_{A\dot{A}\dot{B}\dot{C}}$ introduced in (1.53), which is symmetric in the last two indices, has a totally symmetric part

$$\mu_{A(\dot{A}\dot{B}\dot{C})} = \tfrac{1}{6}(\mu_{A\dot{A}\dot{B}\dot{C}} + \mu_{A\dot{A}\dot{C}\dot{B}} + \mu_{A\dot{B}\dot{A}\dot{C}} + \mu_{A\dot{B}\dot{C}\dot{A}} + \mu_{A\dot{C}\dot{A}\dot{B}} + \mu_{A\dot{C}\dot{B}\dot{A}})$$

$$= \tfrac{1}{3}(\mu_{A\dot{A}\dot{B}\dot{C}} + \mu_{A\dot{B}\dot{A}\dot{C}} + \mu_{A\dot{C}\dot{A}\dot{B}}),$$

where we have made use of the symmetry of $\mu_{A\dot{A}\dot{B}\dot{C}}$ in the last pair of dotted indices. Thus,

$$\mu_{A(\dot{A}\dot{B}\dot{C})} = \mu_{A\dot{A}\dot{B}\dot{C}} + \tfrac{1}{3}(\mu_{A\dot{B}\dot{A}\dot{C}} - \mu_{A\dot{A}\dot{B}\dot{C}} + \mu_{A\dot{C}\dot{A}\dot{B}} - \mu_{A\dot{A}\dot{B}\dot{C}})$$

$$= \mu_{A\dot{A}\dot{B}\dot{C}} + \tfrac{1}{3}\mu_{A}{}^{\dot{R}}{}_{\dot{R}\dot{C}}\varepsilon_{\dot{B}\dot{A}} + \tfrac{1}{3}\mu_{A}{}^{\dot{R}}{}_{\dot{R}\dot{B}}\varepsilon_{\dot{C}\dot{A}},$$

i.e.,

$$\mu_{A\dot{A}\dot{B}\dot{C}} = \mu_{A(\dot{A}\dot{B}\dot{C})} + \tfrac{1}{3}\mu_{A}{}^{\dot{R}}{}_{\dot{R}\dot{C}}\varepsilon_{\dot{A}\dot{B}} + \tfrac{1}{3}\mu_{A}{}^{\dot{R}}{}_{\dot{R}\dot{B}}\varepsilon_{\dot{A}\dot{C}}. \tag{1.226}$$

The image of a totally symmetric spinor under a spin transformation is also totally symmetric, and since the spin transformations have unit determinant, ε_{AB} and $\varepsilon_{\dot{A}\dot{B}}$ are invariant under the spin transformations in the sense that, e.g., $K_A{}^C K_B{}^D \varepsilon_{CD} = \varepsilon_{AB}$ [see (1.30)]; hence, the (complex) vector space formed by the spinors totally symmetric on k undotted indices and totally symmetric on l dotted indices multiplied by a fixed number of ε_{AB}'s and $\varepsilon_{\dot{A}\dot{B}}$'s is invariant under the spin transformations. Furthermore, the action of the group of spin transformations on this vector space is irreducible, in the sense that there is no nontrivial subspace invariant under all the spin transformations. Thus, a decomposition such as the one given in (1.226) corresponds to a decomposition into irreducible parts, and the group of the spin transformations possesses an irreducible representation on the vector space formed by the spinors totally symmetric on k undotted indices and totally symmetric on l dotted indices, denoted by $D(k/2, l/2)$. The dimension of this representation is $(k+1)(l+1)$.

A particularly simple and important case corresponds to the totally symmetric spinors with only one type of indices (undotted or dotted) because they can be expressed as the symmetrized tensor product of one-index spinors (Penrose 1960).

Proposition 1.5. *If $\phi_{AB\ldots L}$ is a k-index totally symmetric spinor, then there exist k one-index spinors (not necessarily distinct), α_A, β_A, ..., ζ_A, such that*

$$\phi_{AB\ldots L} = \alpha_{(A}\beta_B \cdots \zeta_{L)}. \tag{1.227}$$

Proof. Let λ^1, λ^2 be two auxiliary complex variables. Then if $\lambda^2 \neq 0$, the expression

$$(\lambda^2)^{-k}\phi_{AB...L}\lambda^A\lambda^B\cdots\lambda^L = \phi_{11...11}\left(\frac{\lambda^1}{\lambda^2}\right)^k + k\,\phi_{11...12}\left(\frac{\lambda^1}{\lambda^2}\right)^{k-1}$$

$$+\frac{k(k-1)}{2}\phi_{11...22}\left(\frac{\lambda^1}{\lambda^2}\right)^{k-2} + \cdots + \phi_{22...22}$$

is a polynomial in λ^1/λ^2 of degree k provided that $\phi_{11...11}$ does not vanish, and according to the fundamental theorem of algebra, there exist k complex numbers (the roots of the polynomial) that can be written as ratios of complex numbers α^1/α^2, $\beta^1/\beta^2, \ldots, \zeta^1/\zeta^2$ such that

$$\begin{aligned}(\lambda^2)^{-k}&\phi_{AB...L}\lambda^A\lambda^B\cdots\lambda^L\\ &= \phi_{11...11}\left(\frac{\lambda^1}{\lambda^2}-\frac{\alpha^1}{\alpha^2}\right)\left(\frac{\lambda^1}{\lambda^2}-\frac{\beta^1}{\beta^2}\right)\cdots\left(\frac{\lambda^1}{\lambda^2}-\frac{\zeta^1}{\zeta^2}\right)\\ &= \frac{\phi_{11...11}}{(\lambda^2)^k\alpha^2\beta^2\cdots\zeta^2}(\alpha^2\lambda^1-\alpha^1\lambda^2)(\beta^2\lambda^1-\beta^1\lambda^2)\cdots(\zeta^2\lambda^1-\zeta^1\lambda^2);\end{aligned}$$

hence

$$\phi_{AB...L}\lambda^A\lambda^B\cdots\lambda^L = \frac{\phi_{11...11}}{\alpha_1\beta_1\cdots\zeta_1}\alpha_A\lambda^A\beta_B\lambda^B\cdots\zeta_L\lambda^L. \tag{1.228}$$

Since only the ratios α^1/α^2, $\beta^1/\beta^2, \ldots, \zeta^1/\zeta^2$ are fixed, we can make $\phi_{11...11} = \alpha_1\beta_1\cdots\zeta_1$, and using the fact that λ^A is arbitrary one obtains (1.227).

If, for instance, $\phi_{11...11} = 0$ but $\phi_{11...12} \neq 0$, then $(\lambda^2)^{-k}\phi_{AB...L}\lambda^A\lambda^B\cdots\lambda^L$ is a polynomial in λ^1/λ^2 of degree $k-1$; hence, expressing the $k-1$ roots of this polynomial in the form $\beta^1/\beta^2, \ldots, \zeta^1/\zeta^2$,

$$\begin{aligned}(\lambda^2)^{-k}\phi_{AB...L}\lambda^A\lambda^B\cdots\lambda^L &= k\phi_{11...12}\left(\frac{\lambda^1}{\lambda^2}-\frac{\beta^1}{\beta^2}\right)\cdots\left(\frac{\lambda^1}{\lambda^2}-\frac{\zeta^1}{\zeta^2}\right)\\ &= \frac{k\phi_{11...12}}{(\lambda^2)^{k-1}\beta^2\cdots\zeta^2}(\beta^2\lambda^1-\beta^1\lambda^2)\cdots(\zeta^2\lambda^1-\zeta^1\lambda^2).\end{aligned}$$

Thus

$$\phi_{AB...L}\lambda^A\lambda^B\cdots\lambda^L = \frac{k\phi_{11...12}}{\beta_1\cdots\zeta_1}\lambda^2\beta_B\lambda^B\cdots\zeta_L\lambda^L = \alpha_A\lambda^A\beta_B\lambda^B\cdots\zeta_L\lambda^L,$$

with

$$\alpha_1 \equiv 0, \qquad \alpha_2 \equiv \frac{k\phi_{11...12}}{\beta_1\cdots\zeta_1},$$

which leads again to (1.227). In a similar way one can verify the validity of (1.227) when the degree of the polynomial $(\lambda^2)^{-k}\phi_{AB...L}\lambda^A\lambda^B\cdots\lambda^L$ is less than $k-1$. □

The one-index spinors α_A, β_A, $\ldots$, ζ_A appearing in (1.227) are called *principal spinors* of $\phi_{AB...L}$. As pointed out above, if $k \geqslant 2$, each principal spinor is defined up to a scalar factor. From (1.228) it follows that $\phi_{AB...L}\lambda^A\lambda^B\cdots\lambda^L = 0$ if and only if λ^A is proportional to one of the principal spinors of $\phi_{AB...L}$.

A first algebraic classification of the totally symmetric k-index spinors comes from the possible multiplicities of the roots of the polynomial appearing in the proof of Proposition 1.5,

$$\phi_{11...11}z^k + k\,\phi_{11...12}z^{k-1} + \frac{k(k-1)}{2}\phi_{11...22}z^{k-2} + \cdots + \phi_{22...22}. \tag{1.229}$$

If one of the roots of this polynomial is repeated m times, e.g., assuming that precisely the first m roots coincide, $\alpha^1/\alpha^2 = \beta^1/\beta^2 = \cdots$, we can take $\alpha_A = \beta_A = \cdots$; hence,

$$\phi_{AB...D...L} = \underbrace{\alpha_{(A}\alpha_B\cdots\alpha_D}_{m}\cdots\zeta_{L)},$$

which is equivalent to the conditions

$$\phi_{AB...D...L}\underbrace{\alpha^A\alpha^B\cdots\alpha^D}_{k-m} \neq 0, \qquad \phi_{AB...D...L}\underbrace{\alpha^A\alpha^B\cdots\alpha^D}_{k-m+1} = 0. \tag{1.230}$$

It should be clear that all the preceding results, especially (1.227) and (1.230), also apply if all the indices are dotted.

When the signature of V is Euclidean or ultrahyperbolic, the mate of a totally symmetric k-index spinor $\phi_{AB...L}$ can be proportional to $\phi_{AB...L}$, that is, $\widehat{\phi}_{AB...L} = \lambda\phi_{AB...L}$, and in that case, if α_A is an m-fold repeated principal spinor of $\phi_{AB...L}$, then so is $\widehat{\alpha}_A$. This conclusion follows from the fact that the relations (1.230) would then be equivalent to

$$\widehat{\phi}_{AB...D...L}\underbrace{\widehat{\alpha}^A\widehat{\alpha}^B\cdots\widehat{\alpha}^D}_{k-m} \neq 0, \qquad \widehat{\phi}_{AB...D...L}\underbrace{\widehat{\alpha}^A\widehat{\alpha}^B\cdots\widehat{\alpha}^D}_{k-m+1} = 0.$$

As we have shown (Section 1.6), if the signature of V is Euclidean, $\widehat{\alpha}_A$ cannot be proportional to α_A (except in the trivial case $\alpha_A = 0$), and therefore a nonvanishing totally symmetric k-index spinor $\phi_{AB...L}$ such that $\widehat{\phi}_{AB...L}$ is proportional to $\phi_{AB...L}$ must be of the form

$$\phi_{ABCD...LM} = e^{i\theta}\alpha_{(A}\widehat{\alpha}_B\beta_C\widehat{\beta}_D\cdots\zeta_L\widehat{\zeta}_{M)}, \tag{1.231}$$

for some $\theta \in \mathbb{R}$, and necessarily k must be an even number (the one-index spinors α_A, β_A, $\ldots$ need not be distinct). An example is provided by any antisymmetric real two-index tensor (or bivector) t_{ab} whose spinor equivalent is of the form $t_{A\dot{A}B\dot{B}} = \tau_{AB}\varepsilon_{\dot{A}\dot{B}} + \tau_{\dot{A}\dot{B}}\varepsilon_{AB}$, with $\tau_{AB} = \tau_{(AB)}$, $\tau_{\dot{A}\dot{B}} = \tau_{(\dot{A}\dot{B})}$ and $\widehat{\tau}_{AB} = \tau_{AB}$, $\widehat{\tau}_{\dot{A}\dot{B}} = \tau_{\dot{A}\dot{B}}$ [see (1.50) and (1.224)]. Thus, there exists γ_A such that $\tau_{AB} = \lambda\gamma_{(A}\widehat{\gamma}_{B)}$. Then, $\widehat{\tau}_{AB} = \overline{\lambda}\widehat{\gamma}_{(A}\widehat{\widehat{\gamma}}_{B)} = -\overline{\lambda}\gamma_{(A}\widehat{\gamma}_{B)}$, which implies that $\lambda = ib$, for some $b \in \mathbb{R}$. Thus, we have $\tau_{AB} = i\alpha_{(A}\widehat{\alpha}_{B)}$, where $\alpha_A \equiv \sqrt{b}\,\gamma_A$ if $b > 0$, and $\alpha_A \equiv \sqrt{-b}\,\widehat{\gamma}_A$ if $b < 0$ (then $\widehat{\alpha}_A = -\sqrt{-b}\,\gamma_A$). Similarly, it follows that there exists $\beta_{\dot{A}}$ such that $\tau_{\dot{A}\dot{B}} = \mathrm{i}\beta_{(\dot{A}}\widehat{\beta}_{\dot{B})}$ and

$$t_{A\dot{A}B\dot{B}} = \mathrm{i}\alpha_{(A}\widehat{\alpha}_{B)}\varepsilon_{\dot{A}\dot{B}} + \mathrm{i}\beta_{(\dot{A}}\widehat{\beta}_{\dot{B})}\varepsilon_{AB}. \tag{1.232}$$

As an application of the expression (1.232) we can prove that if a bivector t_{ab} satisfies ${}^*t_{ab}t^{ab} = 0$, then t_{ab} is simple. Indeed, from (1.67) and (1.232) we see that ${}^*t_{ab}t^{ab} = 0$ implies $\alpha_{(A}\widehat{\alpha}_{B)}\alpha^{(A}\widehat{\alpha}^{B)} = \beta_{(\dot{A}}\widehat{\beta}_{\dot{B})}\beta^{(\dot{A}}\widehat{\beta}^{\dot{B})}$, which, in turn, is equivalent to $\alpha^A\widehat{\alpha}_A = \beta^{\dot{A}}\widehat{\beta}_{\dot{A}}$. Taking

$$v_{A\dot{A}} = \frac{\alpha_A\beta_{\dot{A}} - \widehat{\alpha}_A\widehat{\beta}_{\dot{A}}}{(2\alpha^R\widehat{\alpha}_R)^{1/2}}, \qquad w_{A\dot{A}} = \frac{\mathrm{i}(\alpha_A\beta_{\dot{A}} + \widehat{\alpha}_A\widehat{\beta}_{\dot{A}})}{(2\alpha^R\widehat{\alpha}_R)^{1/2}},$$

we have [see (1.52)]

$$\begin{aligned} v_{A\dot{A}}w_{B\dot{B}} - v_{B\dot{B}}w_{A\dot{A}} &= v_{(A}{}^{\dot{R}}w_{B)\dot{R}}\varepsilon_{\dot{A}\dot{B}} + v^R{}_{(\dot{A}}w_{|R|\dot{B})}\varepsilon_{AB}, \\ &= \frac{1}{2\alpha^S\widehat{\alpha}_S}\left[2\mathrm{i}\alpha_{(A}\widehat{\alpha}_{B)}\beta^{\dot{R}}\widehat{\beta}_{\dot{R}}\varepsilon_{\dot{A}\dot{B}} + 2\mathrm{i}\beta_{(\dot{A}}\widehat{\beta}_{\dot{B})}\alpha^R\widehat{\alpha}_R\varepsilon_{AB}\right] \\ &= \mathrm{i}\alpha_{(A}\widehat{\alpha}_{B)}\varepsilon_{\dot{A}\dot{B}} + \mathrm{i}\beta_{(\dot{A}}\widehat{\beta}_{\dot{B})}\varepsilon_{AB}, \end{aligned}$$

i.e., $t_{ab} = v_aw_b - v_bw_a$, where v_a and w_a are the vector equivalents of $v_{A\dot{A}}$ and $w_{A\dot{A}}$, respectively, thus showing that t_{ab} is simple. It may be verified that the vectors v_a and w_a are real [see (1.224)].

If the signature is ultrahyperbolic, a nonzero one-index spinor can be proportional to its mate, and therefore, in contrast to (1.231), the principal spinors of a totally symmetric k-index spinor proportional to its mate need not appear in conjugate pairs. For instance, when $k = 2$, if $\widehat{\phi}_{AB} = \phi_{AB}$, then ϕ_{AB} is of one of the forms

$$\begin{cases} \pm\alpha_{(A}\widehat{\alpha}_{B)} & \text{with } \widehat{\alpha}^A\alpha_A \neq 0, \\ \alpha_{(A}\beta_{B)} & \text{with } \widehat{\alpha}_A = \alpha_A, \widehat{\beta}_A = \beta_A, \\ \pm\alpha_A\alpha_B & \text{with } \widehat{\alpha}_A = \alpha_A, \end{cases}$$

which correspond to $\phi^{AB}\phi_{AB}$ being positive, negative, or equal to zero, respectively.

When V has Lorentzian signature, the spinor equivalent of a real bivector t_{ab} is of the form

$$t_{A\dot{A}B\dot{B}} = \alpha_{(A}\beta_{B)}\varepsilon_{\dot{A}\dot{B}} + \overline{\alpha}_{(\dot{A}}\overline{\beta}_{\dot{B})}\varepsilon_{AB}. \tag{1.233}$$

Making use of this expression, we can explicitly show that if ${}^*t_{ab}t^{ab} = 0$, then t_{ab} is simple. In fact, from ${}^*t_{ab}t^{ab} = 0$ it follows that $\alpha_{(A}\beta_{B)}\alpha^{(A}\beta^{B)} = \overline{\alpha}_{(\dot{A}}\overline{\beta}_{\dot{B})}\overline{\alpha}^{(\dot{A}}\overline{\beta}^{\dot{B})}$, or equivalently, $\alpha^A\beta_A = \pm\overline{\alpha}^{\dot{A}}\overline{\beta}_{\dot{A}}$, which means that $\alpha^A\beta_A$ is real or pure imaginary. If $\alpha^A\beta_A$ is real and different from zero, we can assume without loss of generality that $\alpha^A\beta_A$ is positive (otherwise, we have only to interchange α_A and β_A, which leaves (1.233) unchanged); then, letting

$$v_{A\dot{A}} = \frac{\alpha_A\overline{\alpha}_{\dot{A}} - \beta_A\overline{\beta}_{\dot{A}}}{(2\alpha^R\beta_R)^{1/2}}, \qquad w_{A\dot{A}} = \frac{\alpha_A\overline{\alpha}_{\dot{A}} + \beta_A\overline{\beta}_{\dot{A}}}{(2\alpha^R\beta_R)^{1/2}},$$

one finds that $t_{ab} = v_a w_b - v_b w_a$. If $\alpha^A \beta_A$ is different from zero and pure imaginary, we can assume that $\mathrm{i}\alpha^A \beta_A$ is positive. Then with

$$v_{A\dot{A}} = \frac{\mathrm{i}(\alpha_A \overline{\beta}_{\dot{A}} - \beta_A \overline{\alpha}_{\dot{A}})}{(2\mathrm{i}\alpha^R \beta_R)^{1/2}}, \qquad w_{A\dot{A}} = \frac{\alpha_A \overline{\beta}_{\dot{A}} + \beta_A \overline{\alpha}_{\dot{A}}}{(2\mathrm{i}\alpha^R \beta_R)^{1/2}},$$

we obtain $t_{ab} = v_a w_b - v_b w_a$.

Finally, if $\alpha^A \beta_A = 0$, then β_A is proportional to α_A and we can assume that $t_{A\dot{A}B\dot{B}}$ is of the form

$$t_{A\dot{A}B\dot{B}} = \alpha_A \alpha_B \varepsilon_{\dot{A}\dot{B}} + \overline{\alpha}_{\dot{A}} \overline{\alpha}_{\dot{B}} \varepsilon_{AB}. \tag{1.234}$$

Taking

$$v_{A\dot{A}} = \alpha_A \overline{\alpha}_{\dot{A}}, \qquad w_{A\dot{A}} = \alpha_A \overline{\gamma}_{\dot{A}} + \gamma_A \overline{\alpha}_{\dot{A}},$$

where γ_A is a one-index spinor such that $\alpha^R \gamma_R = 1$, one finds that $t_{ab} = v_a w_b - v_b w_a$. Since $t_{A\dot{A}B\dot{B}} t^{A\dot{A}B\dot{B}} = -(\alpha^A \beta_A)^2 - (\overline{\alpha}^{\dot{A}} \overline{\beta}_{\dot{A}})^2$ [see (1.233)], these three possible cases correspond to $t_{ab} t^{ab}$ negative, positive, or zero, respectively. In all cases, the vectors v_a and w_a defined above are real (cf. also (3.29) and (3.30)).

Spin

Now we restrict ourselves to the case that V has Lorentzian signature. In elementary quantum mechanics, the spin of an object is defined by its behavior under spatial rotations. A spin-s object has $2s+1$ components that form a basis for an irreducible representation of SU(2), the double covering group of SO(3). As we shall see, the components $\psi_{AB\ldots\dot{C}\dot{D}\ldots}$ of a spinor with n undotted indices and m dotted indices can be decomposed as the sum of objects of spin $\frac{1}{2}(n+m)$, $\frac{1}{2}(n+m)-1, \ldots, \frac{1}{2}|n-m|$.

In order to isolate the spatial rotations, we employ a timelike vector t_a. The set of vectors orthogonal to t_a form a three-dimensional subspace of V, on which the inner product is positive definite. The $\mathrm{SO}^{\uparrow}(3,1)$ transformations that leave t_a invariant (and hence leave invariant the subspace orthogonal to t_a) form a group isomorphic to SO(3) (the group of rotations of a three-dimensional space with a definite inner product). Correspondingly, the $\mathrm{SL}(2,\mathbb{C})$ matrices $(K^A{}_B)$ satisfying

$$K^A{}_B \overline{K}^{\dot{C}}{}_{\dot{D}} t_{A\dot{C}} = t_{B\dot{D}}, \tag{1.235}$$

where $t_{A\dot{B}}$ is the spinor equivalent of t_a, form a group isomorphic to SU(2). Since $\det(t_{A\dot{B}}) = \frac{1}{2} t^{A\dot{B}} t_{A\dot{B}} = -\frac{1}{2} t^a t_a \neq 0$, the 2×2 matrix $(t_{A\dot{B}})$ is invertible, and therefore there is an invertible relation between $\psi_{AB\ldots\dot{C}\dot{D}\ldots}$ and

$$\phi^{AB\ldots CD\ldots} \equiv t^C{}_{\dot{C}} t^D{}_{\dot{D}} \cdots \psi^{AB\ldots\dot{C}\dot{D}\ldots}.$$

Then, under the spin transformation defined by an $\mathrm{SL}(2,\mathbb{C})$ matrix $(K^A{}_B)$ that satisfies the condition (1.235), we have

$$\begin{aligned}\phi^{AB\dots CD\dots} &\mapsto K^A{}_M K^B{}_N \cdots \overline{K}^{\dot{C}}{}_{\dot{P}} \overline{K}^{\dot{D}}{}_{\dot{Q}} \cdots t^C{}_{\dot{C}} t^D{}_{\dot{D}} \psi^{MN\dots\dot{P}\dot{Q}\dots} \\ &= K^A{}_M K^B{}_N \cdots K^C{}_R t^R{}_{\dot{P}} K^D{}_S t^S{}_{\dot{Q}} \cdots \psi^{MN\dots\dot{P}\dot{Q}\dots} \\ &= K^A{}_M K^B{}_N \cdots K^C{}_R K^D{}_S \cdots \phi^{MN\dots RS\dots},\end{aligned}$$

which means that the components $\phi^{AB\dots CD\dots}$ form a basis for a (possibly reducible) representation of SU(2).

By expressing $\phi^{AB\dots CD\dots}$ as the sum of its totally symmetric part plus totally symmetric objects (with $n+m-2$, $n+m-4$, $\dots$, $|n-m|$ indices) multiplied by ε_{AB}'s (though some terms may be equal to zero), as at the beginning of this section, one decomposes $\phi^{AB\dots CD\dots}$ as a sum of objects of spin $\frac{1}{2}(n+m)$, $\frac{1}{2}(n+m)-1, \dots,$ $\frac{1}{2}|n-m|$ (though some values may be absent). Note that this decomposition depends on the timelike vector t_a, except in the case that m or n is equal to zero (actually, when $\psi_{\dot{A}\dot{B}\dots\dot{L}}$ has only dotted indices, the value of the components of $\phi_{AB\dots L}$ do depend on t_a, but the number of independent totally symmetric spinors that can be extracted from $\phi_{AB\dots L}$ does not depend on t_a). (See, e.g., the discussion about spin-3/2 particles in Berestetskii et al. 1982, §31.)

Exercises

1.1. Show that if $t_{A\dot{A}B\dot{B}}$ is the spinor equivalent of a symmetric two-index tensor t_{ab}, then $s_{A\dot{A}B\dot{B}} = t_{B\dot{A}A\dot{B}}$ is the spinor equivalent of $t_{ab} - \frac{1}{2}t^c{}_c g_{ab}$.

1.2. Show that if $t_{AB\dot{C}\dot{D}}$ is the spinor equivalent of a *symmetric* two-index tensor t_{ab}, then $t_{(AB)\dot{C}\dot{D}} = t_{AB(\dot{C}\dot{D})}$ is the spinor equivalent of $t_{ab} - \frac{1}{4}t^c{}_c g_{ab}$ (the traceless part of t_{ab}).

1.3. Show that the spinor equivalent of a three-index tensor H_{abc} such that

$$H_{abc} = -H_{bac}, \qquad H_{ab}{}^b = 0, \qquad H_{abc} + H_{bca} + H_{cab} = 0,$$

is of the form $H_{A\dot{A}B\dot{B}C\dot{C}} = h_{ABC\dot{C}}\varepsilon_{\dot{A}\dot{B}} + h_{\dot{A}\dot{B}\dot{C}C}\varepsilon_{AB}$, with $h_{ABC\dot{C}} = h_{(ABC)\dot{C}}$ and $h_{\dot{A}\dot{B}\dot{C}C} = h_{(\dot{A}\dot{B}\dot{C})C}$.

1.4. Using the spinor formalism, show that if the signature is Lorentzian, then two nonzero real null vectors are orthogonal to each other if and only if they are proportional.

1.5. Show that

$$t_{A\dot{A}} t_{B\dot{B}} = t_{B\dot{A}} t_{A\dot{B}}$$

if and only if $t_{A\dot{A}}$ is the spinor equivalent of a null vector. (If $t_{A\dot{A}}$ is different from zero, there exist α_A and $\beta_{\dot{A}}$ such that $t_{A\dot{A}}\alpha^A\beta^{\dot{A}} \neq 0$; hence, contracting both sides of the equation above with $\alpha^B\beta^{\dot{B}}$, one concludes that if $t_{A\dot{A}}t^{A\dot{A}} = 0$, then $t_{A\dot{A}} = t_{B\dot{A}}\alpha^B t_{A\dot{B}}\beta^{\dot{B}} / (t_{B\dot{B}}\alpha^B\beta^{\dot{B}})$; this constitutes a proof of Proposition 1.3.)

1.6. Using the spinor formalism, show that if t_{ab} and s_{ab} are two bivectors, then

$$t_{ac}s_b{}^c + (-1)^q\, {}^*s_{ac}{}^*t_b{}^c = \tfrac{1}{2}t_{de}s^{de}\, g_{ab}$$

and conclude that $t_{ac}{}^*t_b{}^c = \frac{1}{4}t_{de}{}^*t^{de}\, g_{ab}$. Hence, a bivector t_{ab} is simple if and only if $t_{ac}{}^*t_b{}^c = 0$, or equivalently, $t_{a[b}t_{cd]} = 0$.

1.7. Using the spinor formalism, show that if t_{ab} is a bivector, then

$$\det(t^{ab}) = \tfrac{1}{16}({}^*t_{ab}t^{ab})^2.$$

1.8. Let t_{ab} be an anti-self-dual bivector (i.e., ${}^*t_{ab} = -\mathrm{i}^q t_{ab}$) and u^a an arbitrary nonnull vector. Show, using the spinor formalism, that if we let $v_a \equiv t_{ab}u^b$, then

$$t_{ab} = \frac{1}{u^c u_c}\big(v_a u_b - u_a v_b + (-\mathrm{i})^q \varepsilon_{abcd} u^c v^d\big).$$

1.9. Assume that the signature of the metric is Lorentzian. If the axes of the reference frame S′ are obtained from those of S by means of a spatial rotation through an angle θ about the axis defined by the unit vector u^a, the Cartesian coordinates of a point with respect to these frames are related by $x'^a = L^a{}_b x^b$, with

$$L^a{}_b = \cos\theta\, \delta^a_b + (1-\cos\theta)(u^a u_b - T^a T_b) + \sin\theta\, g^{am}\varepsilon_{mbcd}u^c T^d,$$

where T^a is a timelike vector orthogonal to u^a such that $T^a T_a = -1$. Show that the spinor equivalent of $L^a{}_b$ is of the form $-K^A{}_B K^{\dot{A}}{}_{\dot{B}}$ with

$$K^A{}_B = \pm\,(\mathrm{e}^{\mathrm{i}\theta/2}\alpha^A\beta_B - \mathrm{e}^{-\mathrm{i}\theta/2}\beta^A\alpha_B)$$

and $\alpha^A\beta_A = 1$. Here $\alpha_A\overline{\alpha}_{\dot{A}}$ and $\beta_A\overline{\beta}_{\dot{A}}$ are the spinor equivalents of the null vectors $(T^a + u^a)/\sqrt{2}$ and $(T^a - u^a)/\sqrt{2}$, respectively. (Note that as in the case of the boost (1.135), we are considering *passive* transformations in the sense that the points are not affected by transformations; only the basis vectors are transformed.)

1.10. Assuming that the signature of the metric is Lorentzian, show that any two-dimensional spacelike plane orthogonal to a given null vector l_a is spanned by the vector equivalents of $\alpha_A\overline{\beta}_{\dot{A}} + \beta_A\overline{\alpha}_{\dot{A}}$ and $\mathrm{i}(\alpha_A\overline{\beta}_{\dot{A}} - \beta_A\overline{\alpha}_{\dot{A}})$, where $\pm\alpha_A\overline{\alpha}_{\dot{A}}$ is the spinor equivalent of l_a and β_A is some one-index spinor such that $\alpha^A\beta_A \neq 0$.

1.11. Show that each matrix in (1.173) can be expressed in the form

$$\nu\delta^A_B + \mu v^a w^b S_{ab}{}^A{}_B,$$

where ν, μ are two real scalars and v^a, w^a are two real vectors.

1.12. Show that if the signature of the metric is Euclidean, the expression

$$\langle\Phi,\Psi\rangle = \widehat{\chi}_A\psi^A + \widehat{\eta}_{\dot{A}}\phi^{\dot{A}},$$

where $\varPhi$ and $\varPsi$ are the bispinors $\varPhi = \binom{\chi_A}{\eta^{\dot{A}}}$, $\varPsi = \binom{\psi_A}{\phi^{\dot{A}}}$, defines a positive definite Hermitian inner product. Show that with respect to this inner product, the operators γ_a are Hermitian. Also show that if the signature of the metric is ultrahyperbolic, then

$$\langle \varPhi, \varPsi \rangle = \mathrm{i}(\widehat{\chi}_A \psi^A + \widehat{\eta}_{\dot{A}} \phi^{\dot{A}})$$

defines an indefinite Hermitian inner product. Show that with respect to this inner product, the operators γ_a are anti-Hermitian.

1.13. Show that the eigenvalues of each operator γ_a are $\pm(g_{aa})^{1/2}$ (without summation on a) and find the corresponding eigenvectors.

1.14. Show that

$$\tfrac{1}{2}\varepsilon_{abcd}[\gamma^c, \gamma^d] = -\mathrm{i}^q \gamma_5 [\gamma_a, \gamma_b].$$

1.15. Prove explicitly that if the signature of the metric is ultrahyperbolic, a real bivector t_{ab} such that ${}^*t_{ab}t^{ab} = 0$ is simple.

Chapter 2
Connection and Curvature

In this chapter some applications of the spinor formalism to four-dimensional Riemannian manifolds are considered. By definition, if p is a point of a Riemannian manifold M, the tangent space to M at p, T_pM, is a vector space with a nondegenerate inner product, and therefore, one can construct null tetrads in the complexification of T_pM, thus allowing the use of the spinor algebra discussed in the previous chapter.

It is shown that the connection and curvature of a four-dimensional Riemannian manifold can be conveniently computed and analyzed by making use of the two-component spinor formalism. As examples, the connection and curvature of the standard metric of S^4, the Schwarzschild metric, and the Euclidean Schwarzschild metric are computed. Some further applications that will be employed in the following chapters, such as the conformal rescalings of the metric, the Killing vector fields, and the algebraic classification of the conformal curvature for all signatures of the metric, are studied. The Newman–Penrose and the Geroch–Held–Penrose notations are also introduced.

It is assumed that the reader is acquainted with the basic concepts related to differentiable manifolds such as tangent vectors, vector fields, differential forms, and connections (see, e.g., Conlon 2001, Hall 2004).

Throughout this chapter it will be assumed the Infeld–van der Waerden symbols are given by one of the forms (1.33)–(1.35) and (1.37), depending on the signature of the metric of M.

2.1 Covariant Differentiation

In what follows, M will denote a four-dimensional Riemannian manifold, that is, M will be a differentiable manifold of dimension four endowed with a Riemannian metric (a differentiable tensor field g such that at each point $p \in M$, g_p is a nondegenerate, symmetric, not necessarily definite, bilinear form defined on T_pM, the tangent space to M at p).

G.F.T. del Castillo, *Spinors in Four-Dimensional Spaces*, Progress in Mathematical Physics 59, DOI 10.1007/978-0-8176-4984-5_2,

Each point $p \in M$ possesses a neighborhood U in M such that there exist four differentiable vector fields $\{\partial_1, \partial_2, \partial_3, \partial_4\}$ defined on U, which, evaluated at a point $q \in U$, form an orthonormal basis of T_qM. Such a set is called an *orthonormal tetrad*. Furthermore, we shall assume that the constants $g_{ab} = g(\partial_a, \partial_b)$ have one of the forms (1.1). (In most cases, the vector fields ∂_a will not be the holonomic basis associated with a coordinate system; that is, in most cases there will not exist coordinates x^a such that $\partial_a = \partial/\partial x^a$. See also Appendix A.)

Given an orthonormal tetrad $\{\partial_a\}$, there exists a unique set of four one-forms $\{\theta^a\}$ such that $\theta^a(\partial_b) = \delta^a_b$. Then, the metric tensor can be expressed locally in the form

$$g = g_{ab}\theta^a \otimes \theta^b, \tag{2.1}$$

where $\otimes$ denotes the tensor product, that is, $g(X,Y) = g_{ab}\theta^a(X)\theta^b(Y)$, for any pair of vector fields X, Y on M. Thus, we have

$$g = \begin{cases} \theta^1 \otimes \theta^1 + \theta^2 \otimes \theta^2 + \theta^3 \otimes \theta^3 + \theta^4 \otimes \theta^4, \\ \theta^1 \otimes \theta^1 + \theta^2 \otimes \theta^2 + \theta^3 \otimes \theta^3 - \theta^4 \otimes \theta^4, \\ \theta^1 \otimes \theta^1 + \theta^2 \otimes \theta^2 - \theta^3 \otimes \theta^3 - \theta^4 \otimes \theta^4, \end{cases}$$

if the signature of the metric is $(++++)$, $(+++-)$, or $(++--)$, respectively. By expressing a given metric in this form, one can identify a set of one-forms $\{\theta^a\}$ corresponding to some orthonormal tetrad $\{\partial_a\}$.

Connection One-Forms

With respect to an orthonormal tetrad $\{\partial_a\}$, a connection ∇ on M is represented by the set of (real-valued) functions $\Gamma^a{}_{bc}$, defined by

$$\nabla_a \partial_b = \Gamma^c{}_{ba}\partial_c, \tag{2.2}$$

where ∇_a denotes the covariant derivative along ∂_a, that is, $\nabla_a = \nabla_{\partial_a}$ [note the position of the subscripts of $\Gamma^c{}_{ba}$ in (2.2)]. It follows from (2.2) and the properties of a connection that the *connection one-forms*

$$\Gamma^a{}_b \equiv \Gamma^a{}_{bc}\theta^c \tag{2.3}$$

satisfy the relation

$$\Gamma^a{}_b(X) = \theta^a(\nabla_X \partial_b), \tag{2.4}$$

for any vector field X on M.

With the torsion of the connection ∇ being defined by

$$T(X,Y) = \nabla_X Y - \nabla_Y X - [X,Y], \tag{2.5}$$

for any pair of vector fields X, Y on M, the *torsion two-forms* with respect to the tetrad $\{\partial_a\}$ will be defined by

$$T^a(X,Y) \equiv \tfrac{1}{2}\theta^a\big(T(X,Y)\big), \tag{2.6}$$

for any pair of vector fields X, Y on M. Then, from (2.4)–(2.6), one derives the *Cartan first structure equations*

$$\mathrm{d}\theta^a + \Gamma^a{}_b \wedge \theta^b = T^a \tag{2.7}$$

if the differential (or exterior derivative) of a one-form α is defined by

$$\mathrm{d}\alpha(X,Y) \equiv \tfrac{1}{2}\{X(\alpha(Y)) - Y(\alpha(X)) - \alpha([X,Y])\}, \tag{2.8}$$

for any pair of vector fields X, Y on M, and the exterior product of two one-forms α, β is defined by

$$(\alpha \wedge \beta)(X,Y) \equiv \tfrac{1}{2}\left[\alpha(X)\beta(Y) - \alpha(Y)\beta(X)\right]. \tag{2.9}$$

If the connection ∇ is compatible with the metric, that is,

$$X[g(Y,Z)] = g(\nabla_X Y, Z) + g(Y, \nabla_X Z),$$

for all vector fields X, Y, Z on M, we have $\partial_a g_{bc} = g(\nabla_a \partial_b, \partial_c) + g(\partial_b, \nabla_a \partial_c) = \Gamma^d{}_{ba} g_{dc} + \Gamma^d{}_{ca} g_{bd} = \Gamma_{cba} + \Gamma_{bca}$ (lowering the tetrad indices by means of g_{ab} in the usual way). Thus, with respect to an orthonormal tetrad, a connection compatible with the metric is represented by a set of functions Γ_{abc} (the *Ricci rotation coefficients*) satisfying

$$\Gamma_{abc} = -\Gamma_{bac}. \tag{2.10}$$

As is well known, in a Riemannian manifold there is a unique torsion-free connection compatible with the metric (the Riemannian or Levi-Civita connection) (see, e.g., Conlon 2001). In what follows we shall consider only Riemannian connections.

The Ricci rotation coefficients can be computed by making use of the Cartan first structure equations (2.7) or the commutators of the vector fields ∂_a. In fact, combining (2.5) and (2.2), one finds that when the torsion vanishes,

$$[\partial_a, \partial_b] = (\Gamma^c{}_{ba} - \Gamma^c{}_{ab})\partial_c. \tag{2.11}$$

By virtue of (2.10), the Ricci rotation coefficients are determined by the commutation relations (2.11). In effect, writing

$$[\partial_a, \partial_b] = C^c{}_{ba}\partial_c,$$

we have $C^c{}_{ba} = \Gamma^c{}_{ba} - \Gamma^c{}_{ab}$, and one readily verifies that

$$\Gamma_{abc} = \tfrac{1}{2}(C_{abc} + C_{bca} - C_{cab}). \tag{2.12}$$

As a consequence of (2.10), the connection one-forms (2.3) satisfy $\Gamma_{ab} = -\Gamma_{ba}$; therefore, their spinor equivalents $\Gamma_{A\dot{A}B\dot{B}} = \frac{1}{2}\sigma^a{}_{A\dot{A}}\sigma^b{}_{B\dot{B}}\Gamma_{ab}$ can be expressed in the form

$$\Gamma_{A\dot{A}B\dot{B}} = \mathbf{\Gamma}_{AB}\varepsilon_{\dot{A}\dot{B}} + \mathbf{\Gamma}_{\dot{A}\dot{B}}\varepsilon_{AB}, \tag{2.13}$$

where $\boldsymbol{\Gamma}_{AB} = \boldsymbol{\Gamma}_{(AB)}$ and $\boldsymbol{\Gamma}_{\dot{A}\dot{B}} = \boldsymbol{\Gamma}_{(\dot{A}\dot{B})}$ are one-forms [see (1.50)]. (We are employing the symbol $\boldsymbol{\Gamma}$ in order to avoid confusion with the connection one-forms $\Gamma^a{}_b$, which also possess two indices.) Then, letting

$$\theta^{A\dot{A}} \equiv \frac{1}{\sqrt{2}} \sigma_a{}^{A\dot{A}} \theta^a, \tag{2.14}$$

the Cartan first structure equations (2.7), with vanishing torsion, are equivalent to

$$\mathrm{d}\theta^{A\dot{A}} = \Gamma^{A\dot{A}}{}_{B\dot{B}} \wedge \theta^{B\dot{B}} = \boldsymbol{\Gamma}^A{}_B \wedge \theta^{B\dot{A}} + \boldsymbol{\Gamma}^{\dot{A}}{}_{\dot{B}} \wedge \theta^{A\dot{B}}. \tag{2.15}$$

(The one-forms $\theta^{A\dot{B}}$ are related to the one-forms $g^{A\dot{B}}$ employed by Plebański (1974, 1975) by $g^{A\dot{B}} = \sqrt{2}\,\theta^{A\dot{B}}$. The factor $1/\sqrt{2}$ included in (2.14) is dictated by the rule (1.41).)

The one-forms $\boldsymbol{\Gamma}_{AB}$ and $\boldsymbol{\Gamma}_{\dot{A}\dot{B}}$ can be expressed in the form

$$\boldsymbol{\Gamma}_{AB} = -\Gamma_{ABC\dot{C}} \theta^{C\dot{C}}, \qquad \boldsymbol{\Gamma}_{\dot{A}\dot{B}} = -\Gamma_{\dot{A}\dot{B}\dot{C}C} \theta^{C\dot{C}}, \tag{2.16}$$

where the $\Gamma_{ABC\dot{C}}$ and $\Gamma_{\dot{A}\dot{B}\dot{C}C}$ are some (possibly complex-valued) functions; then, substituting these expressions into the right-hand side of (2.13) and making use of (2.3), we obtain

$$\Gamma_{ABC\dot{C}} \varepsilon_{\dot{A}\dot{B}} + \Gamma_{\dot{A}\dot{B}\dot{C}C} \varepsilon_{AB} = \frac{1}{2\sqrt{2}} \sigma^a{}_{A\dot{A}} \sigma^b{}_{B\dot{B}} \sigma^c{}_{C\dot{C}} \Gamma_{abc}, \tag{2.17}$$

which leads to the expressions

$$\begin{aligned} \Gamma_{ABC\dot{C}} &= \frac{1}{4\sqrt{2}} S^{ab}{}_{AB} \sigma^c{}_{C\dot{C}} \Gamma_{abc}, \\ \Gamma_{\dot{A}\dot{B}\dot{C}C} &= \frac{1}{4\sqrt{2}} S^{ab}{}_{\dot{A}\dot{B}} \sigma^c{}_{C\dot{C}} \Gamma_{abc} \end{aligned} \tag{2.18}$$

[see (1.69)]. Following Newman and Penrose (1962), the functions $\Gamma_{ABC\dot{C}}$ and $\Gamma_{\dot{A}\dot{B}\dot{C}C}$ will be called *spin coefficients*. (The spin coefficients can also be expressed in terms of the Christoffel symbols associated with some coordinate system; see Appendix A.)

From (2.18), (2.16), and (2.14) it follows that the one-forms $\boldsymbol{\Gamma}_{AB}$ and $\boldsymbol{\Gamma}_{\dot{A}\dot{B}}$ are related to the connection one-forms Γ_{ab} by

$$\boldsymbol{\Gamma}_{AB} = \tfrac{1}{4} S^{ab}{}_{AB} \Gamma_{ab}, \qquad \boldsymbol{\Gamma}_{\dot{A}\dot{B}} = \tfrac{1}{4} S^{ab}{}_{\dot{A}\dot{B}} \Gamma_{ab}.$$

The right-hand side of (2.17) is the spinor equivalent of the Ricci rotation coefficients $\Gamma_{A\dot{A}B\dot{B}C\dot{C}}$. Thus

$$\Gamma_{A\dot{A}B\dot{B}C\dot{C}} = \Gamma_{ABC\dot{C}} \varepsilon_{\dot{A}\dot{B}} + \Gamma_{\dot{A}\dot{B}\dot{C}C} \varepsilon_{AB}. \tag{2.19}$$

The fact that the Ricci rotation coefficients are real-valued functions amounts to certain specific relations between the spin coefficients and their complex conjugates, depending on the signature of the metric. From (1.224) and (1.221) one finds that if the signature of the metric is $(++++)$, then

$$\widehat{\Gamma}_{ABC\dot{D}} = -\Gamma_{ABC\dot{D}}, \qquad \widehat{\Gamma}_{\dot{A}\dot{B}\dot{C}D} = -\Gamma_{\dot{A}\dot{B}\dot{C}D} \tag{2.20}$$

(i.e., $\overline{\Gamma^{ABC\dot{D}}} = -\Gamma_{ABC\dot{D}}$ and $\overline{\Gamma^{\dot{A}\dot{B}\dot{C}D}} = -\Gamma_{\dot{A}\dot{B}\dot{C}D}$); if the signature is $(+++-)$, we have

$$\overline{\Gamma_{ABC\dot{D}}} = \Gamma_{\dot{A}\dot{B}\dot{C}D}, \tag{2.21}$$

and if the signature of the metric is $(++--)$, then

$$\widehat{\Gamma}_{ABC\dot{D}} = \Gamma_{ABC\dot{D}}, \qquad \widehat{\Gamma}_{\dot{A}\dot{B}\dot{C}D} = \Gamma_{\dot{A}\dot{B}\dot{C}D}. \tag{2.22}$$

The spin coefficients can be obtained directly from the relations

$$[\partial_{A\dot{B}}, \partial_{C\dot{D}}] = -\Gamma^{R}{}_{CA\dot{B}}\partial_{R\dot{D}} - \Gamma^{\dot{R}}{}_{\dot{D}\dot{B}A}\partial_{C\dot{R}} + \Gamma^{R}{}_{AC\dot{D}}\partial_{R\dot{B}} + \Gamma^{\dot{R}}{}_{\dot{B}\dot{D}C}\partial_{A\dot{R}}, \tag{2.23}$$

which follow from (2.11) and (2.17), with the definition

$$\partial_{A\dot{B}} \equiv \frac{1}{\sqrt{2}}\sigma^{a}{}_{A\dot{B}}\partial_{a} \tag{2.24}$$

[cf. (1.20)] (see Example 2.2 on pp. 86–87). In actual computations, it is convenient to make use of individual symbols for the 24 spin coefficients, owing to the relatively high number of indices of $\Gamma_{ABC\dot{D}}$ and $\Gamma_{\dot{A}\dot{B}\dot{C}D}$. Following Newman and Penrose (1962), we shall denote the spin coefficients $\Gamma_{ABC\dot{D}}$ according to

$$\begin{array}{llll} \Gamma_{111\dot{1}} = \kappa, & \Gamma_{111\dot{2}} = \sigma, & \Gamma_{112\dot{1}} = \rho, & \Gamma_{112\dot{2}} = \tau, \\ \Gamma_{121\dot{1}} = \varepsilon, & \Gamma_{121\dot{2}} = \beta, & \Gamma_{122\dot{1}} = \alpha, & \Gamma_{122\dot{2}} = \gamma, \\ \Gamma_{221\dot{1}} = \pi, & \Gamma_{221\dot{2}} = \mu, & \Gamma_{222\dot{1}} = \lambda, & \Gamma_{222\dot{2}} = \nu, \end{array} \tag{2.25}$$

and, in a similar way,

$$\begin{array}{llll} \Gamma_{\dot{1}\dot{1}\dot{1}1} = \widetilde{\kappa}, & \Gamma_{\dot{1}\dot{1}\dot{1}2} = \widetilde{\sigma}, & \Gamma_{\dot{1}\dot{1}\dot{2}1} = \widetilde{\rho}, & \Gamma_{\dot{1}\dot{1}\dot{2}2} = \widetilde{\tau}, \\ \Gamma_{\dot{1}\dot{2}\dot{1}1} = \widetilde{\varepsilon}, & \Gamma_{\dot{1}\dot{2}\dot{1}2} = \widetilde{\beta}, & \Gamma_{\dot{1}\dot{2}\dot{2}1} = \widetilde{\alpha}, & \Gamma_{\dot{1}\dot{2}\dot{2}2} = \widetilde{\gamma}, \\ \Gamma_{\dot{2}\dot{2}\dot{1}1} = \widetilde{\pi}, & \Gamma_{\dot{2}\dot{2}\dot{1}2} = \widetilde{\mu}, & \Gamma_{\dot{2}\dot{2}\dot{2}1} = \widetilde{\lambda}, & \Gamma_{\dot{2}\dot{2}\dot{2}2} = \widetilde{\nu}. \end{array} \tag{2.26}$$

Then, the relations (2.23) take the explicit form

$$\begin{aligned}
[D,\delta] &= -(\widetilde{\alpha}+\beta-\widetilde{\pi})D - \kappa\Delta + (\widetilde{\rho}+\varepsilon-\widetilde{\varepsilon})\delta + \sigma\widetilde{\delta}, \\
[\Delta,\delta] &= \widetilde{\nu}D - (\tau-\widetilde{\alpha}-\beta)\Delta - (\mu-\gamma+\widetilde{\gamma})\delta - \widetilde{\lambda}\widetilde{\delta}, \\
[D,\Delta] &= -(\gamma+\widetilde{\gamma})D - (\varepsilon+\widetilde{\varepsilon})\Delta + (\widetilde{\tau}+\pi)\delta + (\tau+\widetilde{\pi})\widetilde{\delta}, \\
[\delta,\widetilde{\delta}] &= (\mu-\widetilde{\mu})D + (\rho-\widetilde{\rho})\Delta - (\alpha-\widetilde{\beta})\delta - (\beta-\widetilde{\alpha})\widetilde{\delta}, \\
[D,\widetilde{\delta}] &= -(\alpha+\widetilde{\beta}-\pi)D - \widetilde{\kappa}\Delta + (\rho+\widetilde{\varepsilon}-\varepsilon)\widetilde{\delta} + \widetilde{\sigma}\delta, \\
[\Delta,\widetilde{\delta}] &= \nu D - (\widetilde{\tau}-\alpha-\widetilde{\beta})\Delta - (\widetilde{\mu}-\widetilde{\gamma}+\gamma)\widetilde{\delta} - \lambda\delta,
\end{aligned} \tag{2.27}$$

with the definitions

$$D \equiv \partial_{1\dot{1}}, \qquad \delta \equiv \partial_{1\dot{2}}, \qquad \widetilde{\delta} \equiv \partial_{2\dot{1}}, \qquad \Delta \equiv \partial_{2\dot{2}}. \tag{2.28}$$

When the metric has Lorentzian signature, then D and Δ are real, each symbol with a tilde is the complex conjugate of the corresponding symbol without a tilde [see (2.21)], and the last two lines of (2.27) are the complex conjugates of the first two lines. (Despite several differences in the conventions, the commutation relations (2.27) coincide with those given in Newman and Penrose 1962.)

When the signature of the metric is Euclidean, the spin coefficients are related by

$$\overline{\kappa} = -\nu, \qquad \overline{\sigma} = \lambda, \qquad \overline{\rho} = \mu, \qquad \overline{\tau} = -\pi, \qquad \overline{\varepsilon} = \gamma, \qquad \overline{\beta} = -\alpha, \tag{2.29}$$

with identical relations for the spin coefficients with a tilde (e.g., $\overline{\widetilde{\kappa}} = -\widetilde{\nu}$) and

$$\overline{D} = -\Delta, \qquad \overline{\delta} = \widetilde{\delta}. \tag{2.30}$$

In the case that the signature of the metric is $(+ + - -)$ and one employs the real Infeld–van der Waerden symbols, the vector fields D, δ, $\widetilde{\delta}$, and Δ are real, and the 24 spin coefficients (2.25) and (2.26) are real and independent.

Using the properties of the covariant derivative, one finds that the covariant derivative of a vector field $X = X^a \partial_a$ with respect to a vector field $Y = Y^a \partial_a$ is given by

$$\begin{aligned} \nabla_Y X &= Y^a \nabla_a (X^b \partial_b) \\ &= Y^a [(\partial_a X^b) \partial_b + X^b \nabla_a \partial_b] \\ &= Y^a (\partial_a X^c + \Gamma^c{}_{ba} X^b) \partial_c. \end{aligned}$$

By abuse of notation, the components of the covariant derivative of X along ∂_a are usually denoted by $\nabla_a X^c$, that is,

$$\nabla_a X^c = \partial_a X^c + \Gamma^c{}_{ba} X^b. \tag{2.31}$$

Hence, according to (2.19), the spinor equivalent of $\nabla_a X^c$ is given by

$$\begin{aligned} \nabla_{A\dot{A}} X^{C\dot{C}} &= \partial_{A\dot{A}} X^{C\dot{C}} - \Gamma^{C\dot{C}}{}_{B\dot{B}A\dot{A}} X^{B\dot{B}} \\ &= \partial_{A\dot{A}} X^{C\dot{C}} - \Gamma^C{}_{BA\dot{A}} X^{B\dot{C}} - \Gamma^{\dot{C}}{}_{\dot{B}\dot{A}A} X^{C\dot{B}}. \end{aligned} \tag{2.32}$$

The covariant derivative of an arbitrary tensor field can be defined as usual, assuming that the covariant differentiation is linear, satisfies the Leibniz rule, commutes with contraction, and the covariant derivative of a function coincides with the directional derivative ($\nabla_X f = Xf$, for any vector field X). In this way one finds that the components of the covariant derivative of a tensor field with tetrad components $t^{ab\ldots}_{cd\ldots}$ are

$$\nabla_a t^{bc\ldots}_{de\ldots} = \partial_a t^{bc\ldots}_{de\ldots} + \Gamma^b{}_{sa} t^{sc\ldots}_{de\ldots} + \Gamma^c{}_{sa} t^{bs\ldots}_{de\ldots} + \cdots - \Gamma^s{}_{da} t^{bc\ldots}_{se\ldots} - \Gamma^s{}_{ea} t^{bc\ldots}_{ds\ldots} - \cdots. \tag{2.33}$$

Tetrad Rotations

In the intersection of their domains, any two orthonormal tetrads $\{\partial_a\}$ and $\{\partial'_a\}$ must be related by means of an expression of the form $\partial'_a = L_a{}^b \partial_b$, where the $L_a{}^b$ are differentiable functions such that at each point p belonging to the common domain of $\{\partial_a\}$ and $\{\partial'_a\}$, $(L_a{}^b(p))$ is an orthogonal matrix [see (1.3)]. From (2.2) and the properties of a connection we have

$$\begin{aligned}\nabla_{\partial'_a}\partial'_b &= L_a{}^c \nabla_{\partial_c}(L_b{}^d \partial_d)\\ &= L_a{}^c \left[L_b{}^d \Gamma^s{}_{dc}\partial_s + (\partial_c L_b{}^d)\partial_d\right]\\ &= L_a{}^c \left[L_b{}^d L^r{}_s \Gamma^s{}_{dc} + (\partial_c L_b{}^d) L^r{}_d\right]\partial'_r.\end{aligned}$$

This last expression must coincide with $\Gamma'^r{}_{ba}\partial'_r$, where $\Gamma'^a{}_{bc}$ are the Ricci rotation coefficients for the tetrad $\{\partial'_a\}$; hence,

$$\Gamma'_{rba} = L_a{}^c L_b{}^d L_r{}^s \Gamma_{sdc} + L_a{}^c L_r{}^d \partial_c L_{bd}. \tag{2.34}$$

The transformation law (2.34) is equivalent to some relation between the corresponding spin coefficients. For instance, if the matrices $(L_a{}^b(p))$ have positive determinant, the functions $L_a{}^b$ can be expressed as

$$L_a{}^b = -\frac{1}{2}\sigma_a{}^{A\dot{A}}\sigma^b{}_{B\dot{B}} K_A{}^B M_{\dot{A}}{}^{\dot{B}}, \tag{2.35}$$

where the $K^A{}_B$ and $M^{\dot{A}}{}_{\dot{B}}$ are differentiable functions such that at each point p where they are defined, the matrices $(K^A{}_B(p))$ and $(M^{\dot{A}}{}_{\dot{B}}(p))$ belong to the appropriate subgroup of $\mathrm{SL}(2,\mathbb{C})$, depending on the signature of the metric. Substituting (2.35) into (2.34), making use of (2.17) and (1.30) we obtain

$$\Gamma'{}_{ABC\dot{C}} = K_A{}^R K_B{}^S K_C{}^D M_{\dot{C}}{}^{\dot{D}} \Gamma_{RSD\dot{D}} - K_A{}^R K_C{}^D M_{\dot{C}}{}^{\dot{D}} \partial_{D\dot{D}} K_{BR} \tag{2.36}$$

and

$$\Gamma'{}_{\dot{A}\dot{B}C\dot{C}} = M_{\dot{A}}{}^{\dot{R}} M_{\dot{B}}{}^{\dot{S}} M_{\dot{C}}{}^{\dot{D}} K_C{}^D \Gamma_{\dot{R}\dot{S}D\dot{D}} - M_{\dot{A}}{}^{\dot{R}} M_{\dot{C}}{}^{\dot{D}} K_C{}^D \partial_{D\dot{D}} M_{\dot{B}\dot{R}}. \tag{2.37}$$

These formulas can be expressed in a more compact form in terms of the connection one-forms $\mathbf{\Gamma}_{AB}$ and $\mathbf{\Gamma}_{\dot{A}\dot{B}}$:

$$\begin{aligned}\mathbf{\Gamma}'{}_{AB} &= K_A{}^R K_B{}^S \mathbf{\Gamma}_{RS} - K_A{}^R \mathrm{d}K_{BR},\\ \mathbf{\Gamma}'{}_{\dot{A}\dot{B}} &= M_{\dot{A}}{}^{\dot{R}} M_{\dot{B}}{}^{\dot{S}} \mathbf{\Gamma}_{\dot{R}\dot{S}} - M_{\dot{A}}{}^{\dot{R}} \mathrm{d}M_{\dot{B}\dot{R}}.\end{aligned} \tag{2.38}$$

Spinor Fields

Throughout Chapter 1, and in what follows, spinors are given by means of their components with respect to some basis, which is induced by an orthonormal basis of a vector space V, or of the tangent space to M at some point. A simultaneous rotation through 2π in both spin spaces leaves the corresponding orthonormal basis invariant, but changes the sign of any spinor with an odd number of indices [see (1.112), (1.116), (1.155), and (1.171)]. In order to define spinor fields on M, it is necessary to assign consistently the change of sign under these rotations over M, which imposes some restrictions on the global structure of M (see, e.g., Geroch 1968, 1970, Penrose and Rindler 1984, Wald 1984, Lawson and Michelsohn 1989, Huggett and Tod 1994). We shall assume that M admits a spinor structure, and we shall restrict ourselves to local considerations.

A spinor field will be given by means of its components with respect to bases of the spin spaces (or spin frames); a basis of one of the spin spaces together with a basis of the other spin space defines a unique null tetrad, but the converse is not true.

Covariant Derivatives of Spinor Fields

Since the Riemannian connection of M is compatible with the metric tensor, the parallel transport of tangent vectors to M along a given curve on M is an orthogonal transformation between the tangent spaces at any two points on the curve. Hence, with respect to a (differentiable) null tetrad $\partial_{A\dot{B}}$, this orthogonal transformation must be of the form $K^A{}_B M^{\dot{C}}{}_{\dot{D}}$ (since by continuity, its determinant must be positive), which means that the parallel transport of spinors along a curve can be defined in such a way that undotted [resp. dotted] spinors are mapped onto undotted [resp. dotted] spinors, and therefore, the covariant derivative of any spinor field will be a spinor field of the same type as the original spinor field.

Equation (2.32) is consistent with the assumption that for one-index spinor fields,

$$\nabla_{A\dot{B}}\phi^C = \partial_{A\dot{B}}\phi^C - \Gamma^C{}_{SA\dot{B}}\phi^S, \qquad \nabla_{A\dot{B}}\phi^{\dot{C}} = \partial_{A\dot{B}}\phi^{\dot{C}} - \Gamma^{\dot{C}}{}_{\dot{S}\dot{B}A}\phi^{\dot{S}}, \tag{2.39}$$

and that the covariant derivative of spinor fields also satisfies the Leibniz rule; then, as a consequence of the symmetry $\Gamma_{ABC\dot{D}} = \Gamma_{BAC\dot{D}}$, we have

$$\nabla_{A\dot{B}}\varepsilon^{CD} = 0, \qquad \nabla_{A\dot{B}}\varepsilon^{\dot{C}\dot{D}} = 0,$$

and therefore, assuming that the covariant derivative commutes with the contractions,

$$\nabla_{A\dot{B}}\varepsilon_{CD} = 0, \qquad \nabla_{A\dot{B}}\varepsilon_{\dot{C}\dot{D}} = 0.$$

Thus, for an arbitrary spinor field $\psi^{A\ldots\dot{B}\ldots}_{C\ldots\dot{D}\ldots}$,

$$\begin{aligned}\nabla_{R\dot{S}}\psi^{A\ldots\dot{B}\ldots}_{C\ldots\dot{D}\ldots} &= \partial_{R\dot{S}}\psi^{A\ldots\dot{B}\ldots}_{C\ldots\dot{D}\ldots} - \Gamma^A{}_{PR\dot{S}}\psi^{P\ldots\dot{B}\ldots}_{C\ldots\dot{D}\ldots} - \cdots - \Gamma^{\dot{B}}{}_{\dot{P}\dot{S}R}\psi^{A\ldots\dot{P}\ldots}_{C\ldots\dot{D}\ldots} - \cdots \\ &\quad + \Gamma^P{}_{CR\dot{S}}\psi^{A\ldots\dot{B}\ldots}_{P\ldots\dot{D}\ldots} + \cdots + \Gamma^{\dot{P}}{}_{\dot{B}\dot{S}R}\psi^{A\ldots\dot{B}\ldots}_{C\ldots\dot{P}\ldots} + \cdots .\end{aligned} \tag{2.40}$$

Weighted Quantities

When the metric has Lorentzian signature, any nonvanishing complex-valued function λ defines a tetrad rotation (2.35) with the SL(2, $\mathbb{C}$)-valued matrix

$$(K^A{}_B) = \begin{pmatrix} \lambda^{-1} & 0 \\ 0 & \lambda \end{pmatrix} \tag{2.41}$$

and $M^{\dot{A}}{}_{\dot{B}} = \overline{K^A{}_B}$. Following Geroch, Held, and Penrose (1973), a quantity η is of type (p,q) if under the tetrad rotation defined by (2.41), it transforms into

$$\eta' = \lambda^p \overline{\lambda}^q \eta.$$

Thus, for instance, from $\partial'_{A\dot{B}} = K_A{}^C M_{\dot{B}}{}^{\dot{D}} \partial_{C\dot{D}}$, it follows that the null tetrad vector fields $\partial_{1\dot{1}}$, $\partial_{1\dot{2}}$, $\partial_{2\dot{1}}$, and $\partial_{2\dot{2}}$ are of type $(1,1)$, $(1,-1)$, $(-1,1)$, and $(-1,-1)$, respectively. In a similar way, each component $\psi_{AB\ldots\dot{C}\dot{D}\ldots}$ of a spinor field is of type (p,q), where p is the number of undotted subscripts, $A, B, \ldots$ equal to 1 minus the number of these subscripts equal to 2, while q is the number of dotted subscripts $\dot{C}, \dot{D}, \ldots$ equal to $\dot{1}$ minus the number of these subscripts equal to $\dot{2}$ [see (1.187)]. According to (2.36) and (2.37), the same rules apply to the spin coefficients $\Gamma_{11A\dot{B}}$, $\Gamma_{22A\dot{B}}$, $\Gamma_{\dot{1}\dot{1}\dot{A}B}$, and $\Gamma_{\dot{2}\dot{2}\dot{A}B}$, but

$$\Gamma'_{12C\dot{C}} = K_C{}^D M_{\dot{C}}{}^{\dot{D}} (\Gamma_{12D\dot{D}} + \lambda^{-1} \partial_{D\dot{D}} \lambda), \tag{2.42}$$

which means that the spin coefficients $\Gamma_{12D\dot{D}}$ are not of a definite type. Similarly, even though each member of the null tetrad $\partial_{A\dot{B}}$ is of a definite type, the directional derivative $\partial_{A\dot{B}}\eta$ of a quantity of type (p,q) is not of a definite type. In fact, under the tetrad rotation defined by (2.41), $\partial_{A\dot{B}}\eta$ transforms into

$$\partial'_{A\dot{B}}\eta' = K_A{}^C M_{\dot{B}}{}^{\dot{D}} \partial_{C\dot{D}} (\lambda^p \overline{\lambda}^q \eta). \tag{2.43}$$

Thus, by combining (2.43) with (2.42) and its conjugate, one finds that

$$(\partial'_{A\dot{B}} - p\Gamma'_{12A\dot{B}} - q\Gamma'_{\dot{1}\dot{2}\dot{B}A})\eta' = K_A{}^C M_{\dot{B}}{}^{\dot{D}} \lambda^p \overline{\lambda}^q (\partial_{C\dot{D}} - p\Gamma_{12C\dot{D}} - q\Gamma_{\dot{1}\dot{2}\dot{D}C})\eta, \tag{2.44}$$

which means that $(\partial_{A\dot{B}} - p\Gamma_{12A\dot{B}} - q\Gamma_{\dot{1}\dot{2}\dot{B}A})\eta$ has a well-defined type (which depends on p, q, and the specific values of A and $\dot{B}$). This fact leads to the definition of the Geroch–Held–Penrose (GHP) operators

$$\begin{aligned} \text{Þ}\eta &\equiv (D - p\varepsilon - q\overline{\varepsilon})\eta, \\ \eth\eta &\equiv (\delta - p\beta - q\overline{\alpha})\eta, \\ \eth'\eta &\equiv (\overline{\delta} - p\alpha - q\overline{\beta})\eta, \\ \text{Þ}'\eta &\equiv (\Delta - p\gamma - q\overline{\gamma})\eta, \end{aligned} \tag{2.45}$$

if η is of type (p,q), where we have made use of the notation (2.25) and (2.28). From (2.44) it follows that $\text{Þ}\eta$, $\eth\eta$, $\text{Þ}'\eta$, and $\eth'\eta$ are of type $(p+1,q+1)$, $(p+1,q-1)$, $(p-1,q+1)$, and $(p-1,q-1)$, respectively.

From the definition one finds that if η is of type (p,q), then its complex conjugate $\overline{\eta}$ is of type (q,p). Hence,

$$\overline{\text{Þ}\eta} = \text{Þ}\overline{\eta}, \qquad \overline{\eth\eta} = \eth'\overline{\eta}, \qquad \overline{\text{Þ}'\eta} = \text{Þ}'\overline{\eta}.$$

The spin weight s and the boost weight w of η are defined by

$$s = \frac{p-q}{2}, \qquad w = \frac{p+q}{2}.$$

When the signature of the metric is Euclidean, any pair of complex-valued functions λ, μ such that $|\lambda| = 1 = |\mu|$ defines a tetrad rotation (2.35) with the SU(2)-valued matrices

$$(K^A{}_B) = \begin{pmatrix} \lambda^{-1} & 0 \\ 0 & \lambda \end{pmatrix}, \qquad (M^{\dot{A}}{}_{\dot{B}}) = \begin{pmatrix} \mu^{-1} & 0 \\ 0 & \mu \end{pmatrix}. \tag{2.46}$$

A quantity η is said to be of type (p,q) if under the tetrad rotation given by (2.46) it transforms according to $\eta' = \lambda^p \mu^q \eta$. Following a procedure similar to that employed in the previous case, one finds that if η is of type (p,q), then

$$\begin{aligned} \text{Þ}\eta &\equiv (D - p\varepsilon - q\widetilde{\varepsilon})\eta, \\ \eth\eta &\equiv (\delta - p\beta - q\widetilde{\alpha})\eta, \\ \eth'\eta &\equiv (\widetilde{\delta} - p\alpha - q\widetilde{\beta})\eta, \\ \text{Þ}'\eta &\equiv (\Delta - p\gamma - q\widetilde{\gamma})\eta, \end{aligned} \tag{2.47}$$

are of type $(p+1,q+1)$, $(p+1,q-1)$, $(p-1,q+1)$, and $(p-1,q-1)$, respectively [cf. (2.45)].

Now, if η is of type (p,q), then $\overline{\eta}$ is of type $(-p,-q)$, and as a consequence of (2.29) and (2.30),

$$\overline{\text{Þ}\eta} = -\text{Þ}'\overline{\eta}, \qquad \overline{\eth\eta} = \eth'\overline{\eta}.$$

Making use of the corresponding definitions, one finds that in all cases, if η is of type (p,q) and κ is of type (r,s), then $\eta\kappa$ is of type $(p+r,q+s)$ and

$$\mathscr{O}(\eta\kappa) = \eta\mathscr{O}\kappa + \kappa\mathscr{O}\eta,$$

where $\mathscr{O}$ is any of the weighted operators Þ, $\eth$, $\eth'$, and Þ$'$.

The use of weighted quantities allows one to write many explicit expressions in a compact form, simplifying some computations (see, e.g., Jeffryes 1986). Whereas the commutators between the vector fields $\partial_{A\dot{B}}$ involve only the spin coefficients [(2.23)], the presence of the spin coefficients in the weighted operators (2.44) implies that the commutators of the latter contain some components of the curvature [see (2.91)–(2.94)].

2.2 Curvature

The *curvature tensor* Ω of the connection ∇ is defined by

$$\Omega(X,Y)Z = \nabla_X \nabla_Y Z - \nabla_Y \nabla_X Z - \nabla_{[X,Y]} Z, \tag{2.48}$$

for any vector fields X, Y, Z on M. With respect to an orthonormal tetrad $\{\partial_a\}$, the curvature of a connection is represented by the *curvature two-forms* $\Omega^a{}_b$, defined by

$$\Omega^a{}_b(X,Y) \equiv \tfrac{1}{2}\theta^a\big(\Omega(X,Y)\partial_b\big), \tag{2.49}$$

for any pair of vector fields X, Y on M. One can verify that the $\Omega^a{}_b$ are indeed two-forms and that

$$\Omega^a{}_b = \tfrac{1}{2}R^a{}_{bcd}\theta^c \wedge \theta^d, \tag{2.50}$$

where the $R^a{}_{bcd}$ are the components of the curvature tensor with respect to the tetrad $\{\partial_a\}$, which are defined by $\Omega(\partial_a,\partial_b)\partial_c = R^d{}_{cab}\partial_d$.

Making use of (2.8), (2.9), (2.48), (2.49), (2.4), and the properties of ∇, one obtains the *Cartan second structure equations*

$$\Omega^a{}_b = \mathrm{d}\Gamma^a{}_b + \Gamma^a{}_c \wedge \Gamma^c{}_b. \tag{2.51}$$

The curvature two-forms satisfy the relation $\Omega_{ab} = -\Omega_{ba}$, which can be readily obtained from (2.51). Indeed, from (2.51) and the fact that the connection one-forms satisfy $\Gamma_{ab} = -\Gamma_{ba}$ we have

$$\Omega_{ab} = \mathrm{d}\Gamma_{ab} + \Gamma_{ac} \wedge \Gamma^c{}_b = -\mathrm{d}\Gamma_{ba} - \Gamma_{bc} \wedge \Gamma^c{}_a = -\Omega_{ba}.$$

Hence, the components of the curvature tensor satisfy $R_{abcd} = -R_{bacd}$.

Applying the exterior derivative operator d to both sides of (2.7), using again these equations and (2.51) one obtains the identities

$$\mathrm{d}T^a + \Gamma^a{}_b \wedge T^b = \Omega^a{}_b \wedge \theta^b, \tag{2.52}$$

and therefore, if the torsion of the connection vanishes,

$$\Omega^a{}_b \wedge \theta^b = 0. \tag{2.53}$$

Substituting (2.50) into (2.53), one finds that when the torsion is equal to zero, $R^a{}_{bcd} + R^a{}_{cdb} + R^a{}_{dbc} = 0$.

Similarly, applying the exterior derivative operator to both sides of (2.51), one obtains the *Bianchi identities*

$$\mathrm{d}\Omega^a{}_b + \Gamma^a{}_c \wedge \Omega^c{}_b - \Gamma^c{}_b \wedge \Omega^a{}_c = 0. \tag{2.54}$$

2.2.1 Curvature Spinors

Any two-form can be expressed locally in terms of the exterior products $\theta^a \wedge \theta^b$, or equivalently, in terms of the exterior products $\theta^{A\dot{A}} \wedge \theta^{B\dot{B}}$, which, according to (1.50) and (1.69), can be written in the form

$$\begin{aligned}\theta^{A\dot{A}} \wedge \theta^{B\dot{B}} &= \tfrac{1}{2}\sigma_a{}^{A\dot{A}}\sigma_b{}^{B\dot{B}}\,\theta^a \wedge \theta^b \\ &= \left(\varepsilon^{\dot{A}\dot{B}}\tfrac{1}{4}\sigma_a{}^{(A|\dot{R}|}\sigma_b{}^{B)}{}_{\dot{R}} + \varepsilon^{AB}\tfrac{1}{4}\sigma_a{}^{R(\dot{A}}\sigma_{bR}{}^{\dot{B})}\right)\theta^a \wedge \theta^b \\ &= \left(\tfrac{1}{4}S_{ab}{}^{AB}\varepsilon^{\dot{A}\dot{B}} + \tfrac{1}{4}S_{ab}{}^{\dot{A}\dot{B}}\varepsilon^{AB}\right)\theta^a \wedge \theta^b \\ &\equiv S^{AB}\varepsilon^{\dot{A}\dot{B}} + S^{\dot{A}\dot{B}}\varepsilon^{AB},\end{aligned} \tag{2.55}$$

where we have introduced the six two-forms

$$S^{AB} \equiv \tfrac{1}{4}S_{ab}{}^{AB}\theta^a \wedge \theta^b, \qquad S^{\dot{A}\dot{B}} \equiv \tfrac{1}{4}S_{ab}{}^{\dot{A}\dot{B}}\theta^a \wedge \theta^b. \tag{2.56}$$

Then, from (2.55) we have

$$S^{AB} = \tfrac{1}{2}\theta^{A\dot{A}} \wedge \theta^B{}_{\dot{A}}, \qquad S^{\dot{A}\dot{B}} = \tfrac{1}{2}\theta^{A\dot{A}} \wedge \theta_A{}^{\dot{B}}, \tag{2.57}$$

or in an explicit form,

$$\begin{aligned}&S^{11} = \theta^{1\dot{1}} \wedge \theta^{1\dot{2}}, \quad S^{12} = \tfrac{1}{2}(\theta^{1\dot{1}} \wedge \theta^{2\dot{2}} - \theta^{1\dot{2}} \wedge \theta^{2\dot{1}}), \quad S^{22} = \theta^{2\dot{1}} \wedge \theta^{2\dot{2}}, \\ &S^{\dot{1}\dot{1}} = \theta^{1\dot{1}} \wedge \theta^{2\dot{1}}, \quad S^{\dot{1}\dot{2}} = \tfrac{1}{2}(\theta^{1\dot{1}} \wedge \theta^{2\dot{2}} - \theta^{2\dot{1}} \wedge \theta^{1\dot{2}}), \quad S^{\dot{2}\dot{2}} = \theta^{1\dot{2}} \wedge \theta^{2\dot{2}}.\end{aligned} \tag{2.58}$$

From (1.76) we see that the duals of the two-forms S^{AB} and $S^{\dot{A}\dot{B}}$ are given by

$${}^*S^{AB} = -\mathrm{i}^q S^{AB}, \qquad {}^*S^{\dot{A}\dot{B}} = \mathrm{i}^q S^{\dot{A}\dot{B}}. \tag{2.59}$$

The two-forms S^{AB} [resp. $S^{\dot{A}\dot{B}}$] form a local basis for the anti-self-dual [resp. self-dual] two-forms.

Making use of (2.57), the properties of the exterior derivative, the Cartan first structure equations (2.15), and the symmetry $\mathbf{\Gamma}_{\dot{R}\dot{S}} = \mathbf{\Gamma}_{\dot{S}\dot{R}}$, one finds that

$$\begin{aligned}\mathrm{d}S^{AB} &= \tfrac{1}{2}\mathrm{d}(\theta^{A\dot{R}} \wedge \theta^B{}_{\dot{R}}) \\ &= \mathrm{d}\theta^{(A|\dot{R}|} \wedge \theta^{B)}{}_{\dot{R}} \\ &= (\mathbf{\Gamma}^{(A}{}_C \wedge \theta^{|C\dot{R}|} + \mathbf{\Gamma}^{\dot{R}}{}_{\dot{S}} \wedge \theta^{(A|\dot{S}|}) \wedge \theta^{B)}{}_{\dot{R}} \\ &= 2\mathbf{\Gamma}^{(A}{}_C \wedge S^{B)C},\end{aligned} \tag{2.60}$$

and in a similar way one obtains

$$\mathrm{d}S^{\dot{A}\dot{B}} = 2\mathbf{\Gamma}^{(\dot{A}}{}_{\dot{C}} \wedge S^{\dot{B})\dot{C}}. \tag{2.61}$$

As we shall show below, the relations (2.60) and (2.61) are very useful in obtaining the spin coefficients for a given tetrad. (An equivalent formulation, for the Lorentzian signature, without the spinor notation is given in Israel 1979.)

Since the components of the curvature tensor of a connection compatible with the metric are antisymmetric in the first and the last pairs of indices $R_{abcd} = -R_{bacd} = -R_{abdc}$, making use of (1.50) it follows that the spinor equivalent of the curvature tensor can be written as

$$\begin{aligned} R_{A\dot{A}B\dot{B}C\dot{C}D\dot{D}} &= X_{ABCD}\varepsilon_{\dot{A}\dot{B}}\varepsilon_{\dot{C}\dot{D}} + \widetilde{C}_{\dot{A}\dot{B}CD}\varepsilon_{AB}\varepsilon_{\dot{C}\dot{D}} \\ &\quad + C_{AB\dot{C}\dot{D}}\varepsilon_{\dot{A}\dot{B}}\varepsilon_{CD} + X_{\dot{A}\dot{B}\dot{C}\dot{D}}\varepsilon_{AB}\varepsilon_{CD}, \end{aligned} \tag{2.62}$$

where the sets of functions X_{ABCD}, $\widetilde{C}_{\dot{A}\dot{B}CD}$, $C_{AB\dot{C}\dot{D}}$, and $X_{\dot{A}\dot{B}\dot{C}\dot{D}}$ are symmetric in the first and the last pairs of indices, $X_{ABCD} = X_{(AB)(CD)}$, $\widetilde{C}_{\dot{A}\dot{B}CD} = \widetilde{C}_{(\dot{A}\dot{B})(CD)}$, $C_{AB\dot{C}\dot{D}} = C_{(AB)(\dot{C}\dot{D})}$, and $X_{\dot{A}\dot{B}\dot{C}\dot{D}} = X_{(\dot{A}\dot{B})(\dot{C}\dot{D})}$. If the Ricci tensor is defined by

$$R_{ab} \equiv R^{c}{}_{acb}, \tag{2.63}$$

from (2.62) and (1.46) we find that the spinor equivalent of the Ricci tensor is

$$\begin{aligned} R_{B\dot{B}D\dot{D}} &= -R^{A\dot{A}}{}_{B\dot{B}A\dot{A}D\dot{D}} \\ &= -X^{A}{}_{BAD}\varepsilon_{\dot{B}\dot{D}} + \widetilde{C}_{\dot{B}\dot{D}BD} + C_{BD\dot{B}\dot{D}} - X^{\dot{A}}{}_{\dot{B}\dot{A}\dot{D}}\varepsilon_{BD}, \end{aligned} \tag{2.64}$$

and the scalar curvature $R \equiv R^{a}{}_{a}$ is given by

$$R = -R^{B\dot{B}}{}_{B\dot{B}} = 2(X^{AB}{}_{AB} + X^{\dot{A}\dot{B}}{}_{\dot{A}\dot{B}}). \tag{2.65}$$

The antisymmetry of the curvature two-forms $\Omega_{ab} = -\Omega_{ba}$ implies that their spinor equivalents $\Omega_{A\dot{A}B\dot{B}}$ are of the form

$$\Omega_{A\dot{A}B\dot{B}} = \mathscr{R}_{AB}\varepsilon_{\dot{A}\dot{B}} + \mathscr{R}_{\dot{A}\dot{B}}\varepsilon_{AB}, \tag{2.66}$$

where $\mathscr{R}_{AB} = \mathscr{R}_{(AB)}$ and $\mathscr{R}_{\dot{A}\dot{B}} = \mathscr{R}_{(\dot{A}\dot{B})}$ are two-forms. From (2.50) and (2.55) it follows that $\Omega_{A\dot{A}B\dot{B}} = \frac{1}{2}R_{A\dot{A}B\dot{B}C\dot{C}D\dot{D}}(S^{CD}\varepsilon^{\dot{C}\dot{D}} + S^{\dot{C}\dot{D}}\varepsilon^{CD})$, and substituting (2.66) and (2.62), one concludes that the two-forms $\mathscr{R}_{AB}$ and $\mathscr{R}_{\dot{A}\dot{B}}$ are given by

$$\begin{aligned} \mathscr{R}_{AB} &= X_{ABCD}S^{CD} + C_{AB\dot{C}\dot{D}}S^{\dot{C}\dot{D}}, \\ \mathscr{R}_{\dot{A}\dot{B}} &= \widetilde{C}_{\dot{A}\dot{B}CD}S^{CD} + X_{\dot{A}\dot{B}\dot{C}\dot{D}}S^{\dot{C}\dot{D}}. \end{aligned} \tag{2.67}$$

The Cartan second structure equations (2.51) expressed in terms of the spinor equivalents of the connection one-forms and of the curvature two-forms given by (2.13) and (2.66), respectively, read

$$\begin{aligned} \mathscr{R}_{AB} &= \mathrm{d}\boldsymbol{\Gamma}_{AB} - \boldsymbol{\Gamma}_{AC} \wedge \boldsymbol{\Gamma}^{C}{}_{B}, \\ \mathscr{R}_{\dot{A}\dot{B}} &= \mathrm{d}\boldsymbol{\Gamma}_{\dot{A}\dot{B}} - \boldsymbol{\Gamma}_{\dot{A}\dot{C}} \wedge \boldsymbol{\Gamma}^{\dot{C}}{}_{\dot{B}} \end{aligned} \tag{2.68}$$

[cf. (2.51)].

Similarly, making use of (2.66) and (2.67), one finds that (2.53) is equivalent to

$$\begin{aligned} 0 &= \mathscr{R}_{AB} \wedge \theta^{B}{}_{\dot{A}} + \mathscr{R}_{\dot{A}\dot{B}} \wedge \theta_{A}{}^{\dot{B}} \\ &= (X_{ABCD}S^{CD} + C_{AB\dot{C}\dot{D}}S^{\dot{C}\dot{D}}) \wedge \theta^{B}{}_{\dot{A}} + (\widetilde{C}_{\dot{A}\dot{B}CD}S^{CD} + X_{\dot{A}\dot{B}\dot{C}\dot{D}}S^{\dot{C}\dot{D}}) \wedge \theta_{A}{}^{\dot{B}}. \end{aligned} \tag{2.69}$$

Owing to the antisymmetry of the exterior product of one-forms, according to (1.58),

$$\theta^{A\dot{A}} \wedge \theta^{B\dot{B}} \wedge \theta^{C\dot{C}} = \varepsilon^{AC}\varepsilon^{\dot{B}\dot{C}}\breve{\theta}^{B\dot{A}} - \varepsilon^{BC}\varepsilon^{\dot{A}\dot{C}}\breve{\theta}^{A\dot{B}}, \tag{2.70}$$

where the $\breve{\theta}^{A\dot{A}}$ are some three-forms, which implies that

$$S^{AB} \wedge \theta^{C\dot{C}} = \varepsilon^{C(A}\breve{\theta}^{B)\dot{C}}, \qquad S^{\dot{A}\dot{B}} \wedge \theta^{C\dot{C}} = -\varepsilon^{\dot{C}(\dot{A}}\breve{\theta}^{|C|\dot{B})} \tag{2.71}$$

[see (2.57)]; hence, (2.69) can be rewritten as

$$0 = X_{A}{}^{B}{}_{BD}\breve{\theta}^{D}{}_{\dot{A}} + C_{AB\dot{A}\dot{D}}\breve{\theta}^{B\dot{D}} - \widetilde{C}_{\dot{A}\dot{B}AD}\breve{\theta}^{D\dot{B}} - X_{\dot{A}}{}^{\dot{B}}{}_{\dot{B}\dot{D}}\breve{\theta}_{A}{}^{\dot{D}}.$$

At each point of its domain, the set of three-forms $\{\breve{\theta}^{A\dot{A}}\}$ is linearly independent (see Exercise 2.2), and therefore the last equation is equivalent to

$$0 = X_{A}{}^{B}{}_{BD}\varepsilon_{\dot{A}\dot{D}} + C_{AD\dot{A}\dot{D}} - \widetilde{C}_{\dot{A}\dot{D}AD} - X_{\dot{A}}{}^{\dot{B}}{}_{\dot{B}\dot{D}}\varepsilon_{AD}. \tag{2.72}$$

Hence, contracting the last equation with $\varepsilon^{\dot{A}\dot{D}}$ yields

$$0 = 2X_{A}{}^{B}{}_{BD} - X^{\dot{D}\dot{B}}{}_{\dot{B}\dot{D}}\varepsilon_{AD}, \tag{2.73}$$

while contracting with ε^{AD} gives us

$$0 = X^{DB}{}_{BD}\varepsilon_{\dot{A}\dot{D}} - 2X_{\dot{A}}{}^{\dot{B}}{}_{\dot{B}\dot{D}}. \tag{2.74}$$

From (2.73) or (2.74) we obtain $X^{AB}{}_{AB} = X^{\dot{A}\dot{B}}{}_{\dot{A}\dot{B}}$, and therefore (2.65) implies that

$$X^{AB}{}_{AB} = X^{\dot{A}\dot{B}}{}_{\dot{A}\dot{B}} = \tfrac{1}{4}R. \tag{2.75}$$

Thus, from (2.73) and (2.74) we obtain

$$X_{A}{}^{B}{}_{BD} = \tfrac{1}{8}R\varepsilon_{AD}, \qquad X_{\dot{A}}{}^{\dot{B}}{}_{\dot{B}\dot{D}} = \tfrac{1}{8}R\varepsilon_{\dot{A}\dot{D}}, \tag{2.76}$$

and substituting these expressions into (2.72), it reduces to

$$\widetilde{C}_{\dot{A}\dot{D}AD} = C_{AD\dot{A}\dot{D}}. \tag{2.77}$$

As a consequence of (2.76), X_{ABCD} and $X_{\dot{A}\dot{B}\dot{C}\dot{D}}$ are symmetric not only on the first and second pairs of indices, but also under the exchange of the first and second pairs of indices. Indeed, from (1.27) and (2.76) we see that

$$\begin{aligned}X_{ABCD} - X_{CDAB} &= X_{ABCD} - X_{ACBD} + X_{ACBD} - X_{CDAB} \\ &= \varepsilon_{BC} X_A{}^M{}_{MD} + \varepsilon_{AD} X_C{}^M{}_{MB} \\ &= \tfrac{1}{8} R(\varepsilon_{BC}\varepsilon_{AD} + \varepsilon_{AD}\varepsilon_{CB}) \\ &= 0,\end{aligned}$$

i.e.,

$$X_{ABCD} = X_{CDAB}, \tag{2.78}$$

and in a similar manner one concludes that

$$X_{\dot{A}\dot{B}\dot{C}\dot{D}} = X_{\dot{C}\dot{D}\dot{A}\dot{B}}. \tag{2.79}$$

From (2.62) and (2.77)–(2.79) it follows that $R_{A\dot{A}B\dot{B}C\dot{C}D\dot{D}} = R_{C\dot{C}D\dot{D}A\dot{A}B\dot{B}}$, which is equivalent to the well-known relation $R_{abcd} = R_{cdab}$.

Making use of (2.78) and (2.76), one finds that the totally symmetric part of X_{ABCD}, which will be denoted by C_{ABCD}, is

$$\begin{aligned}C_{ABCD} &= X_{(ABCD)} \\ &= \tfrac{1}{3}(X_{ABCD} + X_{ACBD} + X_{ADBC}) \\ &= X_{ABCD} + \tfrac{1}{3}(X_{ACBD} - X_{ABCD} + X_{ADBC} - X_{ABCD}) \\ &= X_{ABCD} + \tfrac{1}{3}(\varepsilon_{CB} X_A{}^M{}_{MD} + \varepsilon_{DB} X_A{}^M{}_{MC}) \\ &= X_{ABCD} - \tfrac{1}{24} R(\varepsilon_{AC}\varepsilon_{BD} + \varepsilon_{AD}\varepsilon_{BC}).\end{aligned} \tag{2.80}$$

Similarly, denoting by $C_{\dot{A}\dot{B}\dot{C}\dot{D}}$ the totally symmetric part of $X_{\dot{A}\dot{B}\dot{C}\dot{D}}$, one obtains

$$C_{\dot{A}\dot{B}\dot{C}\dot{D}} = X_{\dot{A}\dot{B}\dot{C}\dot{D}} - \tfrac{1}{24} R(\varepsilon_{\dot{A}\dot{C}}\varepsilon_{\dot{B}\dot{D}} + \varepsilon_{\dot{A}\dot{D}}\varepsilon_{\dot{B}\dot{C}}). \tag{2.81}$$

Thus, from (2.67), (2.77), (2.80), and (2.81) we see that the curvature two-forms can be expressed as

$$\begin{aligned}\mathscr{R}_{AB} &= C_{ABCD} S^{CD} + \tfrac{1}{12} R\, S_{AB} + C_{AB\dot{C}\dot{D}} S^{\dot{C}\dot{D}}, \\ \mathscr{R}_{\dot{A}\dot{B}} &= C_{CD\dot{A}\dot{B}} S^{CD} + \tfrac{1}{12} R\, S_{\dot{A}\dot{B}} + C_{\dot{A}\dot{B}\dot{C}\dot{D}} S^{\dot{C}\dot{D}},\end{aligned} \tag{2.82}$$

and from (2.62), the spinor equivalent of the curvature tensor is given by

$$\begin{aligned}R_{A\dot{A}B\dot{B}C\dot{C}D\dot{D}} = {} & C_{ABCD}\varepsilon_{\dot{A}\dot{B}}\varepsilon_{\dot{C}\dot{D}} + C_{\dot{A}\dot{B}\dot{C}\dot{D}}\varepsilon_{AB}\varepsilon_{CD} \\ & + C_{AB\dot{C}\dot{D}}\varepsilon_{\dot{A}\dot{B}}\varepsilon_{CD} + C_{CD\dot{A}\dot{B}}\varepsilon_{AB}\varepsilon_{\dot{C}\dot{D}} \\ & + \tfrac{1}{24} R(\varepsilon_{AC}\varepsilon_{BD}\varepsilon_{\dot{A}\dot{B}}\varepsilon_{\dot{C}\dot{D}} + \varepsilon_{AD}\varepsilon_{BC}\varepsilon_{\dot{A}\dot{B}}\varepsilon_{\dot{C}\dot{D}} \\ & \qquad + \varepsilon_{AB}\varepsilon_{CD}\varepsilon_{\dot{A}\dot{C}}\varepsilon_{\dot{B}\dot{D}} + \varepsilon_{AB}\varepsilon_{CD}\varepsilon_{\dot{A}\dot{D}}\varepsilon_{\dot{B}\dot{C}}).\end{aligned} \tag{2.83}$$

Making use of the relation

$$\varepsilon_{AB}\varepsilon_{CD} + \varepsilon_{AC}\varepsilon_{DB} + \varepsilon_{AD}\varepsilon_{BC} = 0, \tag{2.84}$$

which is equivalent to (1.27), and its dotted version, one can rewrite in various ways the expression between parentheses in the last two lines of (2.83), e.g.,

$$2(\varepsilon_{AB}\varepsilon_{CD}\varepsilon_{\dot{A}\dot{C}}\varepsilon_{\dot{B}\dot{D}}+\varepsilon_{AD}\varepsilon_{BC}\varepsilon_{\dot{A}\dot{B}}\varepsilon_{\dot{C}\dot{D}}),$$

$$2(\varepsilon_{AB}\varepsilon_{CD}\varepsilon_{\dot{A}\dot{D}}\varepsilon_{\dot{B}\dot{C}}+\varepsilon_{AC}\varepsilon_{BD}\varepsilon_{\dot{A}\dot{B}}\varepsilon_{\dot{C}\dot{D}}),$$

and

$$2(\varepsilon_{AC}\varepsilon_{\dot{A}\dot{C}}\varepsilon_{BD}\varepsilon_{\dot{B}\dot{D}}-\varepsilon_{AD}\varepsilon_{\dot{A}\dot{D}}\varepsilon_{BC}\varepsilon_{\dot{B}\dot{C}});$$

according to (1.21), the last expression is the spinor equivalent of $2(g_{ac}g_{bd}-g_{ad}g_{bc})$.

Substituting (2.76) and (2.77) into (2.64), one finds that the spinor equivalent of the Ricci tensor is related to $C_{AB\dot{A}\dot{B}}$ by $R_{B\dot{B}D\dot{D}}=2C_{BD\dot{B}\dot{D}}-\frac{1}{4}R\varepsilon_{BD}\varepsilon_{\dot{B}\dot{D}}$, i.e.,

$$C_{AB\dot{A}\dot{B}}=\tfrac{1}{2}(R_{A\dot{A}B\dot{B}}+\tfrac{1}{4}R\varepsilon_{AB}\varepsilon_{\dot{A}\dot{B}}). \qquad (2.85)$$

Thus, recalling that $-\varepsilon_{AB}\varepsilon_{\dot{A}\dot{B}}$ is the spinor equivalent of g_{ab}, (2.85) shows that $C_{AB\dot{A}\dot{B}}$ is the spinor equivalent of $\frac{1}{2}(R_{ab}-\frac{1}{4}Rg_{ab})$; that is, apart from a factor of $\frac{1}{2}$, $C_{AB\dot{A}\dot{B}}$ is the spinor equivalent of the traceless part of the Ricci tensor.

Making use of (1.27), it can readily be seen that

$$\begin{aligned}C_{AB\dot{C}\dot{D}}\varepsilon_{\dot{A}\dot{B}}\varepsilon_{CD}+C_{CD\dot{A}\dot{B}}\varepsilon_{AB}\varepsilon_{\dot{C}\dot{D}} &= C_{AD\dot{A}\dot{D}}\varepsilon_{BC}\varepsilon_{\dot{B}\dot{C}}-C_{AC\dot{A}\dot{C}}\varepsilon_{BD}\varepsilon_{\dot{B}\dot{D}}\\ &\quad +C_{BC\dot{B}\dot{C}}\varepsilon_{AD}\varepsilon_{\dot{A}\dot{D}}-C_{BD\dot{B}\dot{D}}\varepsilon_{AC}\varepsilon_{\dot{A}\dot{C}},\end{aligned}$$

which means that the second line of (2.83) is the spinor equivalent of the tensor $\frac{1}{2}(R_{ad}-\frac{1}{4}Rg_{ad})(-g_{bc})-\frac{1}{2}(R_{ac}-\frac{1}{4}Rg_{ac})(-g_{bd})+\frac{1}{2}(R_{bc}-\frac{1}{4}Rg_{bc})(-g_{ad})-\frac{1}{2}(R_{bd}-\frac{1}{4}Rg_{bd})(-g_{ac})=-\frac{1}{2}(R_{ad}g_{bc}-R_{ac}g_{bd}+R_{bc}g_{ad}-R_{bd}g_{ac})+\frac{1}{4}(g_{ad}g_{bc}-g_{ac}g_{bd})$; hence, it follows from (2.83) that $C_{ABCD}\varepsilon_{\dot{A}\dot{B}}\varepsilon_{\dot{C}\dot{D}}+C_{\dot{A}\dot{B}\dot{C}\dot{D}}\varepsilon_{AB}\varepsilon_{CD}$ is the spinor equivalent of

$$C_{abcd}\equiv R_{abcd}+\tfrac{1}{2}(R_{ad}g_{bc}-R_{ac}g_{bd}+R_{bc}g_{ad}-R_{bd}g_{ac})+\tfrac{1}{6}R(g_{ac}g_{bd}-g_{ad}g_{bc}). \qquad (2.86)$$

The tensor field C_{abcd} is known as the *Weyl tensor*, and C_{ABCD} and $C_{\dot{A}\dot{B}\dot{C}\dot{D}}$ are called *Weyl spinors*. Owing to its behavior under conformal rescalings, C_{abcd} is also called the *conformal curvature tensor* (see Section 2.3).

The tensor field C_{abcd} possesses all the symmetries of the curvature tensor R_{abcd} (i.e., $C_{abcd}=-C_{bacd}=-C_{abdc}$, $C_{abcd}+C_{acdb}+C_{adbc}=0$, $C_{abcd}=C_{cdab}$), and in addition, $C^a{}_{bad}=0$, as can readily be seen using the total symmetry of C_{ABCD} and $C_{\dot{A}\dot{B}\dot{C}\dot{D}}$ [see (1.29)].

Since the curvature tensor is real, making use of (2.83), (1.224), and (1.221) one finds that when the signature is $(++++)$ or $(++--)$, the curvature spinors must satisfy

$$\widehat{C}_{ABCD}=C_{ABCD}, \qquad \widehat{C}_{\dot{A}\dot{B}\dot{C}\dot{D}}=C_{\dot{A}\dot{B}\dot{C}\dot{D}}, \qquad \widehat{C}_{AB\dot{C}\dot{D}}=C_{AB\dot{C}\dot{D}}, \qquad (2.87)$$

while when the signature is $(+++-)$ one has

$$\overline{C_{ABCD}} = C_{\dot{A}\dot{B}\dot{C}\dot{D}}, \qquad \overline{C_{AB\dot{C}\dot{D}}} = C_{CD\dot{A}\dot{B}} \tag{2.88}$$

[see (1.47)].

Combining (1.41) and (2.62), we see that

$$X_{ABCD}\varepsilon_{\dot{A}\dot{B}}\varepsilon_{\dot{C}\dot{D}} + C_{CD\dot{A}\dot{B}}\varepsilon_{AB}\varepsilon_{\dot{C}\dot{D}} + C_{AB\dot{C}\dot{D}}\varepsilon_{\dot{A}\dot{B}}\varepsilon_{CD} + X_{\dot{A}\dot{B}\dot{C}\dot{D}}\varepsilon_{AB}\varepsilon_{CD} = \frac{1}{4}\sigma^{a}{}_{A\dot{A}}\sigma^{b}{}_{B\dot{B}}\sigma^{c}{}_{C\dot{C}}\sigma^{d}{}_{D\dot{D}}R_{abcd};$$

hence [see (1.69)],

$$\begin{aligned} X_{ABCD} &= \frac{1}{16}S^{ab}{}_{AB}S^{cd}{}_{CD}R_{abcd}, \\ C_{AB\dot{C}\dot{D}} &= \frac{1}{16}S^{ab}{}_{AB}S^{cd}{}_{\dot{C}\dot{D}}R_{abcd}, \\ X_{\dot{A}\dot{B}\dot{C}\dot{D}} &= \frac{1}{16}S^{ab}{}_{\dot{A}\dot{B}}S^{cd}{}_{\dot{C}\dot{D}}R_{abcd} \end{aligned} \tag{2.89}$$

[cf. (2.85)]. Since C_{ABCD}, $C_{AB\dot{C}\dot{D}}$, and R are independent of each other, from (2.89) we have, for instance,

$$C_{ABCD} = \frac{1}{16}S^{ab}{}_{AB}S^{cd}{}_{CD}C_{abcd}.$$

The spinor components of the curvature can be expressed in terms of the spin coefficients by substituting (2.16) into (2.68), making use of (2.15) and (2.55). Since the differential of a scalar function is expressed locally as $\mathrm{d}f = (\partial_a f)\,\theta^a$, equivalently we have

$$\mathrm{d}f = -(\partial_{A\dot{A}}f)\,\theta^{A\dot{A}}. \tag{2.90}$$

In this way we obtain the explicit expressions

$$X_{ABCD} = \partial_{(C}{}^{\dot{D}}\Gamma_{|AB|D)\dot{D}} - \Gamma_{AB(C|\dot{C}}\Gamma^{\dot{C}\dot{R}}{}_{\dot{R}|D)} + \Gamma_{ABR\dot{C}}\Gamma^{R}{}_{(CD)}{}^{\dot{C}} + \Gamma_{A}{}^{R}{}_{(C}{}^{\dot{D}}\Gamma_{|BR|D)\dot{D}} \tag{2.91}$$

and

$$C_{AB\dot{C}\dot{D}} = \partial^{R}{}_{(\dot{C}}\Gamma_{|ABR|\dot{D})} + \Gamma_{ABC\dot{R}}\Gamma^{\dot{R}}{}_{(\dot{C}\dot{D})}{}^{C} - \Gamma_{ABC(\dot{C}}\Gamma^{CR}{}_{|R|\dot{D})} + \Gamma_{A}{}^{RD}{}_{(\dot{C}}\Gamma_{|BRD|\dot{D})}, \tag{2.92}$$

together with

$$C_{CD\dot{A}\dot{B}} = \partial_{(C}{}^{\dot{R}}\Gamma_{|\dot{A}\dot{B}\dot{R}|D)} + \Gamma_{\dot{A}\dot{B}\dot{C}R}\Gamma^{R}{}_{(CD)}{}^{\dot{C}} - \Gamma_{\dot{A}\dot{B}\dot{C}(C}\Gamma^{\dot{C}\dot{R}}{}_{|\dot{R}|D)} + \Gamma_{\dot{A}}{}^{\dot{R}\dot{D}}{}_{(C}\Gamma_{|\dot{B}\dot{R}\dot{D}|D)} \tag{2.93}$$

and

$$X_{\dot{A}\dot{B}\dot{C}\dot{D}} = \partial^{D}{}_{(\dot{C}}\Gamma_{|\dot{A}\dot{B}|\dot{D})D} - \Gamma_{\dot{A}\dot{B}(\dot{C}|C}\Gamma^{CR}{}_{R|\dot{D})} + \Gamma_{\dot{A}\dot{B}\dot{R}C}\Gamma^{\dot{R}}{}_{(\dot{C}\dot{D})}{}^{C} + \Gamma_{\dot{A}}{}^{\dot{R}}{}_{(\dot{C}}{}^{D}\Gamma_{|\dot{B}\dot{R}|\dot{D})D}. \tag{2.94}$$

As pointed out above, the spin coefficients can be obtained by means of the commutators (2.23) or from (2.60) and (2.61). Substitution of (2.16) and (2.71) into (2.60) and (2.61) gives

$$\mathrm{d}S^{AB} = -\Gamma^{(A}{}_{C}{}^{B)}{}_{\dot{D}}\,\breve{\theta}^{C\dot{D}} - \Gamma^{(A}{}_{C}{}^{|C|}{}_{\dot{D}}\,\breve{\theta}^{B)\dot{D}} \tag{2.95}$$

and

$$\mathrm{d}S^{\dot{A}\dot{B}} = \Gamma^{(\dot{A}}{}_{\dot{C}}{}^{\dot{B})}{}_{D}\,\breve{\theta}^{D\dot{C}} + \Gamma^{(\dot{A}}{}_{\dot{C}}{}^{|\dot{C}}{}_{D}\,\breve{\theta}^{D|\dot{B})}. \tag{2.96}$$

The combinations of the spin coefficients appearing in (2.95) and (2.96) allow us to find the spin coefficients individually. Writing, say, $\mathrm{d}S^{AB} = K^{AB}{}_{C\dot{D}}\,\breve{\theta}^{C\dot{D}}$, with $K_{ABC\dot{D}} = K_{(AB)C\dot{D}}$, and comparing with (2.95) one has $K_{ABC\dot{D}} = -\Gamma_{C(AB)\dot{D}} - \Gamma_{(A|R}{}^{R}{}_{\dot{D}|}\varepsilon_{B)C}$. Then, from this last relation one readily obtains

$$\Gamma_{ABC\dot{D}} = -K_{C(AB)\dot{D}} \tag{2.97}$$

[cf. (2.12)] (see the examples below). In a similar way one finds that

$$\Gamma_{\dot{A}\dot{B}\dot{C}D} = K_{\dot{C}(\dot{A}\dot{B})D} \tag{2.98}$$

if the functions $K_{\dot{A}\dot{B}\dot{C}D}$ are defined by $\mathrm{d}S^{\dot{A}\dot{B}} = K^{\dot{A}\dot{B}}{}_{\dot{C}D}\breve{\theta}^{D\dot{C}}$ (cf. Israel 1979 and the references cited therein).

Frequently, the metric tensor of M is given by a local expression in terms of some coordinate system, and in order to find a set of one-forms $\theta^{A\dot{A}}$, it is convenient to notice that from (2.1) we obtain

$$\begin{aligned} g &= -\varepsilon_{AB}\varepsilon_{\dot{A}\dot{B}}\,\theta^{A\dot{A}} \otimes \theta^{B\dot{B}} \\ &= -\theta^{1\dot{1}} \otimes \theta^{2\dot{2}} - \theta^{2\dot{2}} \otimes \theta^{1\dot{1}} + \theta^{1\dot{2}} \otimes \theta^{2\dot{1}} + \theta^{2\dot{1}} \otimes \theta^{1\dot{2}} \\ &= -2\theta^{1\dot{1}}\theta^{2\dot{2}} + 2\theta^{1\dot{2}}\theta^{2\dot{1}}, \end{aligned} \tag{2.99}$$

where the juxtaposition of one-forms means symmetrized tensor product, and

$$\overline{\theta^{A\dot{B}}} = \begin{cases} -\theta_{A\dot{B}} & \text{if the } \sigma^{a}{}_{A\dot{B}} \text{ are given by (1.33),} \\ \theta^{B\dot{A}} & \text{if the } \sigma^{a}{}_{A\dot{B}} \text{ are given by (1.34),} \\ \theta^{A\dot{B}} & \text{if the } \sigma^{a}{}_{A\dot{B}} \text{ are given by (1.35),} \\ \eta^{AC}\eta^{\dot{B}\dot{D}}\theta_{C\dot{D}} & \text{if the } \sigma^{a}{}_{A\dot{B}} \text{ are given by (1.37).} \end{cases} \tag{2.100}$$

Thus, instead of looking for an orthonormal or null tetrad, a suitable set of one-forms $\theta^{A\dot{A}}$ can be obtained directly by expressing the metric tensor in the form (2.99) in terms of one-forms satisfying the appropriate condition (2.100) (see the examples below).

The behavior of the differential forms S^{AB}, $S^{\dot{A}\dot{B}}$, and $\breve{\theta}^{A\dot{B}}$ under complex conjugation can be derived from (2.100) and the relations (2.57) and (2.71).

Example 2.1. The standard metric of S^4.

The metric induced on the sphere $\mathrm{S}^4 \equiv \{(x^1,x^2,x^3,x^4,x^5) \in \mathbb{R}^5 \,|\, (x^1)^2 + (x^2)^2 + (x^3)^2 + (x^4)^2 + (x^5)^2 = 1\}$ by the standard Euclidean metric of $\mathbb{R}^5$ is locally given by

$$g = \mathrm{d}\psi^2 + \sin^2\psi\,[\mathrm{d}\chi^2 + \sin^2\chi\,(\mathrm{d}\theta^2 + \sin^2\theta\,\mathrm{d}\varphi^2)], \tag{2.101}$$

in terms of spherical coordinates ψ, χ, θ, and φ. We can equivalently write

$$g = -2\left[-\frac{1}{\sqrt{2}}(\mathrm{d}\psi - \mathrm{i}\sin\psi\,\mathrm{d}\chi)\right]\left[\frac{1}{\sqrt{2}}(\mathrm{d}\psi + \mathrm{i}\sin\psi\,\mathrm{d}\chi)\right]$$
$$+2\left[-\frac{1}{\sqrt{2}}\sin\psi\sin\chi(\mathrm{d}\theta - \mathrm{i}\sin\theta\,\mathrm{d}\varphi)\right]\left[-\frac{1}{\sqrt{2}}\sin\psi\sin\chi(\mathrm{d}\theta + \mathrm{i}\sin\theta\,\mathrm{d}\varphi)\right],$$

and by comparing with (2.99) one finds that the one-forms $\theta^{A\dot{A}}$ satisfying the condition $\overline{\theta^{A\dot{B}}} = -\theta_{A\dot{B}}$ (i.e., $\overline{\theta^{1\dot{1}}} = -\theta^{2\dot{2}}$ and $\overline{\theta^{1\dot{2}}} = \theta^{2\dot{1}}$), appropriate for a metric with Euclidean signature, can be taken as

$$\theta^{1\dot{1}} = -\frac{1}{\sqrt{2}}(\mathrm{d}\psi - \mathrm{i}\sin\psi\,\mathrm{d}\chi), \qquad \theta^{2\dot{2}} = \frac{1}{\sqrt{2}}(\mathrm{d}\psi + \mathrm{i}\sin\psi\,\mathrm{d}\chi),$$
$$\theta^{1\dot{2}} = -\frac{\sin\psi\sin\chi}{\sqrt{2}}(\mathrm{d}\theta - \mathrm{i}\sin\theta\,\mathrm{d}\varphi), \quad \theta^{2\dot{1}} = -\frac{\sin\psi\sin\chi}{\sqrt{2}}(\mathrm{d}\theta + \mathrm{i}\sin\theta\,\mathrm{d}\varphi). \tag{2.102}$$

(Alternatively, from (2.101) we see that

$$\theta^1 = \sin\psi\sin\chi\,\mathrm{d}\theta, \qquad \theta^2 = \sin\psi\sin\chi\sin\theta\,\mathrm{d}\varphi,$$
$$\theta^3 = \mathrm{d}\psi, \qquad \theta^4 = \sin\psi\,\mathrm{d}\chi,$$

form the dual basis of an orthonormal tetrad. The one-forms (2.102) are then obtained with the aid of (2.14) and (1.33).)

The two-forms S^{AB} are then given by [see (2.58)]

$$S^{11} = \tfrac{1}{2}\sin\psi\sin\chi\,(\mathrm{d}\psi - \mathrm{i}\sin\psi\,\mathrm{d}\chi)\wedge(\mathrm{d}\theta - \mathrm{i}\sin\theta\,\mathrm{d}\varphi),$$
$$S^{12} = -\tfrac{1}{2}\mathrm{i}(\sin\psi\,\mathrm{d}\psi\wedge\mathrm{d}\chi + \sin^2\psi\sin^2\chi\sin\theta\,\mathrm{d}\theta\wedge\mathrm{d}\varphi), \tag{2.103}$$

with $S^{22} = \overline{S^{11}}$. The two-forms $S^{\dot{A}\dot{B}}$ can be readily obtained by noticing that the one-forms (2.102) have the property that under the substitution of φ by $-\varphi$, $\theta^{A\dot{B}}$ is mapped into $\theta^{B\dot{A}}$, and therefore S^{AB} is mapped into $S^{\dot{A}\dot{B}}$ [see (2.58)]. A straightforward computation, making use of (2.71), gives

$$\mathrm{d}S^{11} = -\frac{\cot\theta}{\sqrt{2}\sin\psi\sin\chi}\breve{\theta}^{1\dot{1}} - \frac{1}{\sqrt{2}}\left(2\cot\psi - \mathrm{i}\frac{\cot\chi}{\sin\psi}\right)\breve{\theta}^{1\dot{2}},$$
$$\mathrm{d}S^{12} = -\frac{1}{\sqrt{2}}\left(\cot\psi + \mathrm{i}\frac{\cot\chi}{\sin\psi}\right)\breve{\theta}^{1\dot{1}} - \frac{1}{\sqrt{2}}\left(\cot\psi - \mathrm{i}\frac{\cot\chi}{\sin\psi}\right)\breve{\theta}^{2\dot{2}}.$$

Hence, the only nonvanishing spin coefficients $\Gamma_{ABC\dot{D}}$ are determined by [see (2.97) and (2.20)]

$$\Gamma_{112\dot{1}} = -\frac{1}{\sqrt{2}}\left(\cot\psi + \mathrm{i}\frac{\cot\chi}{\sin\psi}\right), \qquad \Gamma_{121\dot{1}} = \Gamma_{122\dot{2}} = \frac{1}{2\sqrt{2}}\cot\psi,$$

$$\Gamma_{121\dot{2}} = -\Gamma_{122\dot{1}} = \frac{1}{2\sqrt{2}}\frac{\cot\theta}{\sin\psi\sin\chi},$$

and the connection one-forms $\boldsymbol{\Gamma}_{AB}$ for this tetrad are

$$\begin{aligned}
\boldsymbol{\Gamma}_{11} &= -\frac{1}{2}(\cos\psi\sin\chi + \mathrm{i}\cos\chi)(\mathrm{d}\theta + \mathrm{i}\sin\theta\,\mathrm{d}\varphi),\\
\boldsymbol{\Gamma}_{12} &= -\frac{\mathrm{i}}{2}(\cos\psi\,\mathrm{d}\chi + \cos\theta\,\mathrm{d}\varphi),\\
\boldsymbol{\Gamma}_{22} &= -\frac{1}{2}(\cos\psi\sin\chi - \mathrm{i}\cos\chi)(\mathrm{d}\theta - \mathrm{i}\sin\theta\,\mathrm{d}\varphi).
\end{aligned} \tag{2.104}$$

Substituting into (2.68), one finds that the curvature two-forms $\mathscr{R}_{AB}$ are given explicitly by

$$\begin{aligned}
\mathscr{R}_{11} &= \mathrm{d}\boldsymbol{\Gamma}_{11} - 2\boldsymbol{\Gamma}_{12}\wedge\boldsymbol{\Gamma}_{11} = S^{22},\\
\mathscr{R}_{12} &= \mathrm{d}\boldsymbol{\Gamma}_{12} + \boldsymbol{\Gamma}_{11}\wedge\boldsymbol{\Gamma}_{22} = -S^{12},\\
\mathscr{R}_{22} &= \mathrm{d}\boldsymbol{\Gamma}_{22} + 2\boldsymbol{\Gamma}_{12}\wedge\boldsymbol{\Gamma}_{22} = S^{11},
\end{aligned} \tag{2.105}$$

and hence $C_{AB\dot{A}\dot{B}} = 0$, $C_{ABCD} = 0$, and $R = 12$ [see (2.82)]. In a similar way one finds that $C_{\dot{A}\dot{B}\dot{C}\dot{D}} = 0$.

Example 2.2. The Schwarzschild metric.

The Schwarzschild metric, which, according to Einstein's theory of gravitation, represents the exterior gravitational field of a spherically symmetric distribution of matter, is usually expressed in the form

$$g = -\left(1 - \frac{r_g}{r}\right)c^2\mathrm{d}t^2 + \frac{\mathrm{d}r^2}{1 - r_g/r} + r^2\mathrm{d}\theta^2 + r^2\sin^2\theta\,\mathrm{d}\varphi^2, \tag{2.106}$$

where r_g is a constant (called the gravitational radius or Schwarzschild radius, $r_g = 2GM/c^2$, where G is Newton's constant of gravitation, M corresponds to the Newtonian mass of the source, and c is the speed of light in vacuum), and t, r, θ, and φ are real coordinates (see, e.g., Rindler 1977, Wald 1984). When $r_g = 0$, (2.106) is the metric of Minkowski space-time in spherical coordinates. By comparing the metric (2.106), written as

$$\begin{aligned}
g = &-2\left[\frac{1}{2}\left(1 - \frac{r_g}{r}\right)c\mathrm{d}t - \frac{1}{2}\mathrm{d}r\right]\left[c\mathrm{d}t + \left(1 - \frac{r_g}{r}\right)^{-1}\mathrm{d}r\right]\\
&+2\left[-\frac{r}{\sqrt{2}}(\mathrm{d}\theta - \mathrm{i}\sin\theta\,\mathrm{d}\varphi)\right]\left[-\frac{r}{\sqrt{2}}(\mathrm{d}\theta + \mathrm{i}\sin\theta\,\mathrm{d}\varphi)\right],
\end{aligned}$$

with (2.99) one finds that the one-forms $\theta^{A\dot{B}}$ satisfying the condition $\overline{\theta^{A\dot{B}}} = \theta^{B\dot{A}}$, appropriate for Lorentzian signature, can be taken as

$$\theta^{1\dot{1}} = \frac{1}{2}\left(1-\frac{r_g}{r}\right)c\mathrm{d}t - \frac{1}{2}\mathrm{d}r, \qquad \theta^{2\dot{2}} = c\mathrm{d}t + \left(1-\frac{r_g}{r}\right)^{-1}\mathrm{d}r,$$
$$\theta^{1\dot{2}} = -\frac{r}{\sqrt{2}}(\mathrm{d}\theta - \mathrm{i}\sin\theta\,\mathrm{d}\varphi), \qquad \theta^{2\dot{1}} = -\frac{r}{\sqrt{2}}(\mathrm{d}\theta + \mathrm{i}\sin\theta\,\mathrm{d}\varphi), \tag{2.107}$$

and hence,

$$\partial_{1\dot{1}} = -\left(1-\frac{r_g}{r}\right)^{-1}\frac{1}{c}\frac{\partial}{\partial t} + \frac{\partial}{\partial r}, \qquad \partial_{2\dot{2}} = -\frac{1}{2c}\frac{\partial}{\partial t} - \frac{1}{2}\left(1-\frac{r_g}{r}\right)\frac{\partial}{\partial r},$$
$$\partial_{1\dot{2}} = \frac{1}{\sqrt{2}r}\left(\frac{\partial}{\partial\theta} + \frac{\mathrm{i}}{\sin\theta}\frac{\partial}{\partial\varphi}\right), \qquad \partial_{2\dot{1}} = \frac{1}{\sqrt{2}r}\left(\frac{\partial}{\partial\theta} - \frac{\mathrm{i}}{\sin\theta}\frac{\partial}{\partial\varphi}\right). \tag{2.108}$$

(The coefficients of $\partial/\partial t$ are chosen negative, so that $v^{A\dot{A}} = l^A \bar{l}^{\dot{A}}$ is the spinor equivalent of a future-pointing null vector, for any one-index spinor l_A different from zero.)

The commutators of the vector fields (2.108) can be readily computed, and making use of the notation (2.28), we obtain

$$[D,\delta] = -\frac{1}{r}\delta, \qquad [\Delta,\delta] = \frac{1}{2r}\left(1-\frac{r_g}{r}\right)\delta,$$
$$[D,\Delta] = -\frac{r_g}{2r^2}D, \qquad [\delta,\overline{\delta}] = \frac{\cot\theta}{\sqrt{2}r}(\delta - \overline{\delta}),$$

which, compared with (2.27), implies that the only nonvanishing spin coefficients are

$$\rho = -\frac{1}{r}, \quad \beta = \frac{\cot\theta}{2\sqrt{2}r}, \quad \alpha = -\frac{\cot\theta}{2\sqrt{2}r}, \quad \gamma = \frac{r_g}{4r^2}, \quad \mu = -\frac{1}{2r}\left(1-\frac{r_g}{r}\right); \tag{2.109}$$

therefore, the connection one-forms are given by

$$\begin{aligned}
\boldsymbol{\Gamma}_{11} &= -\frac{1}{\sqrt{2}}(\mathrm{d}\theta + \mathrm{i}\sin\theta\,\mathrm{d}\varphi),\\
\boldsymbol{\Gamma}_{12} &= -\frac{\mathrm{i}}{2}\cos\theta\,\mathrm{d}\varphi - \frac{r_g}{4r^2}\left[c\mathrm{d}t + \left(1-\frac{r_g}{r}\right)^{-1}\mathrm{d}r\right],\\
\boldsymbol{\Gamma}_{22} &= -\frac{1}{2\sqrt{2}}\left(1-\frac{r_g}{r}\right)(\mathrm{d}\theta - \mathrm{i}\sin\theta\,\mathrm{d}\varphi),
\end{aligned} \tag{2.110}$$

and, substituting into (2.68), one finds that the curvature two-forms $\mathscr{R}_{AB}$ are given explicitly by

$$\begin{aligned}
\mathscr{R}_{11} &= \mathrm{d}\boldsymbol{\Gamma}_{11} - 2\boldsymbol{\Gamma}_{12}\wedge\boldsymbol{\Gamma}_{11} = -\frac{r_g}{2r^3}S^{22},\\
\mathscr{R}_{12} &= \mathrm{d}\boldsymbol{\Gamma}_{12} + \boldsymbol{\Gamma}_{11}\wedge\boldsymbol{\Gamma}_{22} = -\frac{r_g}{r^3}S^{12},\\
\mathscr{R}_{22} &= \mathrm{d}\boldsymbol{\Gamma}_{22} + 2\boldsymbol{\Gamma}_{12}\wedge\boldsymbol{\Gamma}_{22} = -\frac{r_g}{2r^3}S^{11},
\end{aligned} \tag{2.111}$$

which implies that $C_{AB\dot{A}\dot{B}} = 0$, $R = 0$ (that is, $R_{ab} = 0$, thus showing that the Einstein vacuum field equations are satisfied), and the only nonvanishing components of the Weyl spinor are determined by

$$C_{1122} = -\frac{r_g}{2r^3} \tag{2.112}$$

[see (2.82)].

Example 2.3. The Euclidean Schwarzschild metric.

By reversing the sign of the first term in (2.106) we obtain a metric with Euclidean signature for $r > r_g$

$$g = \left(1 - \frac{r_g}{r}\right) c^2 \mathrm{d}t^2 + \frac{\mathrm{d}r^2}{1 - r_g/r} + r^2 \mathrm{d}\theta^2 + r^2 \sin^2\theta \, \mathrm{d}\varphi^2, \tag{2.113}$$

which can be expressed in the form $-2\theta^{1\dot{1}}\theta^{2\dot{2}} + 2\theta^{1\dot{2}}\theta^{2\dot{1}}$, with $\overline{\theta^{1\dot{1}}} = -\theta^{2\dot{2}}$ and $\overline{\theta^{1\dot{2}}} = \theta^{2\dot{1}}$, choosing

$$\begin{aligned}
\theta^{1\dot{1}} &= \frac{\mathrm{i}}{\sqrt{2}}\sqrt{1 - \frac{r_g}{r}}\, c\mathrm{d}t - \frac{1}{\sqrt{2}} \frac{\mathrm{d}r}{\sqrt{1 - r_g/r}}, \\
\theta^{2\dot{2}} &= \frac{\mathrm{i}}{\sqrt{2}}\sqrt{1 - \frac{r_g}{r}}\, c\mathrm{d}t + \frac{1}{\sqrt{2}} \frac{\mathrm{d}r}{\sqrt{1 - r_g/r}}, \\
\theta^{1\dot{2}} &= -\frac{r}{\sqrt{2}}(\mathrm{d}\theta - \mathrm{i}\sin\theta\, \mathrm{d}\varphi), \qquad \theta^{2\dot{1}} = -\frac{r}{\sqrt{2}}(\mathrm{d}\theta + \mathrm{i}\sin\theta\, \mathrm{d}\varphi).
\end{aligned}$$

A straightforward computation gives

$$\begin{aligned}
S^{11} &= -\frac{r}{2}\left(\mathrm{i}\sqrt{1 - \frac{r_g}{r}}\, c\mathrm{d}t - \frac{\mathrm{d}r}{\sqrt{1 - r_g/r}}\right) \wedge (\mathrm{d}\theta - \mathrm{i}\sin\theta\, \mathrm{d}\varphi), \\
S^{12} &= \frac{\mathrm{i}}{2}\left(c\mathrm{d}t \wedge \mathrm{d}r - r^2 \sin\theta\, \mathrm{d}\theta \wedge \mathrm{d}\varphi\right),
\end{aligned}$$

with $S^{22} = \overline{S^{11}}$, and

$$\begin{aligned}
\mathrm{d}S^{11} &= \frac{\cot\theta}{\sqrt{2}r}\breve{\theta}^{1\dot{1}} - \frac{1}{\sqrt{2}r}\left(\sqrt{1 - \frac{r_g}{r}} + \frac{r_g/r}{2\sqrt{1 - r_g/r}}\right)\breve{\theta}^{1\dot{2}}, \\
\mathrm{d}S^{12} &= -\frac{1}{\sqrt{2}r}\sqrt{1 - \frac{r_g}{r}}\,(\breve{\theta}^{1\dot{1}} + \breve{\theta}^{2\dot{2}}).
\end{aligned}$$

Then, using the fact that when the signature is Euclidean, $\overline{K_{ABC\dot{D}}} = -K^{ABC\dot{D}}$, the foregoing relations and (2.97) give

$$\begin{aligned}
\boldsymbol{\Gamma}_{11} &= -\frac{1}{2}\sqrt{1 - \frac{r_g}{r}}\,(\mathrm{d}\theta + \mathrm{i}\sin\theta\, \mathrm{d}\varphi), \\
\boldsymbol{\Gamma}_{12} &= -\frac{\mathrm{i}}{2}\cos\theta\, \mathrm{d}\varphi - \frac{\mathrm{i}r_g}{4r^2} c\mathrm{d}t,
\end{aligned}$$

with $\mathbf{\Gamma}_{22} = \overline{\mathbf{\Gamma}_{11}}$ [cf. (2.110)]. The curvature two-forms are

$$\mathscr{R}_{11} = -\frac{r_g}{2r^3} S^{22}, \qquad \mathscr{R}_{12} = -\frac{r_g}{r^3} S^{12},$$

with $\mathscr{R}_{22} = \overline{\mathscr{R}_{11}}$, showing that $C_{AB\dot{C}\dot{D}} = 0 = R$ and the only nonvanishing component of the Weyl spinor C_{ABCD} is given by $C_{1122} = -r_g/(2r^3)$ [real, as follows from (2.87)].

Further examples are given in Sections 2.3 and 4.2.

Bianchi and Ricci Identities

The Bianchi identities in spinor form can be derived from (2.54) or directly by applying the exterior derivative operator to both sides of the Cartan second structure equations (2.68), using again these equations in order to eliminate the exterior derivative of $\mathbf{\Gamma}_{AB}$ or $\mathbf{\Gamma}_{\dot{A}\dot{B}}$. Thus

$$\begin{aligned} \mathrm{d}\mathscr{R}_{AB} &= 2\mathbf{\Gamma}_{C(A} \wedge \mathscr{R}^{C}{}_{B)}, \\ \mathrm{d}\mathscr{R}_{\dot{A}\dot{B}} &= 2\mathbf{\Gamma}_{\dot{C}(\dot{A}} \wedge \mathscr{R}^{\dot{C}}{}_{\dot{B})}. \end{aligned} \tag{2.114}$$

Substituting (2.67) and (2.16) into (2.114), with the aid of (2.71), (2.77), (2.60), and (2.61) one obtains

$$\begin{aligned} \nabla^{C}{}_{\dot{R}} X_{ABCD} - \nabla_{D}{}^{\dot{C}} C_{AB\dot{C}\dot{R}} &= 0, \\ \nabla_{R}{}^{\dot{C}} X_{\dot{A}\dot{B}\dot{C}\dot{D}} - \nabla^{C}{}_{\dot{D}} C_{CR\dot{A}\dot{B}} &= 0, \end{aligned} \tag{2.115}$$

and hence, employing the decompositions (2.80) and (2.81), one finds that these last equations are equivalent to

$$\begin{aligned} \nabla^{C}{}_{\dot{R}} C_{CABD} - \nabla_{(A}{}^{\dot{C}} C_{BD)\dot{C}\dot{R}} &= 0, \\ \nabla_{R}{}^{\dot{C}} C_{\dot{C}\dot{A}\dot{B}\dot{D}} - \nabla^{C}{}_{(\dot{A}} C_{|CR|\dot{B}\dot{D})} &= 0, \\ \tfrac{1}{8}\nabla_{D\dot{R}} R + \nabla^{B\dot{C}} C_{BD\dot{C}\dot{R}} &= 0. \end{aligned} \tag{2.116}$$

Since $C_{AB\dot{C}\dot{D}}$ is the spinor equivalent of $\frac{1}{2}(R_{ab} - \frac{1}{4}Rg_{ab})$, the last of these relations amounts to the so-called contracted Bianchi identities.

Applying the decomposition (1.52) to the commutator of covariant derivatives, we have

$$\nabla_{A\dot{A}}\nabla_{B\dot{B}} - \nabla_{B\dot{B}}\nabla_{A\dot{A}} = \varepsilon_{\dot{A}\dot{B}}\square_{AB} + \varepsilon_{AB}\square_{\dot{A}\dot{B}}, \tag{2.117}$$

where

$$\square_{AB} \equiv \nabla_{(A}{}^{\dot{R}}\nabla_{B)\dot{R}}, \qquad \square_{\dot{A}\dot{B}} \equiv \nabla^{R}{}_{(\dot{A}}\nabla_{|R|\dot{B})}, \tag{2.118}$$

and the spinor equivalent of the Ricci identity $(\nabla_a\nabla_b - \nabla_b\nabla_a)t_c = -R^{d}{}_{cab}t_d$, which follows from (2.48), can be written in the form

$$(\varepsilon_{\dot{A}\dot{B}}\square_{AB} + \varepsilon_{AB}\square_{\dot{A}\dot{B}})t_{C\dot{C}} = R^{D\dot{D}}{}_{C\dot{C}A\dot{A}B\dot{B}}t_{D\dot{D}}. \tag{2.119}$$

Then, making use of (2.83) one finds that (2.119) is equivalent to

$$\square_{AB} t_{C\dot{C}} = C^D{}_{CAB} t_{D\dot{C}} + C_{AB}{}^{\dot{D}}{}_{\dot{C}} t_{C\dot{D}} + \tfrac{1}{24} R(\varepsilon_{CB} t_{A\dot{C}} + \varepsilon_{CA} t_{B\dot{C}}) \tag{2.120}$$

and

$$\square_{\dot{A}\dot{B}} t_{C\dot{C}} = C^{\dot{D}}{}_{\dot{C}\dot{A}\dot{B}} t_{C\dot{D}} + C^D{}_{C\dot{A}\dot{B}} t_{D\dot{C}} + \tfrac{1}{24} R(\varepsilon_{\dot{C}\dot{B}} t_{C\dot{A}} + \varepsilon_{\dot{C}\dot{A}} t_{C\dot{B}}). \tag{2.121}$$

Equations (2.120) and (2.121) also follow directly from the definitions (2.118) and the explicit expressions (2.91)–(2.94).

The commutator of covariant derivatives of a scalar function vanishes; in fact, $(\nabla_a\nabla_b - \nabla_b\nabla_a)f = \nabla_a\partial_b f - \nabla_b\partial_a f = \partial_a\partial_b f - \Gamma^c{}_{ba}\partial_c f - \partial_b\partial_a f + \Gamma^c{}_{ab}\partial_c f$, which is equal to zero if and only if (2.11) holds. Thus, $\square_{AB} f = 0$ and $\square_{\dot{A}\dot{B}} f = 0$.

In the case of one-index spinor fields, making use of (2.40), (2.80), (2.81), and (2.91)–(2.94), one finds that

$$\square_{AB}\psi_C = C^D{}_{CAB}\psi_D + \tfrac{1}{24} R(\varepsilon_{CB}\psi_A + \varepsilon_{CA}\psi_B), \tag{2.122}$$

$$\square_{AB}\psi_{\dot{C}} = C_{AB}{}^{\dot{D}}{}_{\dot{C}}\psi_{\dot{D}}, \tag{2.123}$$

and

$$\square_{\dot{A}\dot{B}}\psi_C = C^D{}_{C\dot{A}\dot{B}}\psi_D, \tag{2.124}$$

$$\square_{\dot{A}\dot{B}}\psi_{\dot{C}} = C^{\dot{D}}{}_{\dot{C}\dot{A}\dot{B}}\psi_{\dot{D}} + \tfrac{1}{24} R(\varepsilon_{\dot{C}\dot{B}}\psi_{\dot{A}} + \varepsilon_{\dot{C}\dot{A}}\psi_{\dot{B}}). \tag{2.125}$$

These identities imply the identities (2.120) and (2.121) by virtue of the formulas

$$\square_{AB}(\psi_{C\ldots\dot{D}\ldots}\phi_{E\ldots\dot{F}\ldots}) = \psi_{C\ldots\dot{D}\ldots}\square_{AB}\phi_{E\ldots\dot{F}\ldots} + \phi_{E\ldots\dot{F}\ldots}\square_{AB}\psi_{C\ldots\dot{D}\ldots}$$

and

$$\square_{\dot{A}\dot{B}}(\psi_{C\ldots\dot{D}\ldots}\phi_{E\ldots\dot{F}\ldots}) = \psi_{C\ldots\dot{D}\ldots}\square_{\dot{A}\dot{B}}\phi_{E\ldots\dot{F}\ldots} + \phi_{E\ldots\dot{F}\ldots}\square_{\dot{A}\dot{B}}\psi_{C\ldots\dot{D}\ldots},$$

which follow from the fact that the covariant derivative satisfies the Leibniz rule. In this manner one finds that

$$\begin{aligned}\square_{AB}\psi_{CD\ldots\dot{E}\dot{F}\ldots} &= C^R{}_{CAB}\psi_{RD\ldots\dot{E}\dot{F}\ldots} + C^R{}_{DAB}\psi_{CR\ldots\dot{E}\dot{F}\ldots} + \cdots \\ &\quad + \tfrac{1}{12} R\big(\varepsilon_{C(A}\psi_{B)D\ldots\dot{E}\dot{F}\ldots} + \varepsilon_{D(A}\psi_{|C|B)\ldots\dot{E}\dot{F}\ldots} + \cdots\big) \\ &\quad + C_{AB}{}^{\dot{R}}{}_{\dot{E}}\psi_{CD\ldots\dot{R}\dot{F}\ldots} + C_{AB}{}^{\dot{R}}{}_{\dot{F}}\psi_{CD\ldots\dot{E}\dot{R}\ldots} + \cdots\end{aligned} \tag{2.126}$$

and

$$\begin{aligned}\square_{\dot{A}\dot{B}}\psi_{CD\ldots\dot{E}\dot{F}\ldots} &= C^{\dot{R}}{}_{\dot{E}\dot{A}\dot{B}}\psi_{CD\ldots\dot{R}\dot{F}\ldots} + C^{\dot{R}}{}_{\dot{F}\dot{A}\dot{B}}\psi_{CD\ldots\dot{E}\dot{R}\ldots} + \cdots \\ &\quad + \tfrac{1}{12} R\big(\varepsilon_{\dot{E}(\dot{A}}\psi_{|CD\ldots|\dot{B})\dot{F}\ldots} + \varepsilon_{\dot{F}(\dot{A}}\psi_{|CD\ldots\dot{E}|\dot{B})\ldots} + \cdots\big) \\ &\quad + C^R{}_{C\dot{A}\dot{B}}\psi_{RD\ldots\dot{E}\dot{F}\ldots} + C^R{}_{D\dot{A}\dot{B}}\psi_{CR\ldots\dot{E}\dot{F}\ldots} + \cdots.\end{aligned} \tag{2.127}$$

2.2.2 Algebraic Classification of the Conformal Curvature

Since the conformal curvature spinors C_{ABCD} and $C_{\dot{A}\dot{B}\dot{C}\dot{D}}$ are totally symmetric, they can be classified according to the multiplicities of their principal spinors (see Section 1.7).

If the signature is $(+\,+\,+\,+)$, the reality conditions (2.87) imply that the only possible nontrivial cases are

$$C_{ABCD} = \begin{cases} \alpha_{(A}\widehat{\alpha}_B\beta_C\widehat{\beta}_{D)} & \text{with } \alpha^A\beta_A \neq 0, \\ \pm\alpha_{(A}\alpha_B\widehat{\alpha}_C\widehat{\alpha}_{D)} \end{cases} \tag{2.128}$$

[see (1.231)], with analogous independent expressions for $C_{\dot{A}\dot{B}\dot{C}\dot{D}}$. For instance, the spinor C_{ABCD} of the Euclidean Schwarzschild metric is of the form $C_{ABCD} = -\alpha_{(A}\alpha_B\widehat{\alpha}_C\widehat{\alpha}_{D)}$.

Similarly, if the metric has signature $(+\,+\,-\,-)$, the only nontrivial possible cases are

$$C_{ABCD} = \begin{cases} \alpha_{(A}\beta_B\gamma_C\delta_{D)} & \text{with } \widehat{\alpha}_A = \alpha_A,\ \widehat{\beta}_A = \beta_A,\ \widehat{\gamma}_A = \gamma_A,\ \widehat{\delta}_A = \delta_A, \\ \alpha_{(A}\beta_B\gamma_C\widehat{\gamma}_{D)} & \text{with } \widehat{\alpha}_A = \alpha_A,\ \widehat{\beta}_A = \beta_A, \\ \pm\alpha_{(A}\widehat{\alpha}_B\beta_C\widehat{\beta}_{D)}, \\ \alpha_{(A}\alpha_B\beta_C\gamma_{D)} & \text{with } \widehat{\alpha}_A = \alpha_A,\ \widehat{\beta}_A = \beta_A,\ \widehat{\gamma}_A = \gamma_A, \\ \pm\alpha_{(A}\alpha_B\beta_C\widehat{\beta}_{D)} & \text{with } \widehat{\alpha}_A = \alpha_A, \\ \pm\alpha_{(A}\alpha_B\beta_C\beta_{D)} & \text{with } \widehat{\alpha}_A = \alpha_A,\ \widehat{\beta}_A = \beta_A, \\ \pm\alpha_{(A}\alpha_B\widehat{\alpha}_C\widehat{\alpha}_{D)}, \\ \alpha_{(A}\alpha_B\alpha_C\beta_{D)} & \text{with } \widehat{\alpha}_A = \alpha_A,\ \widehat{\beta}_A = \beta_A, \\ \pm\alpha_A\alpha_B\alpha_C\alpha_D & \text{with } \widehat{\alpha}_A = \alpha_A, \end{cases} \tag{2.129}$$

with analogous independent expressions for $C_{\dot{A}\dot{B}\dot{C}\dot{D}}$. In particular, there exist Riemannian manifolds with ultrahyperbolic signature with $C_{ABCD} = 0$ and $C_{\dot{A}\dot{B}\dot{C}\dot{D}} \neq 0$ (or vice versa), which turn out to possess interesting applications (see, e.g., Dunajski and West 2008). (The algebraic types (2.128) and (2.129) coincide with the possible algebraic types of a traceless two-index symmetric tensor in a manifold of dimension three; Torres del Castillo 2003, Sec. 5.3.)

It should be remarked that when the signature of the metric is $(+\,+\,+\,+)$ or $(+\,+\,-\,-)$, the algebraic type of the Weyl spinors is not completely determined by the multiplicities of their principal spinors, but in some cases there is an overall sign that has to be specified (cf. Goldblatt 1994a,b and the references cited therein).

In the remaining case, in which the metric has signature $(+\,+\,+\,-)$, the reality conditions (2.88) do not impose any restrictions on the principal spinors of C_{ABCD}, and therefore, if the conformal curvature spinor is different from zero, it can be of the form

$$C_{ABCD} = \begin{cases} \alpha_{(A}\beta_B\gamma_C\delta_{D)} & \text{(type G or } \{1,1,1,1\}), \\ \alpha_{(A}\alpha_B\beta_C\gamma_{D)} & \text{(type II or } \{2,1,1\}), \\ \alpha_{(A}\alpha_B\beta_C\beta_{D)} & \text{(type D or } \{2,2\}), \\ \alpha_{(A}\alpha_B\alpha_C\beta_{D)} & \text{(type III or } \{3,1\}), \\ \alpha_A\alpha_B\alpha_C\alpha_D & \text{(type N or } \{4\}), \end{cases} \tag{2.130}$$

where it is understood that all the contractions $\alpha^A\beta_A$, $\alpha^A\gamma_A$, $\alpha^A\delta_A$, $\beta^A\gamma_A$, $\beta^A\delta_A$, and $\gamma^A\delta_A$ are different from zero. This constitutes the Petrov–Penrose classification of the conformal curvature (Penrose 1960, Pirani 1965). Only if the metric has Lorentzian signature are C_{ABCD} and $C_{\dot{A}\dot{B}\dot{C}\dot{D}}$ not independent of each other, being related by $C_{\dot{A}\dot{B}\dot{C}\dot{D}} = \overline{C_{ABCD}}$ [see (2.88)], and therefore it suffices to consider C_{ABCD} only. Furthermore, only in this case is there associated with each one-index spinor α_A another one-index spinor $\overline{\alpha}_{\dot{A}} = \overline{\alpha_A}$, so that α_A is a principal spinor of C_{ABCD} if and only if $\overline{\alpha}_{\dot{A}}$ is a principal spinor of $C_{\dot{A}\dot{B}\dot{C}\dot{D}}$. If α_A is a principal spinor of C_{ABCD}, then $\pm\alpha_A\overline{\alpha}_{\dot{A}}$ is the spinor equivalent of a real null vector referred to as a Debever–Penrose (DP) vector, and its direction is a *principal null direction* of the conformal curvature.

The DP vectors can be defined directly using the conformal curvature tensor. If l_a is a real null vector, then the spinor equivalent of $l^c C_{abcd}$ is

$$\pm\left(\varepsilon_{\dot{A}\dot{B}}\overline{\alpha}_{\dot{D}}C_{ABCD}\alpha^C + \varepsilon_{AB}\alpha_D C_{\dot{A}\dot{B}\dot{C}\dot{D}}\overline{\alpha}^{\dot{C}}\right),$$

assuming that $\pm\alpha^A\overline{\alpha}^{\dot{A}}$ is the spinor equivalent of l^a; hence

$$l^c C_{abcd} = 0 \qquad \Leftrightarrow \qquad C_{ABCD}\alpha^C = 0,$$

which means that α_A is a quadruple principal spinor of C_{ABCD}. Similarly, the spinor equivalent of $l^c C_{abc[d}l_{e]}$ is [see (1.52)]

$$-\tfrac{1}{2}\left(\varepsilon_{DE}\varepsilon_{\dot{A}\dot{B}}\overline{\alpha}_{\dot{D}}\overline{\alpha}_{\dot{E}}C_{ABCR}\alpha^C\alpha^R + \varepsilon_{AB}\varepsilon_{\dot{D}\dot{E}}\alpha_D\alpha_E C_{\dot{A}\dot{B}\dot{C}\dot{R}}\overline{\alpha}^{\dot{C}}\overline{\alpha}^{\dot{R}}\right),$$

and hence

$$l^c C_{abc[d}l_{e]} = 0 \qquad \Leftrightarrow \qquad C_{ABCR}\alpha^C\alpha^R = 0,$$

which means that α_A is at least a triple principal spinor of C_{ABCD}. It can readily be seen that the spinor equivalent of $l^b l^c C_{abc[d}l_{e]}$ is

$$\pm\tfrac{1}{2}\left(\varepsilon_{DE}\overline{\alpha}_{\dot{A}}\overline{\alpha}_{\dot{D}}\overline{\alpha}_{\dot{E}}C_{ABCR}\alpha^B\alpha^C\alpha^R + \varepsilon_{\dot{D}\dot{E}}\alpha_A\alpha_D\alpha_E C_{\dot{A}\dot{B}\dot{C}\dot{R}}\overline{\alpha}^{\dot{B}}\overline{\alpha}^{\dot{C}}\overline{\alpha}^{\dot{R}}\right),$$

and therefore

$$l^b l^c C_{abc[d}l_{e]} = 0 \qquad \Leftrightarrow \qquad C_{ABCR}\alpha^B\alpha^C\alpha^R = 0,$$

which means that α_A is at least a double principal spinor of C_{ABCD}. Finally, the spinor equivalent of $l^b l^c l_{[f}C_{a]bc[d}l_{e]}$ is

$$\tfrac{1}{4}\left(\varepsilon_{DE}\varepsilon_{FA}\overline{\alpha}_{\dot{F}}\overline{\alpha}_{\dot{A}}\overline{\alpha}_{\dot{D}}\overline{\alpha}_{\dot{E}}C_{SBCR}\alpha^S\alpha^B\alpha^C\alpha^R + \varepsilon_{\dot{D}\dot{E}}\varepsilon_{\dot{F}\dot{A}}\alpha_F\alpha_A\alpha_D\alpha_E C_{\dot{S}\dot{B}\dot{C}\dot{R}}\overline{\alpha}^{\dot{S}}\overline{\alpha}^{\dot{B}}\overline{\alpha}^{\dot{C}}\overline{\alpha}^{\dot{R}}\right),$$

and therefore

$$l^b l^c l_{[f} C_{a]bc[d} l_{e]} = 0 \qquad \Leftrightarrow \qquad C_{SBCR} \alpha^S \alpha^B \alpha^C \alpha^R = 0, \tag{2.131}$$

which means that α_A is a principal spinor of C_{ABCD}.

By virtue of the equivalence (2.131), it follows that there exist four (not necessarily distinct) real null directions represented by null vectors l^a satisfying the condition $l^b l^c l_{[f} C_{a]bc[d} l_{e]} = 0$. If the four principal null directions of the conformal curvature are all distinct [which corresponds to the type G or $\{1,1,1,1\}$ in (2.130)], the curvature is said to be *algebraically general*. The conformal curvature is *algebraically special* or *degenerate* if there are fewer than four distinct principal null directions (which corresponds to the existence of a repeated principal spinor of C_{ABCD}).

2.3 Conformal Rescalings

Two metric tensors g and g' on M are *conformally equivalent* if there exists a positive function ϕ such that $g = \phi^2 g'$. Given a null tetrad $\partial_{A\dot{A}}$ for the metric g,

$$\partial'_{A\dot{A}} = \phi\, \partial_{A\dot{A}} \tag{2.132}$$

is a null tetrad for g'. Note that since the values of ϕ are real, the transformation (2.132) preserves conditions (2.100). Each metric tensor defines a Riemannian connection, and in order to find the spin coefficients corresponding to the Riemannian connection defined by g' with respect to the null tetrad $\partial'_{A\dot{A}}$, we note that the two-forms S^{AB} and their analogues S'^{AB} are then related by $S'^{AB} = \phi^{-2} S^{AB}$; therefore, writing $\mathrm{d}S^{AB} = K^{AB}{}_{C\dot{D}}\, \breve{\theta}^{C\dot{D}}$ as in Section 2.2, and similarly, $\mathrm{d}S'^{AB} = K'^{AB}{}_{C\dot{D}}\, \breve{\theta}'^{C\dot{D}}$, we have

$$\begin{aligned}
\mathrm{d}S'^{AB} &= \phi^{-2} \mathrm{d}S^{AB} - 2\phi^{-3} \mathrm{d}\phi \wedge S^{AB} \\
&= \phi^{-2} K^{AB}{}_{C\dot{D}}\, \breve{\theta}^{C\dot{D}} + 2\phi^{-3} (\partial_{R\dot{D}} \phi)\, \theta^{R\dot{D}} \wedge S^{AB} \\
&= (\phi K^{AB}{}_{C\dot{D}} + 2\partial^{(A}{}_{\dot{D}} \phi\, \varepsilon^{B)}{}_{C})\, \breve{\theta}'^{C\dot{D}}
\end{aligned}$$

[see (2.71) and (2.90)], i.e., $K'_{ABC\dot{D}} = \phi K_{ABC\dot{D}} + \varepsilon_{AC} \partial_{B\dot{D}} \phi + \varepsilon_{BC} \partial_{A\dot{D}} \phi$. Thus, according to (2.97), $\Gamma'_{ABC\dot{D}} = -K'_{C(AB)\dot{D}}$, that is,

$$\Gamma'_{ABC\dot{D}} = \phi\, \Gamma_{ABC\dot{D}} - \varepsilon_{C(A}\, \partial_{B)\dot{D}} \phi. \tag{2.133}$$

In a similar manner one finds that

$$\Gamma'_{\dot{A}\dot{B}\dot{C}D} = \phi\, \Gamma_{\dot{A}\dot{B}\dot{C}D} - \varepsilon_{\dot{C}(\dot{A}}\, \partial_{|D|\dot{B})} \phi. \tag{2.134}$$

Equations (2.133) and (2.134) give the spin coefficients of the Riemannian connection corresponding to g' with respect to the null tetrad $\partial'_{A\dot{A}}$.

The spinor components of the curvature of the rescaled metric g' with respect to the tetrad $\partial'_{A\dot{A}}$ can be obtained by substituting (2.132), (2.133), and (2.134) into (2.91)–(2.94); in this way, from (2.91) one obtains

$$X'_{ABCD} = \phi^2 X_{ABCD} + \tfrac{1}{4}(\varepsilon_{AC}\varepsilon_{BD} + \varepsilon_{AD}\varepsilon_{BC})\,\phi^3\,\nabla^{R\dot{R}}\nabla_{R\dot{R}}\phi^{-1},$$

which is equivalent to the relations

$$C'_{ABCD} = \phi^2 C_{ABCD}, \qquad R' = \phi^2 R + 6\,\phi^3\,\nabla^{R\dot{R}}\nabla_{R\dot{R}}\phi^{-1}, \tag{2.135}$$

whereas (2.92) yields

$$C'_{AB\dot{C}\dot{D}} = \phi^2 C_{AB\dot{C}\dot{D}} + \phi\,\nabla_{A(\dot{C}}\nabla_{|B|\dot{D})}\phi. \tag{2.136}$$

Similarly, from (2.93) we obtain

$$C'_{AB\dot{C}\dot{D}} = \phi^2 C_{AB\dot{C}\dot{D}} + \phi\,\nabla_{(A|\dot{C}|}\nabla_{B)\dot{D}}\phi, \tag{2.137}$$

which is equivalent to (2.136), while from (2.94) we obtain

$$C'_{\dot{A}\dot{B}\dot{C}\dot{D}} = \phi^2 C_{\dot{A}\dot{B}\dot{C}\dot{D}}. \tag{2.138}$$

It should be stressed that equations (2.135)–(2.138) give the relation between the components of the curvature spinors of g and g' with respect to two *different* null tetrads related by (2.132). With the aid of (2.135)–(2.138), one finds that the components of the curvature tensors of g and g' (with respect to the basis induced by a coordinate system) are related by

$$R'_{\mu\nu} = R_{\mu\nu} + 2\phi^{-1}\nabla_\mu\nabla_\nu\phi + \phi^{-1}g_{\mu\nu}\nabla^\rho\nabla_\rho\phi - 3\phi^{-2}g_{\mu\nu}(\nabla^\rho\phi)(\nabla_\rho\phi)$$

and

$$C'_{\mu\nu\rho\sigma} = \phi^{-2}C_{\mu\nu\rho\sigma},$$

and therefore, $C'^{\mu}{}_{\nu\rho\sigma} = C^{\mu}{}_{\nu\rho\sigma}$ (see Appendix A).

Only C_{ABCD} and $C_{\dot{A}\dot{B}\dot{C}\dot{D}}$ transform homogeneously under conformal rescalings, and therefore, C_{ABCD} and $C_{\dot{A}\dot{B}\dot{C}\dot{D}}$ vanish for a metric conformally equivalent to a flat one. It can be shown that conversely, if C_{ABCD} and $C_{\dot{A}\dot{B}\dot{C}\dot{D}}$ are equal to zero, then the metric tensor g is locally conformally equivalent to a flat metric.

For instance, the curvature of the standard metric of the sphere S^4 satisfies $C_{ABCD} = 0 = C_{\dot{A}\dot{B}\dot{C}\dot{D}}$, and we can see explicitly that this metric is indeed locally conformally flat. In fact, letting $r \equiv \cot\frac{1}{2}\psi$, from (2.101) we obtain

$$\begin{aligned} &\mathrm{d}\psi^2 + \sin^2\psi\left[\mathrm{d}\chi^2 + \sin^2\chi\,(\mathrm{d}\theta^2 + \sin^2\theta\,\mathrm{d}\varphi^2)\right] \\ &\qquad = \left(\frac{2}{1+r^2}\right)^2\left\{\mathrm{d}r^2 + r^2\left[\mathrm{d}\chi^2 + \sin^2\chi\,(\mathrm{d}\theta^2 + \sin^2\theta\,\mathrm{d}\varphi^2)\right]\right\}. \end{aligned}$$

The expression within braces can be recognized as the standard metric of $\mathbb{R}^4$, which is flat, written in spherical coordinates. (Note that the conformal factor $\phi = 2/(1+r^2) = 2\sin^2\frac{1}{2}\psi$ vanishes at $\psi = 0$.)

Example 2.4. The hyperbolic space.

The hyperbolic space $\mathbb{H}^4 \equiv \{(x^1,x^2,x^3,x^4) \in \mathbb{R}^4 \,|\, x^4 > 0\}$ possesses the metric

$$g' = (x^4)^{-2}\left[(\mathrm{d}x^1)^2 + (\mathrm{d}x^2)^2 + (\mathrm{d}x^3)^2 + (\mathrm{d}x^4)^2\right]$$

conformally equivalent to the flat metric $g = (\mathrm{d}x^1)^2 + (\mathrm{d}x^2)^2 + (\mathrm{d}x^3)^2 + (\mathrm{d}x^4)^2$. Thus, in the notation employed throughout this section, $\phi = x^4$. The Laplacian of $\phi^{-1} = (x^4)^{-1}$, relative to the flat metric g, can be readily computed by making use of Cartesian coordinates x^a, and one obtains $\nabla^a\nabla_a\phi^{-1} = -\nabla^a(\phi^{-2}\delta^4_a) = 2\phi^{-3}$. Similarly, $\nabla_a\nabla_b\phi = 0$, which amounts to $\nabla_{A\dot{B}}\nabla_{C\dot{D}}\phi = 0$. Then, from (2.135)–(2.138) we see that the only nonvanishing component of the curvature of g' is the scalar curvature

$$R' = -12.$$

2.4 Killing Vectors. Lie Derivative of Spinors

The infinitesimal generator K of a one-parameter group of isometries satisfies the condition $\pounds_K g = 0$, where $\pounds_K$ denotes the Lie derivative with respect to the vector field K (see, e.g., Kobayashi and Nomizu 1963, Petersen 1998), or equivalently, in terms of the Levi-Civita connection

$$\nabla_a K_b + \nabla_b K_a = 0. \tag{2.139}$$

These equations are referred to as the Killing equations, and their solutions are called Killing vector fields, or simply *Killing vectors*. In a four-dimensional manifold, the Killing equations constitute a set of ten independent linear partial differential equations, and the set of Killing vector fields is a real Lie algebra, with the commutator or Lie bracket [,], of dimension less than or equal to ten. Thus, a vector field K is a Killing vector field if and only if $\nabla_a K_b$ is antisymmetric and therefore its spinor equivalent is of the form (1.50),

$$\nabla_{A\dot{A}} K_{B\dot{B}} = L_{AB}\varepsilon_{\dot{A}\dot{B}} + L_{\dot{A}\dot{B}}\varepsilon_{AB}, \tag{2.140}$$

with L_{AB} and $L_{\dot{A}\dot{B}}$ being symmetric spinor fields. From (2.140) we obtain

$$L_{AB} = \tfrac{1}{2}\nabla_A{}^{\dot{R}}K_{B\dot{R}}, \qquad L_{\dot{A}\dot{B}} = \tfrac{1}{2}\nabla^R{}_{\dot{A}}K_{R\dot{B}}. \tag{2.141}$$

The spinor components L_{AB} and $L_{\dot{A}\dot{B}}$ can be conveniently calculated by expressing the differential of the one-form $K_a\theta^a$ in terms of the two-forms S^{AB} and $S^{\dot{A}\dot{B}}$, since

$$\mathrm{d}(K_a\theta^a) = \nabla_{(A}{}^{\dot{R}}K_{B)\dot{R}}\,S^{AB} + \nabla^R{}_{(\dot{A}}K_{|R|\dot{B})}\,S^{\dot{A}\dot{B}} \tag{2.142}$$

(see Exercise 2.5). In this way, the spin coefficients need not be known in order to find L_{AB} and $L_{\dot{A}\dot{B}}$.

The Lie derivative $\pounds_K X$ of an arbitrary vector field $X = X^a \partial_a = -X^{A\dot{A}} \partial_{A\dot{A}}$ with respect to a Killing vector $K = K^a \partial_a = -K^{A\dot{A}} \partial_{A\dot{A}}$ coincides with their commutator, or Lie bracket, $[K,X]$; therefore, if the torsion of the connection vanishes, then $\pounds_K X = [K,X] = \nabla_K X - \nabla_X K$ [see (2.5)], and making use of (2.140), we find that the spinor equivalent of $\pounds_K X$ is determined by

$$\begin{aligned}\pounds_K X &= (K^{A\dot{A}} \nabla_{A\dot{A}} X^{B\dot{B}} - X^{A\dot{A}} \nabla_{A\dot{A}} K^{B\dot{B}})\, \partial_{B\dot{B}} \\ &= (K^{A\dot{A}} \nabla_{A\dot{A}} X^{B\dot{B}} + L^B{}_A X^{A\dot{B}} + L^{\dot{B}}{}_{\dot{A}} X^{B\dot{A}})\, \partial_{B\dot{B}} \\ &\equiv -(\pounds_K X^{B\dot{B}})\, \partial_{B\dot{B}}.\end{aligned}$$

This result suggests the following definition for the Lie derivative of an arbitrary spinor field with respect to a Killing vector:

$$\begin{aligned}\pounds_K \psi^{A\ldots\dot{B}\ldots}_{C\ldots\dot{D}\ldots} = &-K^{R\dot{S}} \nabla_{R\dot{S}} \psi^{A\ldots\dot{B}\ldots}_{C\ldots\dot{D}\ldots} - L^A{}_S \psi^{S\ldots\dot{B}\ldots}_{C\ldots\dot{D}\ldots} - \cdots - L^{\dot{B}}{}_{\dot{S}} \psi^{A\ldots\dot{S}\ldots}_{C\ldots\dot{D}\ldots} \\ &- \cdots + L^S{}_C \psi^{A\ldots\dot{B}\ldots}_{S\ldots\dot{D}\ldots} + \cdots + L^{\dot{S}}{}_{\dot{D}} \psi^{A\ldots\dot{B}\ldots}_{C\ldots\dot{S}\ldots} + \cdots .\end{aligned} \tag{2.143}$$

Then, for instance, $\pounds_K \varepsilon_{AB} = 0$ and $\pounds_K \varepsilon_{\dot{A}\dot{B}} = 0$.

The Killing equations (2.140) have some integrability conditions. From (2.141) and (2.117) we obtain

$$\begin{aligned}\nabla_{A\dot{A}} L_{BC} &= \tfrac{1}{2} \nabla_{A\dot{A}} \nabla_B{}^{\dot{R}} K_{C\dot{R}} \\ &= \tfrac{1}{2} (\nabla_{A\dot{A}} \nabla_B{}^{\dot{R}} - \nabla_B{}^{\dot{R}} \nabla_{A\dot{A}} + \nabla_B{}^{\dot{R}} \nabla_{A\dot{A}}) K_{C\dot{R}} \\ &= \tfrac{1}{2} (\varepsilon_{AB} \square_{\dot{A}}{}^{\dot{R}} + \varepsilon_{\dot{A}}{}^{\dot{R}} \square_{AB} + \nabla_B{}^{\dot{R}} \nabla_{A\dot{A}}) K_{C\dot{R}} \\ &= \tfrac{1}{2} \varepsilon_{AB} \square_{\dot{A}}{}^{\dot{R}} K_{C\dot{R}} - \tfrac{1}{2} \square_{AB} K_{C\dot{A}} - \tfrac{1}{2} \nabla_B{}^{\dot{R}} \nabla_{C\dot{R}} K_{A\dot{A}},\end{aligned}$$

where we have made use of the fact that $\nabla_{A\dot{A}} K_{C\dot{R}} = -\nabla_{C\dot{R}} K_{A\dot{A}}$ [see (2.139)]. Since the left-hand side of the last equation is symmetric on the indices BC, we have

$$\begin{aligned}\nabla_{A\dot{A}} L_{BC} &= \tfrac{1}{2} \varepsilon_{A(B} \square_{|\dot{A}|}{}^{\dot{R}} K_{C)\dot{R}} - \tfrac{1}{2} \square_{A(B} K_{C)\dot{A}} - \tfrac{1}{2} \square_{BC} K_{A\dot{A}} \\ &= -C^D{}_{ABC} K_{D\dot{A}} - C_{BC}{}^{\dot{D}}{}_{\dot{A}} K_{A\dot{D}} - \tfrac{1}{12} R \varepsilon_{A(B} K_{C)\dot{A}},\end{aligned} \tag{2.144}$$

by virtue of the Ricci identities (2.120) and (2.121). Analogously, we obtain

$$\nabla_{A\dot{A}} L_{\dot{B}\dot{C}} = -C^{\dot{D}}{}_{\dot{A}\dot{B}\dot{C}} K_{A\dot{D}} - C^D{}_{A\dot{B}\dot{C}} K_{D\dot{A}} - \tfrac{1}{12} R \varepsilon_{\dot{A}(\dot{B}} K_{|A|\dot{C})}. \tag{2.145}$$

Equations (2.144) together with (2.145) are equivalent to the well-known relation $\nabla_a \nabla_b K_c = R^d{}_{abc} K_d$, which follows from the Killing equations. From (2.144) and (2.145) one finds that the spinor fields L_{AB} and $L_{\dot{A}\dot{B}}$ are covariantly constant along the Killing vector K,

$$K^{A\dot{A}} \nabla_{A\dot{A}} L_{BC} = 0, \qquad K^{A\dot{A}} \nabla_{A\dot{A}} L_{\dot{B}\dot{C}} = 0,$$

or equivalently [see (2.143)],

$$\pounds_K L_{BC} = 0, \qquad \pounds_K L_{\dot{B}\dot{C}} = 0. \tag{2.146}$$

Combining (2.118) and (2.144), one can compute $\Box_{AD} L_{BC}$ and $\Box_{\dot{A}\dot{D}} L_{BC}$; then, making use of (2.140) and the Ricci and Bianchi identities, one finds that

$$\begin{aligned} K^{R\dot{S}} \nabla_{R\dot{S}} C_{ABCD} - 4L^R{}_{(A} C_{BCD)R} &= 0, \\ K^{R\dot{S}} \nabla_{R\dot{S}} R &= 0, \\ K^{R\dot{S}} \nabla_{R\dot{S}} C_{AB\dot{C}\dot{D}} - 2L^R{}_{(A} C_{B)R\dot{C}\dot{D}} - 2L^{\dot{R}}{}_{(\dot{C}} C_{|AB|\dot{D})\dot{R}} &= 0, \end{aligned} \tag{2.147}$$

which can also be written as $\pounds_K C_{ABCD} = 0$, $\pounds_K R = 0$, and $\pounds_K C_{AB\dot{C}\dot{D}} = 0$, respectively. In a similar way one finds that $\pounds_K C_{\dot{A}\dot{B}\dot{C}\dot{D}} = 0$.

It is also possible to define the Lie derivative of spinor fields with respect to a *conformal Killing vector*. A vector field K is a conformal Killing vector if $\nabla_a K_b + \nabla_b K_a = 2\chi g_{ab}$, where χ is a function different from zero (if χ is a nonzero constant, K is a *homothetic Killing vector*); hence, the spinor equivalent of a conformal Killing vector satisfies

$$\nabla_{A\dot{A}} K_{B\dot{B}} = L_{AB}\varepsilon_{\dot{A}\dot{B}} + L_{\dot{A}\dot{B}}\varepsilon_{AB} - \chi\varepsilon_{AB}\varepsilon_{\dot{A}\dot{B}},$$

with L_{AB} and $L_{\dot{A}\dot{B}}$ symmetric. Proceeding as in the case of a Killing vector, one obtains the formula

$$\begin{aligned} \pounds_K \psi^{A\ldots\dot{B}\ldots}_{C\ldots\dot{D}\ldots} = {} & -K^{R\dot{S}} \nabla_{R\dot{S}} \psi^{A\ldots\dot{B}\ldots}_{C\ldots\dot{D}\ldots} - (L^A{}_S + \tfrac{1}{2}\chi\delta^A_S)\psi^{S\ldots\dot{B}\ldots}_{C\ldots\dot{D}\ldots} - \cdots \\ & - (L^{\dot{B}}{}_{\dot{S}} + \tfrac{1}{2}\chi\delta^{\dot{B}}_{\dot{S}})\psi^{A\ldots\dot{S}\ldots}_{C\ldots\dot{D}\ldots} - \cdots + (L^S{}_C + \tfrac{1}{2}\chi\delta^S_C)\psi^{A\ldots\dot{B}\ldots}_{S\ldots\dot{D}\ldots} \\ & + \cdots + (L^{\dot{S}}{}_{\dot{D}} + \tfrac{1}{2}\chi\delta^{\dot{S}}_{\dot{D}})\psi^{A\ldots\dot{B}\ldots}_{C\ldots\dot{S}\ldots} + \cdots, \end{aligned} \tag{2.148}$$

which reduces to (2.143) when $\chi = 0$. Now we have, for instance, $\pounds_K \varepsilon^{AB} = -\chi\varepsilon^{AB}$ and $\pounds_K \varepsilon_{AB} = \chi\varepsilon_{AB}$.

Example 2.5. Total angular momentum.

A flat four-dimensional Riemannian manifold possesses ten linearly independent Killing vectors. In fact, from (2.144) and (2.145) we obtain $\nabla_{A\dot{A}} L_{BC} = 0$, $\nabla_{A\dot{A}} L_{\dot{B}\dot{C}} = 0$. On the other hand, if x^a denote Cartesian coordinates, the vector fields $\partial_a = \partial/\partial x^a$ form an orthonormal tetrad with $\Gamma_{abc} = 0$; therefore, $\nabla_{A\dot{A}} L_{BC} = 0$ and $\nabla_{A\dot{A}} L_{\dot{B}\dot{C}} = 0$ reduce to $\partial_{A\dot{A}} L_{BC} = 0$ and $\partial_{A\dot{A}} L_{\dot{B}\dot{C}} = 0$, which means that the components L_{BC} and $L_{\dot{B}\dot{C}}$ with respect to the null tetrad $\partial_{A\dot{A}} = (1/\sqrt{2})\,\sigma^a{}_{A\dot{A}}\,\partial/\partial x^a$ are constant. Letting $x^{A\dot{A}} \equiv (1/\sqrt{2})\,\sigma_a{}^{A\dot{A}} x^a$, we have

$$\partial_{A\dot{A}} x^{B\dot{B}} = \frac{1}{2}\sigma^a{}_{A\dot{A}} \frac{\partial}{\partial x^a} \sigma_b{}^{B\dot{B}} x^b = -\delta^B_A \delta^{\dot{B}}_{\dot{A}},$$

which means that

$$\partial_{A\dot{A}} = -\frac{\partial}{\partial x^{A\dot{A}}}. \tag{2.149}$$

Then, from (2.140) we obtain

$$-\frac{\partial K_{B\dot{B}}}{\partial x^{A\dot{A}}} = L_{AB}\varepsilon_{\dot{A}\dot{B}} + L_{\dot{A}\dot{B}}\varepsilon_{AB},$$

and the solution is

$$K_{B\dot{B}} = L_{BS}x^{S}{}_{\dot{B}} + L_{\dot{B}\dot{S}}x_{B}{}^{\dot{S}} + b_{B\dot{B}},$$

where the $b_{A\dot{A}}$ are constants (which correspond to rigid translations). The Killing vectors of the form

$$K_{A\dot{A}} = L_{AB}x^{B}{}_{\dot{A}} + L_{\dot{A}\dot{B}}x_{A}{}^{\dot{B}} = (L_{AB}\varepsilon_{\dot{A}\dot{B}} + L_{\dot{A}\dot{B}}\varepsilon_{AB})x^{B\dot{B}} \tag{2.150}$$

generate a group of isometries isomorphic to SO(p,q).

In the specific case of the Minkowski space-time (Lorentzian signature), the Killing vector given by (2.150) with $L_{AB} = -\mathrm{i}\alpha_{(A}\beta_{B)}$, where α_A and β_A are constant with $\alpha^A\beta_A = 1$, generates rotations in the plane spanned by the orthogonal spacelike unit vectors given by

$$v_{A\dot{A}} = \frac{1}{\sqrt{2}}(\alpha_A\overline{\beta}_{\dot{A}} + \beta_A\overline{\alpha}_{\dot{A}}), \qquad w_{A\dot{A}} = -\frac{\mathrm{i}}{\sqrt{2}}(\alpha_A\overline{\beta}_{\dot{A}} - \beta_A\overline{\alpha}_{\dot{A}}).$$

Then $v_{(A}{}^{\dot{R}}w_{B)\dot{R}} = L_{AB}$, and from (1.52) and (1.69) we have

$$v_{A\dot{A}}w_{B\dot{B}} - v_{B\dot{B}}w_{A\dot{A}} = L_{AB}\varepsilon_{\dot{A}\dot{B}} + L_{\dot{A}\dot{B}}\varepsilon_{AB}$$

and

$$v^a w^b S_{abAB} = v^a\sigma_{a(A}{}^{\dot{R}}w^b\sigma_{|b|B)\dot{R}} = 2v_{(A}{}^{\dot{R}}w_{B)\dot{R}} = 2L_{AB}.$$

Therefore, for instance, the Lie derivative of a bispinor, defined as the bispinor formed by the Lie derivatives of the two-component spinors, is [see (2.143), (2.150), and (1.216)]

$$\begin{aligned}
\begin{pmatrix} \pounds_K\psi_A \\ \pounds_K\phi^{\dot{A}} \end{pmatrix} &= -K^{B\dot{B}}\nabla_{B\dot{B}}\begin{pmatrix} \psi_A \\ \phi^{\dot{A}} \end{pmatrix} + \begin{pmatrix} L^{C}{}_{A}\psi_C \\ -L^{\dot{A}}{}_{\dot{C}}\phi^{\dot{C}} \end{pmatrix} \\
&= -(v^{B\dot{B}}w^{C\dot{C}} - v^{C\dot{C}}w^{B\dot{B}})x_{C\dot{C}}\nabla_{B\dot{B}}\begin{pmatrix} \psi_A \\ \phi^{\dot{A}} \end{pmatrix} + \frac{1}{2}v^a w^b\begin{pmatrix} S_{abA}{}^{C}\psi_C \\ -S_{ab}{}^{\dot{A}}{}_{\dot{C}}\phi^{\dot{C}} \end{pmatrix} \\
&= -(v^b w_c - v_c w^b)x^c\partial_b\begin{pmatrix} \psi_A \\ \phi^{\dot{A}} \end{pmatrix} + \frac{1}{4}v^a w^b[\gamma_a,\gamma_b]\begin{pmatrix} \psi_A \\ \phi^{\dot{A}} \end{pmatrix} \\
&= v^a w^b\left(x_a\frac{\partial}{\partial x^b} - x_b\frac{\partial}{\partial x^a}\right)\begin{pmatrix} \psi_A \\ \phi^{\dot{A}} \end{pmatrix} + \frac{1}{4}v^a w^b[\gamma_a,\gamma_b]\begin{pmatrix} \psi_A \\ \phi^{\dot{A}} \end{pmatrix}.
\end{aligned}$$

Apart from a factor $-\mathrm{i}\hbar$, this Lie derivative represents the component of the total angular momentum along the spatial direction orthogonal to the plane spanned by v^a and w^a.

The fact that the components of the metric tensor (2.106) do not depend on the coordinate φ implies that $\partial/\partial\varphi$ is a Killing vector of the Schwarzschild metric. The nonvanishing components of the vector field $K = \partial/\partial\varphi$ with respect to the null tetrad (2.108) are $K^{1\dot{2}} = \mathrm{i}r\sin\theta/\sqrt{2}$ and $K^{2\dot{1}} = -\mathrm{i}r\sin\theta/\sqrt{2}$; thus, according to (2.141) and (2.109) [or using (2.142) and (2.107)], the components of the corresponding field L_{AB} are

$$L_{11} = \frac{\mathrm{i}}{\sqrt{2}}\sin\theta, \qquad L_{12} = \frac{\mathrm{i}}{2}\cos\theta, \qquad L_{22} = -\frac{\mathrm{i}}{2\sqrt{2}}\left(1 - \frac{r_g}{r}\right)\sin\theta.$$

Then, from (2.143) and (2.109) one obtains the remarkably simple expression

$$\pounds_{\partial/\partial\varphi}\psi^A = \frac{\partial}{\partial\varphi}\psi^A.$$

The Schwarzschild metric possesses the two additional spacelike Killing vectors

$$K_{(1)} \equiv -\sin\varphi\frac{\partial}{\partial\theta} - \cot\theta\cos\varphi\frac{\partial}{\partial\varphi}, \qquad K_{(2)} \equiv \cos\varphi\frac{\partial}{\partial\theta} - \cot\theta\sin\varphi\frac{\partial}{\partial\varphi},$$

which, together with $K_{(3)} \equiv \partial/\partial\varphi$, generate a group of isometries isomorphic to SO(3) (one can verify that $[K_{(i)}, K_{(j)}] = -\varepsilon_{ijk}K_{(k)}$). Following the steps outlined above, one finds that

$$\pounds_{K_{(1)}}\psi^A = \left(-\sin\varphi\frac{\partial}{\partial\theta} - \cot\theta\cos\varphi\frac{\partial}{\partial\varphi} - \mathrm{i}s\frac{\cos\varphi}{\sin\theta}\right)\psi^A,$$

$$\pounds_{K_{(2)}}\psi^A = \left(\cos\varphi\frac{\partial}{\partial\theta} - \cot\theta\sin\varphi\frac{\partial}{\partial\varphi} - \mathrm{i}s\frac{\sin\varphi}{\sin\theta}\right)\psi^A,$$

where s is the spin weight of ψ^A ($s = -1/2$ for ψ^1, $s = 1/2$ for ψ^2) (see also Torres del Castillo 2003, Chap. 3). It should be noted that all these expressions do not involve the parameter r_g, and therefore, they are also valid when $r_g = 0$, in which case the Schwarzschild metric (2.106) reduces to the metric of Minkowski spacetime in spherical coordinates.

Exercises

2.1. Show that $\mathrm{d}\breve{\theta}^{A\dot{A}} = \mathbf{\Gamma}^A{}_B \wedge \breve{\theta}^{B\dot{A}} + \mathbf{\Gamma}^{\dot{A}}{}_{\dot{B}} \wedge \breve{\theta}^{A\dot{B}}$.

2.2. Show that $\theta^{A\dot{A}} \wedge \breve{\theta}^{B\dot{B}} = \varepsilon^{AB}\varepsilon^{\dot{A}\dot{B}}\,(\theta^{1\dot{1}} \wedge \theta^{1\dot{2}} \wedge \theta^{2\dot{1}} \wedge \theta^{2\dot{2}})$.

2.3. Show that

$$\begin{aligned}
S^{AB} \wedge S^{CD} &= \tfrac{1}{2}(\varepsilon^{AC}\varepsilon^{BD} + \varepsilon^{AD}\varepsilon^{BC})\,\theta^{1\dot{1}} \wedge \theta^{1\dot{2}} \wedge \theta^{2\dot{1}} \wedge \theta^{2\dot{2}},\\
S^{AB} \wedge S^{\dot{C}\dot{D}} &= 0,\\
S^{\dot{A}\dot{B}} \wedge S^{\dot{C}\dot{D}} &= -\tfrac{1}{2}(\varepsilon^{\dot{A}\dot{C}}\varepsilon^{\dot{B}\dot{D}} + \varepsilon^{\dot{A}\dot{D}}\varepsilon^{\dot{B}\dot{C}})\,\theta^{1\dot{1}} \wedge \theta^{1\dot{2}} \wedge \theta^{2\dot{1}} \wedge \theta^{2\dot{2}}.
\end{aligned}$$

(The second of these relations amounts to the fact that with respect to the inner product induced on the two-forms, the self-dual two-forms are orthogonal to the anti-self-dual two-forms [cf. (1.71)].)

2.4. Show that

$$\breve{\theta}^{A\dot{B}} = \frac{1}{6\sqrt{2}} \breve{\sigma}_{abc}{}^{A\dot{B}} \theta^a \wedge \theta^b \wedge \theta^c,$$

which, together with (1.82), means that the three-form $\breve{\theta}^{A\dot{A}}$ is proportional to the dual of the one-form $\theta^{A\dot{A}}$.

2.5. Show that

$$\begin{aligned} \mathrm{d}(-K_{A\dot{B}}\theta^{A\dot{B}}) &= (\nabla_A{}^{\dot{C}} K_{B\dot{C}}) S^{AB} + (\nabla^C{}_{\dot{A}} K_{C\dot{B}}) S^{\dot{A}\dot{B}}, \\ \mathrm{d}(\phi_{AB} S^{AB}) &= -(\nabla^C{}_{\dot{B}} \phi_{CA}) \breve{\theta}^{A\dot{B}}, \\ \mathrm{d}(\phi_{\dot{A}\dot{B}} S^{\dot{A}\dot{B}}) &= (\nabla_A{}^{\dot{C}} \phi_{\dot{C}\dot{B}}) \breve{\theta}^{A\dot{B}}. \end{aligned}$$

2.6. Find the curvature of the metric (2.113) in the region $r < r_g$, where it has ultra-hyperbolic signature. Find the algebraic type of the Weyl spinors.

2.7. Show that if K is a Killing vector, then

$$\begin{aligned} \nabla_{A\dot{A}} \pounds_K f &= \pounds_K \nabla_{A\dot{A}} f, \\ \nabla_{A\dot{A}} \pounds_K \psi_B &= \pounds_K \nabla_{A\dot{A}} \psi_B, \\ \nabla_{A\dot{A}} \pounds_K \psi_{\dot{B}} &= \pounds_K \nabla_{A\dot{A}} \psi_{\dot{B}}, \end{aligned}$$

where f is a differentiable function, and ψ_B and $\psi_{\dot{B}}$ are one-index spinors.

2.8. Show that if K_1 and K_2 are two Killing vectors, then $[\pounds_{K_1}, \pounds_{K_2}]\psi_A = \pounds_{[K_1,K_2]}\psi_A$ and $[\pounds_{K_1}, \pounds_{K_2}]\psi_{\dot{A}} = \pounds_{[K_1,K_2]}\psi_{\dot{A}}$.

2.9. A two-index Killing tensor is a symmetric tensor field K_{ab} such that $\nabla_{(a}K_{bc)} = 0$ [cf. (2.139)]. Show that this equation is equivalent to

$$\nabla_{(A}{}^{(\dot{D}} P_{BC)}{}^{\dot{E}\dot{F})} = 0$$

and

$$\nabla^{C\dot{D}} P_{AC\dot{B}\dot{D}} = \tfrac{3}{4} \partial_{A\dot{B}} K,$$

where $P_{AB\dot{C}\dot{D}} = P_{(AB)(\dot{C}\dot{D})}$ is the spinor equivalent of the traceless part of K_{ab} and $K = K^a{}_a$.

2.10. The three-index tensor field H_{abc} is a Lanczos potential if

$$H_{abc} = -H_{bac}, \qquad H_{ab}{}^b = 0, \qquad H_{abc} + H_{bca} + H_{cab} = 0,$$

and the conformal curvature tensor can be expressed in the form

$$C_{abcd} = \nabla_d H_{abc} - \nabla_c H_{abd} + \nabla_b H_{cda} - \nabla_a H_{cdb} \\ + H_{(bc)} g_{ad} - H_{(ac)} g_{bd} + H_{(ad)} g_{bc} - H_{(bd)} g_{ac},$$

where $H_{ab} \equiv \nabla^c H_{acb}$ (Lanczos 1962). The spinor equivalent of H_{abc} is of the form $H_{A\dot{A}B\dot{B}C\dot{C}} = h_{ABC\dot{C}} \varepsilon_{\dot{A}\dot{B}} + h_{\dot{A}\dot{B}\dot{C}C} \varepsilon_{AB}$, with $h_{ABC\dot{C}} = h_{(ABC)\dot{C}}$ and $h_{\dot{A}\dot{B}\dot{C}C} = h_{(\dot{A}\dot{B}\dot{C})C}$ (see Exercise 1.3). Show that the Weyl spinors are given by

$$C_{ABCD} = -2\nabla_{(A}{}^{\dot{R}} h_{BCD)\dot{R}}, \qquad C_{\dot{A}\dot{B}\dot{C}\dot{D}} = -2\nabla^{R}{}_{(\dot{A}} h_{\dot{B}\dot{C}\dot{D})R}.$$

(Cf. also Maher and Zund 1968, Zund 1975, Bampi and Caviglia 1983, Ares de Parga et al. 1989, O'Donnell 2003.)

Chapter 3
Applications to General Relativity

In this chapter some applications of the spinor formalism to special and general relativity are considered. Some of the examples given here are related to tensor fields (such as the electromagnetic and the gravitational fields), which can be studied using the traditional tensor formalism; in these cases one can compare and appreciate the advantages of the use of the spinor formalism. Other examples involve spinor fields in an essential manner (such as those of the Dirac field and the Weyl neutrino field) and also serve to show the usefulness of the two-component spinor formalism.

Even though one can always construct the spinor equivalent of any vector or tensor, in some cases it is more convenient to use the original objects than their spinor equivalents. (For instance, the four-momentum of a particle with zero rest mass, being a null vector, can be represented by a single one-index spinor, but the four-momentum of a particle with nonzero rest mass, which is a timelike vector, does not have a particularly simple spinor equivalent.) The examples considered here have been selected in such a way that the advantages of the spinor formalism can be more clearly exhibited. These examples include the Maxwell equations, the self-dual electromagnetic fields, the energy–momentum tensor, and the principal null directions of the electromagnetic field. The Dirac equation is expressed in terms of two-component spinors, as well as the zero-rest-mass field equations for any spin.

This chapter contains proofs of the Mariot–Robinson and the Goldberg–Sachs theorems. In these theorems the shear-free congruences of null geodesics, which are represented by one-index spinor fields, play a central role. These congruences also appear in most of the examples considered in the rest of the book, such as the Killing spinors. The Ernst equations for solutions of the Einstein equations or the Einstein–Maxwell equations that admit a Killing vector are also derived by means of the spinor formalism.

Throughout this chapter it is assumed that M is a possibly curved space-time manifold with Lorentzian signature and that the Infeld–van der Waerden symbols satisfy the condition $\overline{\sigma_{aA\dot{B}}} = \sigma_{aB\dot{A}}$.

G.F.T. del Castillo, *Spinors in Four-Dimensional Spaces*, Progress in Mathematical Physics 59, DOI 10.1007/978-0-8176-4984-5_3,

3.1 Maxwell's Equations

In the framework of special or general relativity, the electromagnetic field is represented by an antisymmetric two-index tensor field F_{ab} (or differential two-form) on the space-time manifold. It is assumed that the electromagnetic field in vacuum satisfies the Maxwell equations

$$\nabla_b F^{ab} = \frac{4\pi}{c} J^a, \qquad \nabla_b {}^*F^{ab} = 0, \tag{3.1}$$

where the F_{ab} are the components of the electromagnetic field tensor with respect to an orthonormal tetrad ∂_a, J^a are the components of the four-current density, and c is the velocity of light in vacuum (see, e.g., Wald 1984). (The effect of the electromagnetic field on a charged particle is given by $\mathrm{d}U^a/\mathrm{d}\tau = (q/m_0 c) F^a{}_b U^b$, where τ, q, m_0, and U^a are the proper time, electric charge, rest mass, and four-velocity of the particle, respectively; if the signature of the metric is chosen as $(- - - +)$, instead of the first equation in (3.1) we have $\nabla_b F^{ba} = (4\pi/c) J^a$.)

According to (1.50), the spinor equivalent of F_{ab} is of the form

$$F_{A\dot{A}B\dot{B}} = f_{AB}\varepsilon_{\dot{A}\dot{B}} + f_{\dot{A}\dot{B}}\varepsilon_{AB}, \tag{3.2}$$

with $f_{AB} = \frac{1}{2} F_A{}^{\dot{R}}{}_{B\dot{R}}$ and $f_{\dot{A}\dot{B}} = \frac{1}{2} F^R{}_{\dot{A}R\dot{B}}$ being symmetric spinor fields, and since F_{ab} is real, $\overline{F_{A\dot{B}C\dot{D}}} = F_{B\dot{A}D\dot{C}}$ [see (1.47)]. Hence

$$f_{\dot{A}\dot{B}} = \overline{f_{AB}}. \tag{3.3}$$

Thus, the spinor equivalent of the Maxwell equations (3.1) is given by

$$-\nabla_{B\dot{B}}(f^{AB}\varepsilon^{\dot{A}\dot{B}} + f^{\dot{A}\dot{B}}\varepsilon^{AB}) = \frac{4\pi}{c} J^{A\dot{A}}$$

and

$$\mathrm{i}\nabla_{B\dot{B}}(f^{AB}\varepsilon^{\dot{A}\dot{B}} - f^{\dot{A}\dot{B}}\varepsilon^{AB}) = 0$$

[see (1.65)]. By combining these equations one obtains

$$\nabla_B{}^{\dot{A}} f^{AB} = \frac{2\pi}{c} J^{A\dot{A}}, \qquad \nabla^A{}_{\dot{B}} f^{\dot{A}\dot{B}} = \frac{2\pi}{c} J^{A\dot{A}}. \tag{3.4}$$

Owing to (3.3) and to the fact that $J^{A\dot{B}}$ is the spinor equivalent of a real vector field, the second equation is the complex conjugate of the first one.

Making use of (1.72) and the second line of (1.77) we find an explicit relation between the components of the electromagnetic spinor f_{AB} and the tetrad components F_{ab}:

$$\begin{aligned}(f^A{}_B) &= \tfrac{1}{4}(F^{ab} S_{ab}{}^A{}_B) \\ &= \tfrac{1}{2}\big[(-F^{14} + \mathrm{i}F^{23})\,\sigma_1 + (F^{24} - \mathrm{i}F^{31})\,\sigma_2 + (-F^{34} + \mathrm{i}F^{12})\,\sigma_3\big];\end{aligned}$$

hence,

$$(f_{AB}) = \frac{1}{2}\begin{pmatrix} -F^{14}+\mathrm{i}F^{23}+\mathrm{i}(F^{24}-\mathrm{i}F^{31}) & F^{34}-\mathrm{i}F^{12} \\ F^{34}-\mathrm{i}F^{12} & F^{14}-\mathrm{i}F^{23}+\mathrm{i}(F^{24}-\mathrm{i}F^{31}) \end{pmatrix}$$
$$= \frac{1}{2}\begin{pmatrix} E_x+\mathrm{i}B_x-\mathrm{i}(E_y+\mathrm{i}B_y) & -(E_z+\mathrm{i}B_z) \\ -(E_z+\mathrm{i}B_z) & -(E_x+\mathrm{i}B_x)-\mathrm{i}(E_y+\mathrm{i}B_y) \end{pmatrix}, \tag{3.5}$$

where (E_x, E_y, E_z) and (B_x, B_y, B_z) are the Cartesian components of the electric and magnetic field, respectively, with respect to the orthonormal tetrad ∂_a.

The Maxwell equations $\nabla_b {}^*F^{ab} = 0$, which can also be written as $\nabla_a F_{bc} + \nabla_b F_{ca} + \nabla_c F_{ab} = 0$, are locally equivalent to the existence of a vector field A_a (the four-potential) such that

$$F_{ab} = \nabla_a A_b - \nabla_b A_a. \tag{3.6}$$

Thus, if $A_{B\dot{B}}$ is the spinor equivalent of A_b, then

$$f_{BC} = \nabla_{(B}{}^{\dot{B}} A_{C)\dot{B}}, \qquad f_{\dot{B}\dot{C}} = \nabla^B{}_{(\dot{B}} A_{|B|\dot{C})}. \tag{3.7}$$

For a given electromagnetic field, the four-potential A_a is defined by (3.6) up to a gauge transformation

$$A_a \mapsto A_a + \partial_a \xi, \tag{3.8}$$

where ξ is an arbitrary differentiable function.

In a source-free region, $\nabla_b F^{ab} = 0$; therefore, by analogy with (3.6), there exists locally a second (real) vector field $\check{A}_a$ such that

$${}^*F_{ab} = \nabla_a \check{A}_b - \nabla_b \check{A}_a,$$

or equivalently [see (1.65)],

$$f_{BC} = \mathrm{i}\nabla_{(B}{}^{\dot{B}} \check{A}_{C)\dot{B}}, \qquad f_{\dot{B}\dot{C}} = -\mathrm{i}\nabla^B{}_{(\dot{B}} \check{A}_{|B|\dot{C})}, \tag{3.9}$$

where $\check{A}_{B\dot{B}}$ is the spinor equivalent of $\check{A}_b$. Then, the *complex* vector field

$$\Phi_a \equiv \tfrac{1}{2}(A_a - \mathrm{i}\check{A}_a)$$

is a potential for the self-dual field $\frac{1}{2}(F_{ab} - \mathrm{i}^*F_{ab})$ [i.e., ${}^*(F_{ab} - \mathrm{i}^*F_{ab}) = \mathrm{i}(F_{ab} - \mathrm{i}^*F_{ab})$]; in fact, from (3.7) and (3.9) we obtain

$$\nabla_{(B}{}^{\dot{B}} \Phi_{C)\dot{B}} = \tfrac{1}{2}(\nabla_{(B}{}^{\dot{B}} A_{C)\dot{B}} - \mathrm{i}\nabla_{(B}{}^{\dot{B}} \check{A}_{C)\dot{B}}) = 0,$$

that is,

$$\nabla_{(B}{}^{\dot{B}} \Phi_{C)\dot{B}} = 0. \tag{3.10}$$

Thus, we have proved the following result.

Proposition 3.1. *If F_{ab} is a (real) solution of the source-free Maxwell equations, then there exists locally a complex vector field Φ_a satisfying the self-duality condition (3.10) such that $F_{ab} = \nabla_a A_b - \nabla_b A_a$, where $A_a = 2\,\mathrm{Re}\,\Phi_a$.*

The self-duality condition (3.10) has the advantage of being a set of first-order partial differential equations for Φ_a, in contrast to the original source-free Maxwell equations $\nabla_b F^{ab} = 0$, which, expressed in terms of the four-potential, are second-order equations. Making use of (3.7) and the complex conjugate of (3.10), one finds that the self-dual part of the *real* electromagnetic field generated by the four-potential $A_a = 2\,\mathrm{Re}\,\Phi_a$ is given by

$$f_{\dot{A}\dot{B}} = \nabla^{B}{}_{(\dot{A}}\Phi_{|B|\dot{B})}. \tag{3.11}$$

In terms of the Newman–Penrose notation, the self-duality condition (3.10) is equivalent to the explicit expressions

$$\begin{aligned}
&(D-\varepsilon+\overline{\varepsilon}-\overline{\rho})\Phi_{1\dot{2}} - (\delta-\beta-\overline{\alpha}+\overline{\pi})\Phi_{1\dot{1}} + \kappa\Phi_{2\dot{2}} - \sigma\Phi_{2\dot{1}} = 0,\\
&\quad (D+\varepsilon+\overline{\varepsilon}+\rho-\overline{\rho})\Phi_{2\dot{2}} - (\delta+\beta-\overline{\alpha}+\tau+\overline{\pi})\Phi_{2\dot{1}}\\
&\quad +(\overline{\delta}-\alpha+\overline{\beta}-\pi-\overline{\tau})\Phi_{1\dot{2}} - (\Delta-\gamma-\overline{\gamma}-\mu+\overline{\mu})\Phi_{1\dot{1}} = 0,\\
&(\overline{\delta}+\alpha+\overline{\beta}-\overline{\tau})\Phi_{2\dot{2}} - (\Delta+\gamma-\overline{\gamma}+\overline{\mu})\Phi_{2\dot{1}} + \nu\Phi_{1\dot{1}} - \lambda\Phi_{1\dot{2}} = 0.
\end{aligned} \tag{3.12}$$

As an application of the last proposition, we can find the solution of the source-free Maxwell equations assuming that the metric of the space-time is the Schwarzschild metric. Making use of (2.108) and (2.109), after some rearrangement, one finds that equations (3.12) take the form

$$\frac{1}{r}\left[-\left(1-\frac{r_g}{r}\right)^{-1}\frac{1}{c}\frac{\partial}{\partial t}+\frac{\partial}{\partial r}\right] r\Phi_{1\dot{2}} - \frac{1}{\sqrt{2}r}\left(\frac{\partial}{\partial\theta}+\frac{\mathrm{i}}{\sin\theta}\frac{\partial}{\partial\varphi}\right)\Phi_{1\dot{1}} = 0,$$

$$\begin{aligned}
\left[-\left(1-\frac{r_g}{r}\right)^{-1}\frac{1}{c}\frac{\partial}{\partial t}+\frac{\partial}{\partial r}\right]\Phi_{2\dot{2}} &- \frac{1}{\sqrt{2}r}\left(\frac{\partial}{\partial\theta}+\frac{\mathrm{i}}{\sin\theta}\frac{\partial}{\partial\varphi}+\cot\theta\right)\Phi_{2\dot{1}}\\
&+\frac{1}{\sqrt{2}r}\left(\frac{\partial}{\partial\theta}-\frac{\mathrm{i}}{\sin\theta}\frac{\partial}{\partial\varphi}+\cot\theta\right)\Phi_{1\dot{2}}\\
&+\frac{1}{2}\left[\left(1-\frac{r_g}{r}\right)^{-1}\frac{1}{c}\frac{\partial}{\partial t}+\frac{\partial}{\partial r}\right]\left(1-\frac{r_g}{r}\right)\Phi_{1\dot{1}} = 0,
\end{aligned}$$

$$\frac{1}{\sqrt{2}r}\left(\frac{\partial}{\partial\theta}-\frac{\mathrm{i}}{\sin\theta}\frac{\partial}{\partial\varphi}\right)\Phi_{2\dot{2}} + \frac{1}{2r}\left[\frac{1}{c}\frac{\partial}{\partial t}+\left(1-\frac{r_g}{r}\right)\frac{\partial}{\partial r}\right] r\Phi_{2\dot{1}} = 0.$$

These equations admit separable solutions of the form

$$
\begin{aligned}
\Phi_{1\dot{1}} &= 2\left(1-\frac{r_g}{r}\right)^{-1} F(r)\, Y_{jm}(\theta,\varphi)\,\mathrm{e}^{-\mathrm{i}\omega t},\\
\Phi_{1\dot{2}} &= \sqrt{2}\,G(r)\,{}_1Y_{jm}(\theta,\varphi)\,\mathrm{e}^{-\mathrm{i}\omega t},\\
\Phi_{2\dot{1}} &= \sqrt{2}\,f(r)\,{}_{-1}Y_{jm}(\theta,\varphi)\,\mathrm{e}^{-\mathrm{i}\omega t},\\
\Phi_{2\dot{2}} &= g(r)\,Y_{jm}(\theta,\varphi)\,\mathrm{e}^{-\mathrm{i}\omega t},
\end{aligned}
\tag{3.13}
$$

where the ${}_sY_{jm}$ are spin-weighted spherical harmonics (Newman and Penrose 1966, Torres del Castillo 2003), j and m are integers, $j=1,2,\ldots,\ -j \leqslant m \leqslant j$, and ω is a constant. The factor $2(1-r_g/r)^{-1}$ is introduced for convenience and corrects the asymmetry introduced in the definition of the null tetrad (2.108). Following the conventions of Newman and Penrose (1966), we have

$$
\begin{aligned}
\left(\frac{\partial}{\partial\theta}+\frac{\mathrm{i}}{\sin\theta}\frac{\partial}{\partial\varphi}-s\cot\theta\right){}_sY_{jm} &= -\sqrt{j(j+1)-s(s+1)}\,{}_{s+1}Y_{jm},\\
\left(\frac{\partial}{\partial\theta}-\frac{\mathrm{i}}{\sin\theta}\frac{\partial}{\partial\varphi}+s\cot\theta\right){}_sY_{jm} &= \sqrt{j(j+1)-s(s-1)}\,{}_{s-1}Y_{jm},
\end{aligned}
\tag{3.14}
$$

and the ${}_0Y_{jm}$ are the ordinary spherical harmonics Y_{jm}. Thus, the one-variable functions F, G, f, and g must obey the system of ordinary differential equations

$$
\begin{aligned}
&\left[\frac{\mathrm{i}\omega}{c}+\left(1-\frac{r_g}{r}\right)\frac{\mathrm{d}}{\mathrm{d}r}\right] rG+\sqrt{j(j+1)}\,F=0,\\
&\left[\frac{\mathrm{i}\omega}{c}+\left(1-\frac{r_g}{r}\right)\frac{\mathrm{d}}{\mathrm{d}r}\right] g+\frac{\sqrt{j(j+1)}}{r}\left(1-\frac{r_g}{r}\right)(f+G)\\
&\qquad +\left[-\frac{\mathrm{i}\omega}{c}+\left(1-\frac{r_g}{r}\right)\frac{\mathrm{d}}{\mathrm{d}r}\right] F=0,\\
&\sqrt{j(j+1)}\,g+\left[-\frac{\mathrm{i}\omega}{c}+\left(1-\frac{r_g}{r}\right)\frac{\mathrm{d}}{\mathrm{d}r}\right] rf=0.
\end{aligned}
\tag{3.15}
$$

These equations imply that $f+G$ must satisfy the decoupled equation

$$
\left(1-\frac{r_g}{r}\right) j(j+1)\frac{f+G}{r}=\left[\frac{\omega^2}{c^2}+\left(1-\frac{r_g}{r}\right)\frac{\mathrm{d}}{\mathrm{d}r}\left(1-\frac{r_g}{r}\right)\frac{\mathrm{d}}{\mathrm{d}r}\right] r(f+G), \tag{3.16}
$$

leaving $f-G$ unspecified, which is related to the gauge freedom (3.8) [see equation (3.17) below].

Then, letting

$$
\xi \equiv \frac{r(f-G)}{\sqrt{j(j+1)}}\,Y_{jm}\mathrm{e}^{-\mathrm{i}\omega t}, \qquad \chi \equiv \frac{f+G}{\sqrt{j(j+1)}}\,Y_{jm}\mathrm{e}^{-\mathrm{i}\omega t},
$$

from (2.108) and (3.13)–(3.16) one obtains

$$\begin{aligned}
\Phi_{1\dot{1}} &= \partial_{1\dot{1}}\xi + \partial_{1\dot{1}}(r\chi), & \Phi_{2\dot{2}} &= \partial_{2\dot{2}}\xi - \partial_{2\dot{2}}(r\chi), \\
\Phi_{1\dot{2}} &= \partial_{1\dot{2}}\xi + \partial_{1\dot{2}}(r\chi), & \Phi_{2\dot{1}} &= \partial_{2\dot{1}}\xi - \partial_{2\dot{1}}(r\chi),
\end{aligned} \tag{3.17}$$

where ξ is an arbitrary function [cf. (3.8)] and χ is a solution of the second-order linear partial differential equation

$$\left[\left(1-\frac{r_g}{r}\right)\left(\frac{\partial}{\partial r}\left(1-\frac{r_g}{r}\right)\frac{\partial}{\partial r}+\frac{1}{\sin\theta}\frac{\partial}{\partial\theta}\sin\theta\frac{\partial}{\partial\theta}+\frac{1}{\sin^2\theta}\frac{\partial^2}{\partial\varphi^2}\right)-\frac{1}{c^2}\frac{\partial^2}{\partial t^2}\right] r\chi = 0. \tag{3.18}$$

(When $r_g = 0$, this equation reduces to the usual scalar wave equation for χ.) Note that (3.17) and (3.18) do not contain the separation constants j, m, and ω. By choosing the function ξ equal to $r\chi$ or to $-r\chi$, one makes $\Phi_{2\dot{A}} = 0$ or $\Phi_{1\dot{A}} = 0$, respectively (see (3.76) below).

A similar result holds in other space-times; the solution of the self-duality condition (3.10) can be expressed locally in terms of a single complex function if there exists a repeated principal spinor of the conformal curvature l_A such that $l^A l^B \nabla_{A\dot{A}} l_B = 0$ (see Proposition 3.6 below).

When the Ricci tensor vanishes, each Killing vector K is the four-potential of a (possibly trivial) solution of the source-free Maxwell equations. From (2.144) and (2.85) we have $\nabla^A{}_{\dot{A}} L_{BC} = C_{ACD\dot{A}} K^{AD} - \frac{1}{8} R K_{C\dot{A}} = \frac{1}{2} R_{C\dot{A}A\dot{D}} K^{A\dot{D}}$, which is equal to zero if $R_{ab} = 0$ or under the weaker condition $R_{ab} K^b = 0$. In any of these cases, $f_{AB} = L_{AB}$ satisfies the source-free Maxwell equations (3.4). Furthermore, by virtue of (2.146), the Lie derivative with respect to K of this electromagnetic field is equal to zero. In the case of Minkowski space-time one obtains only uniform electromagnetic fields in this manner (see Example 2.5). However, the timelike Killing vector $\partial/\partial t$ of the Schwarzschild metric gives rise to a nontrivial spherically symmetric field.

Energy–Momentum Tensor of the Electromagnetic Field

In special relativity, the four-vector $\frac{1}{c}F^a{}_b J^b$ is the Lorentz force density on the charge and current distribution represented by J^a, in the presence of the electromagnetic field F_{ab} (see, e.g., Jackson 1975, Sec. 12.10). The spinor equivalent of $\frac{1}{c}F^a{}_b J^b$, expressed in terms of the electromagnetic field alone, is

$$\begin{aligned}
-\frac{1}{c}F^{A\dot{A}}{}_{B\dot{B}}J^{B\dot{B}} &= -\frac{1}{c}(f^A{}_B \varepsilon^{\dot{A}}{}_{\dot{B}} + f^{\dot{A}}{}_{\dot{B}} \varepsilon^A{}_B) J^{B\dot{B}} \\
&= -\frac{1}{c}(f^A{}_B J^{B\dot{A}} + f^{\dot{A}}{}_{\dot{B}} J^{A\dot{B}}) \\
&= -\frac{1}{2\pi}(f^A{}_B \nabla^B{}_{\dot{C}} f^{\dot{A}\dot{C}} + f^{\dot{A}}{}_{\dot{B}} \nabla_C{}^{\dot{B}} f^{AC}) \\
&= \frac{1}{2\pi}\nabla_{B\dot{B}}(f^{AB} f^{\dot{A}\dot{B}}),
\end{aligned}$$

where we have made use of (3.2) and (3.4). The spinor field

$$T_{AB\dot{A}\dot{B}} \equiv \frac{1}{2\pi} f_{AB} f_{\dot{A}\dot{B}} \tag{3.19}$$

thus obtained is the spinor equivalent of the energy–momentum tensor of the electromagnetic field T_{ab}. As a consequence of the symmetries $T_{AB\dot{A}\dot{B}} = T_{(AB)(\dot{A}\dot{B})}$, T_{ab} is symmetric and traceless.

In order to find the expression for T_{ab} in terms of the electromagnetic field tensor F_{ab}, we rewrite (3.19) in the form

$$\begin{aligned} 4\pi T_{AB\dot{A}\dot{B}} &= -(f_{AC}\varepsilon_{\dot{A}\dot{C}} + f_{\dot{A}\dot{C}}\varepsilon_{AC})(f_B{}^C \varepsilon_{\dot{B}}{}^{\dot{C}} + f_{\dot{B}}{}^{\dot{C}} \varepsilon_B{}^C) \\ &\quad - f_{AC} f_B{}^C \varepsilon_{\dot{A}\dot{B}} - f_{\dot{A}\dot{C}} f_{\dot{B}}{}^{\dot{C}} \varepsilon_{AB}. \end{aligned}$$

Owing to (1.23), $f_{AC} f_B{}^C = -f_A{}^C f_{BC} = -f_{BC} f_A{}^C$; hence, by virtue of (1.26), $f_{AC} f_B{}^C = -\frac{1}{2} f_{RC} f^{RC} \varepsilon_{AB}$, and therefore, by complex conjugation, $f_{\dot{A}\dot{C}} f_{\dot{B}}{}^{\dot{C}} = -\frac{1}{2} f_{\dot{R}\dot{C}} f^{\dot{R}\dot{C}} \varepsilon_{\dot{A}\dot{B}}$. Thus

$$\begin{aligned} 4\pi T_{AB\dot{A}\dot{B}} &= -(f_{AC}\varepsilon_{\dot{A}\dot{C}} + f_{\dot{A}\dot{C}}\varepsilon_{AC})(f_B{}^C \varepsilon_{\dot{B}}{}^{\dot{C}} + f_{\dot{B}}{}^{\dot{C}} \varepsilon_B{}^C) \\ &\quad + \tfrac{1}{2}(f_{RC} f^{RC} + f_{\dot{R}\dot{C}} f^{\dot{R}\dot{C}})\, \varepsilon_{AB}\varepsilon_{\dot{A}\dot{B}} \\ &= -(f_{AC}\varepsilon_{\dot{A}\dot{C}} + f_{\dot{A}\dot{C}}\varepsilon_{AC})(f_B{}^C \varepsilon_{\dot{B}}{}^{\dot{C}} + f_{\dot{B}}{}^{\dot{C}} \varepsilon_B{}^C) \\ &\quad + \tfrac{1}{4}(f_{RC}\varepsilon_{\dot{R}\dot{C}} + f_{\dot{R}\dot{C}}\varepsilon_{RC})(f^{RC}\varepsilon^{\dot{R}\dot{C}} + f^{\dot{R}\dot{C}}\varepsilon^{RC})\, \varepsilon_{AB}\varepsilon_{\dot{A}\dot{B}}, \end{aligned}$$

and this amounts to

$$4\pi T_{ab} = F_{ac} F_b{}^c - \tfrac{1}{4} F_{cd} F^{cd} g_{ab}. \tag{3.20}$$

Equivalently, we can start from

$$\begin{aligned} 4\pi T_{AB\dot{A}\dot{B}} &= (f_{AC}\varepsilon_{\dot{A}\dot{C}} - f_{\dot{A}\dot{C}}\varepsilon_{AC})(f_B{}^C \varepsilon_{\dot{B}}{}^{\dot{C}} - f_{\dot{B}}{}^{\dot{C}} \varepsilon_B{}^C) \\ &\quad + f_{AC} f_B{}^C \varepsilon_{\dot{A}\dot{B}} + f_{\dot{A}\dot{C}} f_{\dot{B}}{}^{\dot{C}} \varepsilon_{AB}, \end{aligned}$$

which leads to the alternative form [see (1.65)]

$$4\pi T_{ab} = {}^*F_{ac}{}^*F_b{}^c + \tfrac{1}{4} F_{cd} F^{cd} g_{ab}.$$

By combining the two expressions found for T_{ab} we also have

$$4\pi T_{ab} = \tfrac{1}{2}(F_{ac} F_b{}^c + {}^*F_{ac}{}^*F_b{}^c). \tag{3.21}$$

In addition to the algebraic properties $T_{ab} = T_{(ab)}$ and $T_a{}^a = 0$, the energy–momentum tensor of the electromagnetic field satisfies the so-called *dominant energy condition*: for every timelike vector t^a, $T_{ab}t^a t^b \geqslant 0$ and $T_{ab}t^b$ is a nonspacelike vector, i.e., $(T_{ab}t^b)(T^a{}_c t^c) \leqslant 0$. Physically, this means that for any observer, the energy density of the electromagnetic field is always positive and the energy flux vector is nonspacelike. We can readily prove the validity of this condition by making use of the fact that f_{AB} can be expressed as the symmetrized tensor product of its principal spinors (see Section 1.7).

If the principal spinors α_A, β_A of f_{AB} are not proportional to each other, the electromagnetic field is called *algebraically general*; then, writing $f_{AB} = \alpha_{(A}\beta_{B)}$, we have

$$\begin{aligned}
T_{ab}t^a t^b &= T_{AB\dot{A}\dot{B}}t^{A\dot{A}}t^{B\dot{B}} \\
&= \frac{1}{2\pi}\alpha_{(A}\beta_{B)}\overline{\alpha}_{(\dot{A}}\overline{\beta}_{\dot{B})}t^{A\dot{A}}t^{B\dot{B}} \\
&= \frac{1}{4\pi}(t^{A\dot{A}}\alpha_A\overline{\alpha}_{\dot{A}}t^{B\dot{B}}\beta_B\overline{\beta}_{\dot{B}} + t^{A\dot{A}}\beta_A\overline{\alpha}_{\dot{A}}t^{B\dot{B}}\alpha_B\overline{\beta}_{\dot{B}}).
\end{aligned} \tag{3.22}$$

On the other hand,

$$\begin{aligned}
t^a t_a &= -\varepsilon_{AB}\varepsilon_{\dot{A}\dot{B}}t^{A\dot{A}}t^{B\dot{B}} \\
&= -\frac{1}{\alpha^R\beta_R\overline{\alpha}^{\dot{S}}\overline{\beta}_{\dot{S}}}(\alpha_A\beta_B - \alpha_B\beta_A)(\overline{\alpha}_{\dot{A}}\overline{\beta}_{\dot{B}} - \overline{\alpha}_{\dot{B}}\overline{\beta}_{\dot{A}})t^{A\dot{A}}t^{B\dot{B}} \\
&= -\frac{2}{|\alpha^R\beta_R|^2}(t^{A\dot{A}}\alpha_A\overline{\alpha}_{\dot{A}}t^{B\dot{B}}\beta_B\overline{\beta}_{\dot{B}} - t^{A\dot{A}}\beta_A\overline{\alpha}_{\dot{A}}t^{B\dot{B}}\alpha_B\overline{\beta}_{\dot{B}});
\end{aligned}$$

hence, $t^a t_a < 0$ if and only if $t^{A\dot{A}}\alpha_A\overline{\alpha}_{\dot{A}}\, t^{B\dot{B}}\beta_B\overline{\beta}_{\dot{B}} > |t^{A\dot{A}}\alpha_A\overline{\beta}_{\dot{A}}|^2$, and from (3.22) we then obtain $T_{ab}t^a t^b > \frac{1}{2\pi}|t^{A\dot{A}}\alpha_A\overline{\beta}_{\dot{A}}|^2 \geqslant 0$.

When the two principal spinors of the electromagnetic spinor f_{AB} are proportional to each other, the electromagnetic field is called *algebraically special* or *null*, and by absorbing the proportionality factor into α_A, we can write $f_{AB} = \alpha_A\alpha_B$; then $T_{ab}t^a t^b = \frac{1}{2\pi}(t^{A\dot{A}}\alpha_A\overline{\alpha}_{\dot{A}})^2 \geqslant 0$. As a matter of fact, for $t_a t^a < 0$, $t^{A\dot{A}}\alpha_A\overline{\alpha}_{\dot{A}}$ vanishes only if $\alpha_A = 0$ (see Exercise 3.1).

In order to prove that for every timelike vector t^a, $T_{ab}t^b$ is a nonspacelike vector, we first note that the spinor equivalent of $T_{ab}T^a{}_c$ is given by

$$\begin{aligned}
-\frac{1}{(2\pi)^2}f_{AB}f_{\dot{A}\dot{B}}f^A{}_C f^{\dot{A}}{}_{\dot{C}} &= -\frac{1}{16\pi^2}f_{AR}f^{AR}\varepsilon_{BC}f_{\dot{A}\dot{S}}f^{\dot{A}\dot{S}}\varepsilon_{\dot{B}\dot{C}} \\
&= -\frac{1}{16\pi^2}\left|f_{AR}f^{AR}\right|^2\varepsilon_{BC}\varepsilon_{\dot{B}\dot{C}},
\end{aligned}$$

which means that $T_{ab}T^a{}_c$ is proportional to g_{bc}. In terms of the electromagnetic field tensor and its dual, $f^{AB}f_{AB}$ can be expressed as [see (1.65)]

$$\begin{aligned}
f^{AB}f_{AB} &= \tfrac{1}{2}f^{AB}\varepsilon^{\dot{A}\dot{B}}f_{AB}\varepsilon_{\dot{A}\dot{B}} \\
&= \tfrac{1}{2}(f^{AB}\varepsilon^{\dot{A}\dot{B}} + f^{\dot{A}\dot{B}}\varepsilon^{AB})f_{AB}\varepsilon_{\dot{A}\dot{B}} \\
&= \tfrac{1}{2}F^{AB\dot{A}\dot{B}}\,\tfrac{1}{2}(F_{AB\dot{A}\dot{B}} + \mathrm{i}\,{}^*F_{AB\dot{A}\dot{B}}) \\
&= \tfrac{1}{4}(F^{ab}F_{ab} + \mathrm{i}\,{}^*F^{ab}F_{ab});
\end{aligned} \tag{3.23}$$

hence,

$$T_{ab}T^a{}_c = \frac{1}{(16\pi)^2}\left[(F^{rs}F_{rs})^2 + ({}^*F^{rs}F_{rs})^2\right]g_{bc}, \tag{3.24}$$

and from this last expression we readily see that $(T_{ab}t^b)(T^a{}_c t^c) \leqslant 0$ for every timelike vector t^a. As a matter of fact, $T_{ab}t^a T^a{}_c t^c = 0$ for any timelike vector t^a, that is, the energy flux vector is null for any observer if and only if $f^{AB} f_{AB} = 0$.

The proportionality of $T_{ab}T^a{}_c$ to g_{bc} is a characteristic of the energy–momentum tensor of the electromagnetic field. More precisely, if $T_{ab} = T_{(ab)}$, $T^a{}_a = 0$, and $T_{ab}T^a{}_c$ is proportional to g_{bc}, then there exists $F_{ab} = -F_{ba}$ such that $4\pi T_{ab} = \pm(F_{ac}F_b{}^c - \frac{1}{4}F_{cd}F^{cd}g_{ab})$. In order to prove this assertion we note that for an object μ_{ABCD} such that $\mu_{ABCD} = -\mu_{CDAB}$, the identity

$$\begin{aligned}
\mu_{ABCD} &= \tfrac{1}{2}(\mu_{ABCD} - \mu_{CDAB}) \\
&= \tfrac{1}{2}(\mu_{ABCD} - \mu_{CBAD} + \mu_{CBAD} - \mu_{CDAB}) \\
&= \tfrac{1}{2}(\varepsilon_{AC}\mu^R{}_{BRD} + \varepsilon_{BD}\mu_C{}^R{}_{AR}) \\
&= \tfrac{1}{2}\varepsilon_{AC}(\mu^R{}_{(B|R|D)} + \tfrac{1}{2}\varepsilon_{BD}\mu^{RS}{}_{RS}) \\
&\quad + \tfrac{1}{2}\varepsilon_{BD}(\mu_{(C}{}^R{}_{A)R} + \tfrac{1}{2}\varepsilon_{CA}\mu^{SR}{}_{SR}) \\
&= \tfrac{1}{2}(\varepsilon_{AC}\mu^R{}_{(B|R|D)} + \varepsilon_{BD}\mu_{(C}{}^R{}_{A)R})
\end{aligned} \tag{3.25}$$

holds. Since we are assuming that T_{ab} is symmetric and traceless, it follows that $T_{AB\dot{A}\dot{B}} = T_{(AB)(\dot{A}\dot{B})}$, and from (3.25) we have

$$T^{AB}{}_{\dot{A}\dot{B}}T^{CD}{}_{\dot{C}\dot{D}} - T^{CD}{}_{\dot{A}\dot{B}}T^{AB}{}_{\dot{C}\dot{D}} = \varepsilon^{AC}T^{R(B}{}_{\dot{A}\dot{B}}T_R{}^{D)}{}_{\dot{C}\dot{D}} + \varepsilon^{BD}T^{R(C}{}_{\dot{A}\dot{B}}T_R{}^{A)}{}_{\dot{C}\dot{D}}.$$

Now $T^{R(B}{}_{\dot{A}\dot{B}}T_R{}^{D)}{}_{\dot{C}\dot{D}} = -T^{R(B}{}_{\dot{C}\dot{D}}T_R{}^{D)}{}_{\dot{A}\dot{B}}$, and applying the analogue of (3.25) for objects with dotted indices, we find that

$$\begin{aligned}
&T^{AB}{}_{\dot{A}\dot{B}}T^{CD}{}_{\dot{C}\dot{D}} - T^{CD}{}_{\dot{A}\dot{B}}T^{AB}{}_{\dot{C}\dot{D}} \\
&\quad = \tfrac{1}{2}\varepsilon^{AC}(\varepsilon_{\dot{A}\dot{C}}T^{R(B|\dot{R}|}{}_{(\dot{B}}T_{|R}{}^{D)}{}_{\dot{R}|\dot{D})} + \varepsilon_{\dot{B}\dot{D}}T^{R(B|\dot{R}|}{}_{(\dot{A}}T_{|R}{}^{D)}{}_{\dot{R}|\dot{C})}) \\
&\qquad + \tfrac{1}{2}\varepsilon^{BD}(\varepsilon_{\dot{A}\dot{C}}T^{R(A|\dot{R}|}{}_{(\dot{B}}T_{|R}{}^{C)}{}_{\dot{R}|\dot{D})} + \varepsilon_{\dot{B}\dot{D}}T^{R(A|\dot{R}|}{}_{(\dot{A}}T_{|R}{}^{C)}{}_{\dot{R}|\dot{C})}),
\end{aligned}$$

which is equal to zero, since by hypothesis, $T_{AB\dot{A}\dot{B}}T^A{}_C{}^{\dot{A}}{}_{\dot{C}}$ is proportional to $\varepsilon_{BC}\varepsilon_{\dot{B}\dot{C}}$. Hence,

$$T_{AB\dot{A}\dot{B}}T_{CD\dot{C}\dot{D}} = T_{CD\dot{A}\dot{B}}T_{AB\dot{C}\dot{D}}, \tag{3.26}$$

and contracting both sides of this last equation with $\alpha^C\overline{\alpha}^{\dot{C}}\alpha^D\overline{\alpha}^{\dot{D}}$, assuming that $T_{CD\dot{C}\dot{D}}\alpha^C\overline{\alpha}^{\dot{C}}\alpha^D\overline{\alpha}^{\dot{D}}$ is different from zero, we obtain

$$T_{AB\dot{A}\dot{B}} = \frac{T_{AB\dot{C}\dot{D}}\overline{\alpha}^{\dot{C}}\overline{\alpha}^{\dot{D}}\, T_{CD\dot{A}\dot{B}}\alpha^C\alpha^D}{T_{CD\dot{C}\dot{D}}\alpha^C\overline{\alpha}^{\dot{C}}\alpha^D\overline{\alpha}^{\dot{D}}},$$

thus showing that $T_{AB\dot{A}\dot{B}}$ is of the form

$$T_{AB\dot{A}\dot{B}} = \pm\frac{1}{2\pi}f_{AB}f_{\dot{A}\dot{B}}. \tag{3.27}$$

The spinor f_{AB} appearing in this last expression is defined by $T_{AB\dot{A}\dot{B}}$ up to a *duality rotation* $f_{AB} \mapsto e^{i\chi} f_{AB}$, where χ is an arbitrary real-valued function. According to the preceding discussion, the sign on the right-hand side must be positive if $T_{ab}t^a t^b \geqslant 0$ for every timelike vector t^a.

Moreover, it may be noticed that by contracting both sides of (3.26) with $\phi^{CD}\overline{\phi}^{\dot{C}\dot{D}}$, where ϕ^{AB} is a symmetric spinor, we obtain

$$T_{AB\dot{A}\dot{B}} = \frac{T_{AB\dot{C}\dot{D}}\overline{\phi}^{\dot{C}\dot{D}}\, T_{CD\dot{A}\dot{B}}\phi^{CD}}{T_{CD\dot{C}\dot{D}}\phi^{CD}\overline{\phi}^{\dot{C}\dot{D}}},$$

which is again of the form (3.27), with f_{AB} proportional to $T_{AB\dot{C}\dot{D}}\overline{\phi}^{\dot{C}\dot{D}}$ and $f_{\dot{A}\dot{B}}$ proportional to $T_{CD\dot{A}\dot{B}}\phi^{CD}$, provided that $T_{CD\dot{C}\dot{D}}\phi^{CD}\overline{\phi}^{\dot{C}\dot{D}}$ is different from zero. Then, using the fact that $T_{AB\dot{R}\dot{S}}\overline{\phi}^{\dot{R}\dot{S}}\varepsilon_{\dot{A}\dot{B}} + T_{RS\dot{A}\dot{B}}\phi^{RS}\varepsilon_{AB}$ is the spinor equivalent of $T_{ac}\phi^c{}_b - T_{bc}\phi^c{}_a$, where ϕ_{ab} is the tensor equivalent of $\phi_{AB}\varepsilon_{\dot{A}\dot{B}} + \overline{\phi}_{\dot{A}\dot{B}}\varepsilon_{AB}$, one concludes that up to duality rotations, the electromagnetic field tensor corresponding to a given energy–momentum tensor T_{ab} is given by

$$F_{ab} = \sqrt{4\pi}\,\frac{T_{ac}\phi^c{}_b - T_{bc}\phi^c{}_a}{\sqrt{T_{cd}\phi^c{}_s\phi^{ds}}}, \tag{3.28}$$

where ϕ_{ab} is any (real) antisymmetric tensor field such that $T_{cd}\phi^c{}_s\phi^{ds}$ is different from zero.

Principal Null Directions of the Electromagnetic Field

If the electromagnetic field is different from zero, the electromagnetic spinor has one of the forms

$$f_{AB} = \begin{cases} \alpha_{(A}\beta_{B)} \text{ with } \alpha^A\beta_A \neq 0, \\ \alpha_A\alpha_B. \end{cases}$$

In the first case (where the electromagnetic field is algebraically general) $f^{AB}f_{AB} = -\frac{1}{2}(\alpha^A\beta_A)^2 \neq 0$, and in the second case (where the electromagnetic field is algebraically special) $f^{AB}f_{AB} = 0$. Since the two well-known invariants of the electromagnetic field $F^{ab}F_{ab}$ and ${}^*F^{ab}F_{ab}$ are both real, from (3.23) we see that the electromagnetic field is algebraically special if and only if $F^{ab}F_{ab}$ and ${}^*F^{ab}F_{ab}$ are both equal to zero. Furthermore, when the electromagnetic field is algebraically special, from (3.19) we see that

$$T_{ab} = \frac{1}{2\pi}l_a l_b,$$

where l_a is the vector equivalent of $\alpha_A\overline{\alpha}_{\dot{A}}$, which is a real null vector. Conversely, if T_{ab} is proportional to $l_a l_b$, where l_a is a real vector, l_a must be null (since $T^a{}_a = 0$), and from (3.19) or (3.24) it follows that the electromagnetic field is algebraically special.

Each principal spinor α_A of an algebraically general electromagnetic field f_{AB} gives rise to a (real) null vector $l^a = -(1/\sqrt{2})\sigma^a{}_{A\dot{A}}\alpha^A\overline{\alpha}^{\dot{A}}$. Since the principal spinors of f_{AB} are defined up to a complex factor, l^a is defined up to a (positive) real factor. Hence, at each point of the space-time, an algebraically general electromagnetic field defines two real null directions, called *principal null directions*, of the electromagnetic field. When the electromagnetic field is algebraically special, these two principal null directions coincide. (Note that if $f_{AB} = \alpha_A\alpha_B$, then α_A is defined up to sign.)

The principal null directions of the electromagnetic field can be characterized as those null directions l^a along which $T_{ab}l^a l^b$ attains its minimum value (zero). Indeed, the spinor equivalent of a real null vector l^a is of the form $l^{A\dot{A}} = \pm\gamma^A\overline{\gamma}^{\dot{A}}$. Then $T_{ab}l^a l^b = \frac{1}{2\pi}|\gamma^A\alpha_A|^2\,|\gamma^B\beta_B|^2$, which is equal to zero if and only if γ_A is proportional to α_A or to β_A.

The principal null directions of the electromagnetic field are the only real null eigenvectors of the electromagnetic field tensor F_{ab}. If l_a is a real null eigenvector of F_{ab}, then its spinor equivalent is of the form $\pm\gamma_A\overline{\gamma}_{\dot{A}}$, and the condition $F_{ab}l^b = \lambda l_a$ is equivalent to

$$(f_{AB}\varepsilon_{\dot{A}\dot{B}} + f_{\dot{A}\dot{B}}\varepsilon_{AB})\gamma^B\overline{\gamma}^{\dot{B}} = -\lambda\gamma_A\overline{\gamma}_{\dot{A}}.$$

Hence, $f_{AB}\gamma^B\overline{\gamma}_{\dot{A}} + f_{\dot{A}\dot{B}}\gamma_A\overline{\gamma}^{\dot{B}} = -\lambda\gamma_A\overline{\gamma}_{\dot{A}}$. Contracting this last equation with γ^A, we obtain $f_{AB}\gamma^A\gamma^B = 0$, and therefore γ_A is a principal spinor of f_{AB} [see (1.230)]. Conversely, if γ_A is a principal spinor of f_{AB}, then $f_{AB}\gamma^A = \mu\gamma_B$, for some μ, and $(f_{AB}\varepsilon_{\dot{A}\dot{B}} + f_{\dot{A}\dot{B}}\varepsilon_{AB})\gamma^B\overline{\gamma}^{\dot{B}} = (\mu+\overline{\mu})\gamma_A\overline{\gamma}_{\dot{A}}$, which means that $\gamma_A\overline{\gamma}_{\dot{A}}$ is the spinor equivalent of a real null eigenvector of F_{ab}. The condition $F_{ab}l^b = \lambda l_a$ is usually expressed in the equivalent form

$$l^a F_{a[b}l_{c]} = 0$$

[cf. (2.131)].

In the case of an algebraically general electromagnetic field, from (3.2) and (1.27) we obtain

$$\begin{aligned}
F_{A\dot{A}B\dot{B}} &= f_{AB}\varepsilon_{\dot{A}\dot{B}} + f_{\dot{A}\dot{B}}\varepsilon_{AB} \\
&= \frac{1}{2}(\alpha_A\beta_B + \alpha_B\beta_A)\frac{1}{\overline{\alpha}^{\dot{S}}\overline{\beta}_{\dot{S}}}(\overline{\alpha}_{\dot{A}}\overline{\beta}_{\dot{B}} - \overline{\alpha}_{\dot{B}}\overline{\beta}_{\dot{A}}) + \text{c.c.} \\
&= \frac{1}{2\overline{\alpha}^{\dot{S}}\overline{\beta}_{\dot{S}}}(\alpha_A\overline{\alpha}_{\dot{A}}\beta_B\overline{\beta}_{\dot{B}} - \alpha_B\overline{\alpha}_{\dot{B}}\beta_A\overline{\beta}_{\dot{A}} + \alpha_B\overline{\alpha}_{\dot{A}}\beta_A\overline{\beta}_{\dot{B}} - \alpha_A\overline{\alpha}_{\dot{B}}\beta_B\overline{\beta}_{\dot{A}}) + \text{c.c.},
\end{aligned}$$

where c.c. denotes complex conjugate. If l_a and n_a are the vector equivalents of $\alpha_A\overline{\alpha}_{\dot{A}}$ and $\beta_A\overline{\beta}_{\dot{A}}$, respectively (that is, l_a and n_a point along the two principal null directions of the electromagnetic field), the spinor equivalent of $\varepsilon_{abcd}l^c n^d$ is [see (1.64)]

$$\mathrm{i}(\varepsilon_{AC}\varepsilon_{BD}\varepsilon_{\dot{A}\dot{D}}\varepsilon_{\dot{B}\dot{C}} - \varepsilon_{AD}\varepsilon_{BC}\varepsilon_{\dot{A}\dot{C}}\varepsilon_{\dot{B}\dot{D}})\,\alpha^C\overline{\alpha}^{\dot{C}}\beta^D\overline{\beta}^{\dot{D}} = \mathrm{i}(\alpha_A\overline{\alpha}_{\dot{B}}\beta_B\overline{\beta}_{\dot{A}} - \alpha_B\overline{\alpha}_{\dot{A}}\beta_A\overline{\beta}_{\dot{B}}),$$

and hence

$$F_{ab} = (\mathrm{Re}\,\mathscr{F})\,(l_a n_b - l_b n_a) + (\mathrm{Im}\,\mathscr{F})\,\varepsilon_{abcd} l^c n^d, \tag{3.29}$$

with $\mathscr{F} \equiv 1/(\alpha^S \beta_S)$ [cf. (1.157)]. The function $\mathscr{F}$ is related to the complex invariant $f^{AB} f_{AB}$ through $f^{AB} f_{AB} = -\frac{1}{2}\mathscr{F}^{-2}$. Directly from (3.29) we have

$$^*F_{ab} = -(\mathrm{Im}\,\mathscr{F})\,(l_a n_b - l_b n_a) + (\mathrm{Re}\,\mathscr{F})\,\varepsilon_{abcd} l^c n^d.$$

The energy–momentum tensor can also be expressed in terms of the null vectors l_a, n_a; the result is

$$T_{ab} = \frac{1}{4\pi}(l_a n_b + n_a l_b - \tfrac{1}{2} l^c n_c g_{ab}),$$

which shows that the subspace spanned by l_a, n_a is formed by eigenvectors of $T^a{}_b$, with eigenvalue $l^c n_c/8\pi$, and the vectors orthogonal to l_a and n_a are eigenvectors of $T^a{}_b$, with eigenvalue $-l^c n_c/8\pi$.

When the electromagnetic field is algebraically special one can find γ_A such that $\alpha^A \gamma_A = 1$; then

$$\begin{aligned} F_{A\dot{A}B\dot{B}} &= f_{AB}\varepsilon_{\dot{A}\dot{B}} + f_{\dot{A}\dot{B}}\varepsilon_{AB} \\ &= \alpha_A \alpha_B (\overline{\alpha}_{\dot{A}} \overline{\gamma}_{\dot{B}} - \overline{\alpha}_{\dot{B}} \overline{\gamma}_{\dot{A}}) + \overline{\alpha}_{\dot{A}} \overline{\alpha}_{\dot{B}} (\alpha_A \gamma_B - \alpha_B \gamma_A) \\ &= \alpha_A \overline{\alpha}_{\dot{A}} (\alpha_B \overline{\gamma}_{\dot{B}} + \gamma_B \overline{\alpha}_{\dot{B}}) - \alpha_B \overline{\alpha}_{\dot{B}} (\alpha_A \overline{\gamma}_{\dot{A}} + \gamma_A \overline{\alpha}_{\dot{A}}). \end{aligned}$$

Thus, if s_a is the vector equivalent of $\alpha_A \overline{\gamma}_{\dot{A}} + \gamma_A \overline{\alpha}_{\dot{A}}$, we have

$$F_{ab} = l_a s_b - l_b s_a, \tag{3.30}$$

and $l^a s_a = 0$, $s^a s_a = 2$. Furthermore,

$$\begin{aligned} ^*F_{A\dot{A}B\dot{B}} &= -\mathrm{i}\alpha_A \alpha_B (\overline{\alpha}_{\dot{A}} \overline{\gamma}_{\dot{B}} - \overline{\alpha}_{\dot{B}} \overline{\gamma}_{\dot{A}}) + \mathrm{i}\overline{\alpha}_{\dot{A}} \overline{\alpha}_{\dot{B}} (\alpha_A \gamma_B - \alpha_B \gamma_A) \\ &= \alpha_A \overline{\alpha}_{\dot{A}} (\mathrm{i}\gamma_B \overline{\alpha}_{\dot{B}} - \mathrm{i}\alpha_B \overline{\gamma}_{\dot{B}}) - \alpha_B \overline{\alpha}_{\dot{B}} (\mathrm{i}\gamma_A \overline{\alpha}_{\dot{A}} - \mathrm{i}\alpha_A \overline{\gamma}_{\dot{A}}), \end{aligned}$$

that is,

$$^*F_{ab} = l_a r_b - l_b r_a,$$

where r_a is the vector equivalent of $\mathrm{i}(\gamma_A \overline{\alpha}_{\dot{A}} - \alpha_A \overline{\gamma}_{\dot{A}})$, which is also orthogonal to l_a. Hence, we conclude that when the electromagnetic field is algebraically special, F_{ab} and $^*F_{ab}$ are simple and $F_{ab} l^b = 0 = {}^*F_{ab} l^b$, where l_a is a null vector that points along the repeated principal null direction of the electromagnetic field.

For example, in Minkowski space-time, the electric and magnetic fields with Cartesian components

$$\begin{aligned} \mathbf{E} &= A\big(\cos(kz - \omega t), \pm \sin(kz - \omega t), 0\big), \\ \mathbf{B} &= A\big(\mp \sin(kz - \omega t), \cos(kz - \omega t), 0\big), \end{aligned}$$

where A is a positive real constant, correspond to a monochromatic, circularly polarized, electromagnetic plane wave propagating along the z-axis, with angular frequency ω and wave number $k = \omega/c$ (the upper or the lower sign determines the sense of rotation of the fields). According to (3.5), the components of the electromagnetic spinor are simply

$$f_{11} = A\mathrm{e}^{\pm\mathrm{i}(kz-\omega t)}, \qquad f_{12} = 0 = f_{22}. \tag{3.31}$$

Therefore $f^{AB} f_{AB} = 0$, which means that this field is algebraically special. By inspection, one finds that $f_{AB} = \alpha_A \alpha_B$ with

$$\begin{pmatrix} \alpha_1 \\ \alpha_2 \end{pmatrix} = \pm A^{1/2} \mathrm{e}^{\pm\mathrm{i}(kz-\omega t)/2} \begin{pmatrix} 1 \\ 0 \end{pmatrix}.$$

The only principal null direction of this field is proportional to $(0,0,1,1)$.

A simple example of an algebraically general electromagnetic field is provided by the field of a static electric point charge $\mathbf{E} = q(x,y,z)/r^3$, $\mathbf{B} = \mathbf{0}$, with $r = (x^2 + y^2 + z^2)^{1/2}$. Making use of (3.5) one obtains the spinor components

$$(f_{AB}) = \frac{q}{2r^3} \begin{pmatrix} x - \mathrm{i}y & -z \\ -z & -x - \mathrm{i}y \end{pmatrix} = \frac{q}{2r^2} \begin{pmatrix} \mathrm{e}^{-\mathrm{i}\varphi} \sin\theta & -\cos\theta \\ -\cos\theta & -\mathrm{e}^{\mathrm{i}\varphi} \sin\theta \end{pmatrix},$$

where we have replaced the Cartesian coordinates by spherical coordinates. The ratios α^1/α^2 and β^1/β^2 of the components of the principal spinors of f_{AB} are the roots of the polynomial [see (1.229)]

$$f_{11} z^2 + 2 f_{12} z + f_{22} = \frac{q}{2r^2} (\mathrm{e}^{-\mathrm{i}\varphi} \sin\theta\, z^2 - 2\cos\theta\, z - \mathrm{e}^{\mathrm{i}\varphi} \sin\theta),$$

which are $-\mathrm{e}^{\mathrm{i}\varphi} \tan\frac{1}{2}\theta$ and $\mathrm{e}^{\mathrm{i}\varphi} \cot\frac{1}{2}\theta$. Hence, the principal spinors of f_{AB} can be chosen as

$$\begin{pmatrix} \alpha^1 \\ \alpha^2 \end{pmatrix} = \frac{q}{r} \begin{pmatrix} -\mathrm{e}^{\mathrm{i}\varphi/2} \sin\frac{1}{2}\theta \\ \mathrm{e}^{-\mathrm{i}\varphi/2} \cos\frac{1}{2}\theta \end{pmatrix}, \qquad \begin{pmatrix} \beta^1 \\ \beta^2 \end{pmatrix} = \frac{1}{r} \begin{pmatrix} \mathrm{e}^{\mathrm{i}\varphi/2} \cos\frac{1}{2}\theta \\ \mathrm{e}^{-\mathrm{i}\varphi/2} \sin\frac{1}{2}\theta \end{pmatrix},$$

and $\mathscr{F}^{-1} = \alpha^S \beta_S = q/r^2$. The first of these spinors corresponds to a null vector whose spatial part points radially away from the origin [see (1.163)], and the second spinor corresponds to a null vector with a spatial part pointing radially toward the origin.

3.2 Dirac's Equation

Another illustrative example of the applications of the spinor formalism is given by the Dirac equation. As a matter of fact, in the study of this equation, Infeld and van der Waerden (1933) developed the two-component spinor formalism for curved space-times.

The Dirac equation in flat space-time, expressed in terms of Cartesian coordinates $(x^1, x^2, x^3, x^4) = (x, y, z, ct)$, has the form

$$\gamma^a \frac{\partial \Psi}{\partial x^a} = \frac{m_0 c}{\hbar} \Psi, \tag{3.32}$$

where m_0 is the rest mass of the particle and the γ^a must satisfy the relations (1.194), that is,

$$\gamma^a \gamma^b + \gamma^b \gamma^a = 2g^{ab}$$

(see, e.g., Messiah 1962, Davydov 1988, Merzbacher 1998). As shown in Section 1.5, we can find a linear representation of the γ^a in a complex vector space of dimension four and with respect to a suitable basis [see (1.211)]

$$\gamma_a \begin{pmatrix} \psi_A \\ \phi^{\dot{A}} \end{pmatrix} = \begin{pmatrix} \sigma_{aA\dot{B}} \phi^{\dot{B}} \\ -\sigma_a{}^{B\dot{A}} \psi_B \end{pmatrix},$$

where the $\sigma_{aA\dot{B}}$ are Infeld–van der Waerden symbols. Hence, the Dirac equation (3.32) is equivalent to

$$\partial_{A\dot{B}} \phi^{\dot{B}} = \frac{m_0 c}{\sqrt{2}\hbar} \psi_A, \qquad \partial_{B\dot{A}} \psi^B = \frac{m_0 c}{\sqrt{2}\hbar} \phi_{\dot{A}}, \tag{3.33}$$

with

$$\partial_{A\dot{B}} = \frac{1}{\sqrt{2}} \sigma^a{}_{A\dot{B}} \frac{\partial}{\partial x^a}.$$

Since the spin coefficients for this null tetrad are all equal to zero, equations (3.33) are equivalent to

$$\nabla_{A\dot{B}} \phi^{\dot{B}} = \frac{m_0 c}{\sqrt{2}\hbar} \psi_A, \qquad \nabla_{B\dot{A}} \psi^B = \frac{m_0 c}{\sqrt{2}\hbar} \phi_{\dot{A}}. \tag{3.34}$$

This set of equations is form-invariant under null tetrad transformations, and it is assumed that it also holds in a curved space-time.

The combination of the Dirac equations (3.34) and their complex conjugates yields the continuity equation

$$\nabla_{A\dot{A}} (\overline{\phi}^A \phi^{\dot{A}} + \psi^A \overline{\psi}^{\dot{A}}) = \frac{m_0 c}{\sqrt{2}\hbar} (\overline{\psi}_{\dot{A}} \phi^{\dot{A}} + \overline{\phi}^A \psi_A + \psi^A \overline{\phi}_A + \overline{\psi}^{\dot{A}} \phi_{\dot{A}}) = 0.$$

The fact that $\overline{\phi}^A \phi^{\dot{A}}$ and $\psi^A \overline{\psi}^{\dot{A}}$ are the spinor equivalents of two future-pointing null vectors implies that $J^{A\dot{A}} \equiv \overline{\phi}^A \phi^{\dot{A}} + \psi^A \overline{\psi}^{\dot{A}}$ is the spinor equivalent of a future-pointing timelike current density J^a; hence, J^4 is nonnegative and can be interpreted as a probability density.

Making use of the definitions (2.25), (2.26), and (2.28), the Dirac equation (3.34) takes the form

$$
\begin{aligned}
(\delta-\overline{\alpha}+\overline{\pi})\phi_{\dot{1}}-(D+\overline{\varepsilon}-\overline{\rho})\phi_{\dot{2}} &= \frac{m_0 c}{\sqrt{2}\hbar}\psi_1, \\
(\Delta-\overline{\gamma}+\overline{\mu})\phi_{\dot{1}}-(\overline{\delta}+\overline{\beta}-\overline{\tau})\phi_{\dot{2}} &= \frac{m_0 c}{\sqrt{2}\hbar}\psi_2, \\
(\overline{\delta}-\alpha+\pi)\psi_1-(D+\varepsilon-\rho)\psi_2 &= \frac{m_0 c}{\sqrt{2}\hbar}\phi_{\dot{1}}, \\
(\Delta-\gamma+\mu)\psi_1-(\delta+\beta-\tau)\psi_2 &= \frac{m_0 c}{\sqrt{2}\hbar}\phi_{\dot{2}}.
\end{aligned}
\tag{3.35}
$$

As in the case of the Maxwell equations, the Dirac equation is integrable in flat space-time or in a curved space-time without restriction on the curvature.

Example 3.1. Solution of the Dirac equation in the Schwarzschild background.

The Dirac equation can be solved by taking as the background metric that of the Schwarzschild space-time, assuming that the energy–momentum tensor of the Dirac field is small enough and that the space-time metric is not affected by the presence of the Dirac field. Substituting (2.108) and (2.109) into (3.35), we obtain

$$
\begin{aligned}
\frac{m_0 c}{\sqrt{2}\hbar}\psi_1 &= \frac{1}{\sqrt{2}r}\left(\frac{\partial}{\partial\theta}+\frac{\mathrm{i}}{\sin\theta}\frac{\partial}{\partial\varphi}+\frac{\cot\theta}{2}\right)\phi_{\dot{1}} \\
&\quad -\frac{1}{r}\left[-\left(1-\frac{r_g}{r}\right)^{-1}\frac{1}{c}\frac{\partial}{\partial t}+\frac{\partial}{\partial r}\right] r\phi_{\dot{2}}, \\
\frac{m_0 c}{\sqrt{2}\hbar}\psi_2 &= -\frac{1}{2r}\left[\frac{1}{c}\frac{\partial}{\partial t}+\left(1-\frac{r_g}{r}\right)\frac{\partial}{\partial r}+\frac{r_g}{2r^2}\right] r\phi_{\dot{1}} \\
&\quad -\frac{1}{\sqrt{2}r}\left(\frac{\partial}{\partial\theta}-\frac{\mathrm{i}}{\sin\theta}\frac{\partial}{\partial\varphi}+\frac{\cot\theta}{2}\right)\phi_{\dot{2}}, \\
\frac{m_0 c}{\sqrt{2}\hbar}\phi_{\dot{1}} &= \frac{1}{\sqrt{2}r}\left(\frac{\partial}{\partial\theta}-\frac{\mathrm{i}}{\sin\theta}\frac{\partial}{\partial\varphi}+\frac{\cot\theta}{2}\right)\psi_1 \\
&\quad -\frac{1}{r}\left[-\left(1-\frac{r_g}{r}\right)^{-1}\frac{1}{c}\frac{\partial}{\partial t}+\frac{\partial}{\partial r}\right] r\psi_2, \\
\frac{m_0 c}{\sqrt{2}\hbar}\phi_{\dot{2}} &= -\frac{1}{2r}\left[\frac{1}{c}\frac{\partial}{\partial t}+\left(1-\frac{r_g}{r}\right)\frac{\partial}{\partial r}+\frac{r_g}{2r^2}\right] r\psi_1 \\
&\quad -\frac{1}{\sqrt{2}r}\left(\frac{\partial}{\partial\theta}+\frac{\mathrm{i}}{\sin\theta}\frac{\partial}{\partial\varphi}+\frac{\cot\theta}{2}\right)\psi_2.
\end{aligned}
$$

(When $r_g = 0$, the background space-time is flat.) These equations admit separable solutions of the form

$$
\begin{aligned}
\psi_1 &= \sqrt{2}\left(1-\frac{r_g}{r}\right)^{-1/2} F(r)\,{}_{\frac{1}{2}}Y_{jm}(\theta,\varphi)\,\mathrm{e}^{-\mathrm{i}Et/\hbar},\\
\psi_2 &= G(r)\,{}_{-\frac{1}{2}}Y_{jm}(\theta,\varphi)\,\mathrm{e}^{-\mathrm{i}Et/\hbar},\\
\phi_1 &= \sqrt{2}\left(1-\frac{r_g}{r}\right)^{-1/2} f(r)\,{}_{-\frac{1}{2}}Y_{jm}(\theta,\varphi)\,\mathrm{e}^{-\mathrm{i}Et/\hbar},\\
\phi_2 &= g(r)\,{}_{\frac{1}{2}}Y_{jm}(\theta,\varphi)\,\mathrm{e}^{-\mathrm{i}Et/\hbar},
\end{aligned}
\tag{3.36}
$$

where the ${}_sY_{jm}$ are spin-weighted spherical harmonics (Newman and Penrose 1966, Torres del Castillo 2003), j and m are half-integers with $-j \leqslant m \leqslant j$, and E is a real constant. The factor $\sqrt{2}(1-r_g/r)^{-1/2}$ is introduced for later convenience and corrects the asymmetry in the definition of the null tetrad (2.108) [cf. (3.13)]. (The spin weight of the functions on the right-hand side of (3.36) is determined by the type of the corresponding component appearing on the left-hand side as defined in Section 2.1.) According to (3.14) we have

$$
\left(\frac{\partial}{\partial\theta} \pm \frac{\mathrm{i}}{\sin\theta}\frac{\partial}{\partial\varphi} + \frac{1}{2}\cot\theta\right){}_{\mp\frac{1}{2}}Y_{jm} = \mp(j+\tfrac{1}{2})\,{}_{\pm\frac{1}{2}}Y_{jm},
\tag{3.37}
$$

and therefore the one-variable functions F, G, f, and g must obey the system of ordinary differential equations

$$
\begin{aligned}
\frac{m_0c}{\hbar}F &= -\frac{1}{r}(j+\tfrac{1}{2})f - \frac{1}{r}\left(1-\frac{r_g}{r}\right)^{1/2}\left[\frac{\mathrm{i}E}{\hbar c}\left(1-\frac{r_g}{r}\right)^{-1} + \frac{\mathrm{d}}{\mathrm{d}r}\right] rg,\\
\frac{m_0c}{\hbar}G &= -\frac{1}{r}\left(1-\frac{r_g}{r}\right)^{1/2}\left[-\frac{\mathrm{i}E}{\hbar c}\left(1-\frac{r_g}{r}\right)^{-1} + \frac{\mathrm{d}}{\mathrm{d}r}\right] rf - \frac{1}{r}(j+\tfrac{1}{2})g,\\
\frac{m_0c}{\hbar}f &= \frac{1}{r}(j+\tfrac{1}{2})F - \frac{1}{r}\left(1-\frac{r_g}{r}\right)^{1/2}\left[\frac{\mathrm{i}E}{\hbar c}\left(1-\frac{r_g}{r}\right)^{-1} + \frac{\mathrm{d}}{\mathrm{d}r}\right] rG,\\
\frac{m_0c}{\hbar}g &= -\frac{1}{r}\left(1-\frac{r_g}{r}\right)^{1/2}\left[-\frac{\mathrm{i}E}{\hbar c}\left(1-\frac{r_g}{r}\right)^{-1} + \frac{\mathrm{d}}{\mathrm{d}r}\right] rF + \frac{1}{r}(j+\tfrac{1}{2})G.
\end{aligned}
$$

These equations can be partially decoupled by expressing them in the equivalent form

$$
\begin{aligned}
\frac{m_0c}{\hbar}A &= \mathrm{i}\frac{j+\frac{1}{2}}{r}A - \frac{\mathrm{i}}{r}\left(1-\frac{r_g}{r}\right)^{1/2}\left[\frac{\mathrm{d}}{\mathrm{d}r} + \frac{\mathrm{i}E}{\hbar c}\left(1-\frac{r_g}{r}\right)^{-1}\right] rB,\\
\frac{m_0c}{\hbar}B &= -\mathrm{i}\frac{j+\frac{1}{2}}{r}B + \frac{\mathrm{i}}{r}\left(1-\frac{r_g}{r}\right)^{1/2}\left[\frac{\mathrm{d}}{\mathrm{d}r} - \frac{\mathrm{i}E}{\hbar c}\left(1-\frac{r_g}{r}\right)^{-1}\right] rA,\\
\frac{m_0c}{\hbar}C &= -\mathrm{i}\frac{j+\frac{1}{2}}{r}C + \frac{\mathrm{i}}{r}\left(1-\frac{r_g}{r}\right)^{1/2}\left[\frac{\mathrm{d}}{\mathrm{d}r} + \frac{\mathrm{i}E}{\hbar c}\left(1-\frac{r_g}{r}\right)^{-1}\right] rD,\\
\frac{m_0c}{\hbar}D &= \mathrm{i}\frac{j+\frac{1}{2}}{r}D - \frac{\mathrm{i}}{r}\left(1-\frac{r_g}{r}\right)^{1/2}\left[\frac{\mathrm{d}}{\mathrm{d}r} - \frac{\mathrm{i}E}{\hbar c}\left(1-\frac{r_g}{r}\right)^{-1}\right] rC,
\end{aligned}
$$

with $A \equiv F + \mathrm{i}f$, $B \equiv G - \mathrm{i}g$, $C \equiv F - \mathrm{i}f$, and $D \equiv G + \mathrm{i}g$.

The fact that the pair of functions A, B is decoupled from the pair C, D implies that one can consider *particular* solutions of the Dirac equation such that either $A = 0 = B$ or $C = 0 = D$; that is, there exist solutions (3.36) such that $f = \pm \mathrm{i}F$ and $g = \mp \mathrm{i}G$, which contain only two radial functions instead of four. As we shall show below (Section 3.4), this possibility is related to the existence of a symmetry operator for the Dirac equation in any type-D vacuum space-time [see (3.128)]. (The solution of the Dirac equation in the Kerr background given in Chandrasekhar 1983, §104, belongs to this class of particular solutions that involve two radial functions only.) When $r_g = 0$ the radial functions are linear combinations of spherical Bessel functions of order $j \pm \frac{1}{2}$ (recall that j is a half-integer).

The Dirac equation in a possibly curved space-time (3.34) can be written in terms of the Dirac matrices γ_a and bispinors. Indeed, from (3.34), (2.39), (1.211), (2.18), and (1.216), we obtain

$$
\begin{aligned}
\frac{m_0 c}{\hbar}\begin{pmatrix}\psi_A\\ \phi^{\dot A}\end{pmatrix} &= \sqrt{2}\begin{pmatrix}\partial_{A\dot B}\phi^{\dot B} - \Gamma^{\dot B}{}_{\dot S\dot B A}\phi^{\dot S}\\ -\partial^{B\dot A}\psi_B - \Gamma^{xS}{}_{B}{}^{B\dot A}\psi_S\end{pmatrix}\\
&= \gamma^a\partial_a\begin{pmatrix}\psi_A\\ \phi^{\dot A}\end{pmatrix} - \frac{1}{4}\Gamma_{abc}\begin{pmatrix}S^{ab\dot B}{}_{\dot S}\sigma^c{}_{A\dot B}\phi^{\dot S}\\ S^{abS}{}_{B}\sigma^{cB\dot A}\psi_S\end{pmatrix}\\
&= \gamma^a\partial_a\begin{pmatrix}\psi_A\\ \phi^{\dot A}\end{pmatrix} + \frac{1}{8}\Gamma_{abc}\gamma^c[\gamma^a,\gamma^b]\begin{pmatrix}\psi_A\\ \phi^{\dot A}\end{pmatrix},
\end{aligned}
$$

or equivalently,

$$\gamma^a\nabla_a\Psi = \frac{m_0 c}{\hbar}\Psi \tag{3.38}$$

[cf. (3.32)], where

$$\nabla_a\Psi \equiv \partial_a\Psi + \frac{1}{8}\Gamma_{bca}[\gamma^b,\gamma^c]\Psi \tag{3.39}$$

is the covariant derivative of a bispinor (see also Lichnerowicz 1964).

Letting $m_0 = 0$ in (3.34), one obtains the decoupled pair of equations

$$\nabla_{A\dot A}\phi^{\dot A} = 0, \qquad \nabla_{A\dot A}\psi^{A} = 0. \tag{3.40}$$

Each of these equations is known as the Weyl equation.

Zero-Rest-Mass Fields

The Weyl equation and the source-free Maxwell equations [see (3.4)] are two particular cases of the so-called spin-s zero-rest-mass field equations

$$\nabla_{A\dot A}\Phi^{AB\ldots L} = 0 \tag{3.41}$$

or

$$\nabla_{A\dot A}\Phi^{\dot A\dot B\ldots\dot L} = 0, \tag{3.42}$$

where $\Phi^{AB\ldots L}$ and $\Phi^{\dot{A}\dot{B}\ldots\dot{L}}$ are totally symmetric $2s$-index spinor fields. (See the discussion about spin at the end of Chapter 1.)

In flat space-time the zero-rest-mass field equations admit plane-wave solutions. The Ricci rotation coefficients for the orthonormal basis $\partial_a = \partial/\partial x^a$, induced by a set of Cartesian coordinates x^a, are equal to zero; therefore $\Gamma_{ABC\dot{D}}$ and $\Gamma_{\dot{A}\dot{B}\dot{C}D}$ are also equal to zero, and substituting, for instance, $\Phi^{AB\ldots L} = M^{AB\ldots L}\exp(-\mathrm{i}k_{R\dot{R}}x^{R\dot{R}})$ into (3.41), where $M^{AB\ldots L} = M^{(AB\ldots L)}$ and $k_{A\dot{A}}$ are constants, and $x^{A\dot{A}}$ denotes the spinor equivalent of x^a [cf. (3.31)], we obtain

$$\mathrm{i}k_{A\dot{A}}M^{AB\ldots L}\exp(-\mathrm{i}k_{R\dot{R}}x^{R\dot{R}}) = 0.$$

Hence, $k_{A\dot{A}}M^{AB\ldots L} = 0$, and by contracting this last equation with $k^{S\dot{A}}$ it follows that $k^{A\dot{A}}k_{A\dot{A}}M^{SB\ldots L} = 0$, which implies that in order to have a nontrivial solution, $k_{A\dot{A}}$ must be null and therefore $k_{A\dot{A}} = \pm\alpha_A\overline{\alpha}_{\dot{A}}$, for some (constant) one-index spinor α_A. Thus we have $\alpha_A M^{AB\ldots L} = 0$, which implies that $M^{AB\ldots L}$ is proportional to $\alpha^A\alpha^B\cdots\alpha^L$ [see (1.230)]. Thus

$$\Phi^{AB\ldots L} = C\alpha^A\alpha^B\cdots\alpha^L\exp(\mp\mathrm{i}\alpha_R\overline{\alpha}_{\dot{R}}x^{R\dot{R}}), \tag{3.43}$$

where C is some constant [cf. (3.31)].

Similarly, one finds that the equations (3.42) admit plane wave solutions of the form

$$\Phi^{\dot{A}\dot{B}\ldots\dot{L}} = \overline{C}\overline{\alpha}^{\dot{A}}\overline{\alpha}^{\dot{B}}\cdots\overline{\alpha}^{\dot{L}}\exp(\mp\mathrm{i}\alpha_R\overline{\alpha}_{\dot{R}}x^{R\dot{R}}), \tag{3.44}$$

where C is a complex constant and α_A is a constant one-index spinor.

As in the case of the circularly polarized electromagnetic wave (3.31), the fields (3.43) and (3.44) possess a definite helicity. Let β_A be a constant one-index spinor such that $\alpha^A\beta_A = 1$. The vector equivalents of $\alpha_A\overline{\beta}_{\dot{A}} + \beta_A\overline{\alpha}_{\dot{A}}$ and $\mathrm{i}(\alpha_A\overline{\beta}_{\dot{A}} - \beta_A\overline{\alpha}_{\dot{A}})$ span a two-dimensional spacelike plane orthogonal to the wave four-vector k^a, and conversely, any two-dimensional spacelike plane orthogonal to k^a can be constructed in this manner, with a suitable choice of β_A (see Exercise 1.10). The $SL(2,\mathbb{C})$ matrix with entries $K^A{}_B = \mathrm{e}^{\mathrm{i}\theta/2}\alpha^A\beta_B - \mathrm{e}^{-\mathrm{i}\theta/2}\beta^A\alpha_B$ corresponds to a rotation through an angle θ in this plane (see Exercise 1.8); under this rotation $k_{A\dot{A}}$ is invariant and $\Phi^{AB\ldots L}$, given by (3.43), is mapped into $\mathrm{e}^{\mathrm{i}s\theta}\Phi^{AB\ldots L}$. At a point with fixed spatial coordinates (x^1, x^2, x^3), the effect of this rotation is the same as that produced in the course of the time $t = \mp s\theta/\omega$, where ω is the angular frequency of the wave ($k^4 = \omega/c$). Thus, at a point with fixed spatial coordinates, the field (3.43) rotates with angular velocity ω/s in any two-dimensional spacelike plane orthogonal to the wave four-vector k^a (the sense of the rotation depends on the sign of k^4).

In an analogous way, under the rotation corresponding to $K^A{}_B = \mathrm{e}^{\mathrm{i}\theta/2}\alpha^A\beta_B - \mathrm{e}^{-\mathrm{i}\theta/2}\beta^A\alpha_B$, $\Phi^{\dot{A}\dot{B}\ldots\dot{L}}$ is mapped into $\mathrm{e}^{-\mathrm{i}s\theta}\Phi^{\dot{A}\dot{B}\ldots\dot{L}}$, and therefore, for a given sign of ω, at a point with fixed spatial coordinates the fields (3.43) and (3.44) rotate in opposite senses.

The zero-rest-mass field equations (3.41) and (3.42) maintain their form under conformal rescalings of the metric. For example, making use of (2.40), (2.132), and (2.133), one finds that if $\Phi^{AB\dots L}$ obeys the zero-rest-mass field equations (3.41), then

$$\begin{aligned}
0 &= \partial_{A\dot{A}}\Phi^{ABC\dots L} - \Gamma^{A}{}_{SA\dot{A}}\Phi^{SBC\dots L} - (2s-1)\Gamma^{(B}{}_{SA\dot{A}}\Phi^{C\dots L)SA} \\
&= \phi^{-1}\partial'_{A\dot{A}}\Phi^{ABC\dots L} - (\phi^{-1}\Gamma'^{A}{}_{SA\dot{A}} + \tfrac{3}{2}\partial'_{S\dot{A}}\phi^{-1})\Phi^{SBC\dots L} \\
&\quad - (2s-1)(\phi^{-1}\Gamma'^{(B}{}_{SA\dot{A}} + \tfrac{1}{2}\delta_{A}^{(B}\partial'_{S\dot{A}}\phi^{-1})\Phi^{C\dots L)SA} \\
&= \phi^{-s-2}\nabla'_{A\dot{A}}(\phi^{s+1}\Phi^{ABC\dots L}),
\end{aligned}$$

which shows that the zero-rest-mass field equations are form-invariant under conformal rescalings if we take $\Phi'^{AB\dots L} = \phi^{s+1}\Phi^{AB\dots L}$, provided that the null tetrads are related as in (2.132).

A conformally invariant equation for a spin-0 zero-rest-mass field is given by

$$\nabla_{A\dot{A}}\nabla^{A\dot{A}}\Phi + \tfrac{1}{6}R\Phi = 0. \tag{3.45}$$

In effect, making use of (2.40), (2.132), and (2.133)–(2.135), one finds that

$$\begin{aligned}
\nabla_{A\dot{A}}\nabla^{A\dot{A}}\Phi &= \partial_{A\dot{A}}\partial^{A\dot{A}}\Phi - \Gamma^{A}{}_{SA\dot{A}}\partial^{S\dot{A}}\Phi - \Gamma^{\dot{A}}{}_{\dot{S}\dot{A}A}\partial^{A\dot{S}}\Phi \\
&= \phi^{-1}\partial'_{A\dot{A}}(\phi^{-1}\partial'^{A\dot{A}}\Phi) - (\phi^{-1}\Gamma'^{A}{}_{SA\dot{A}} + \tfrac{3}{2}\partial'_{S\dot{A}}\phi^{-1})\phi^{-1}\partial'^{S\dot{A}}\Phi \\
&\quad - (\phi^{-1}\Gamma'^{\dot{A}}{}_{\dot{S}\dot{A}A} + \tfrac{3}{2}\partial'_{A\dot{S}}\phi^{-1})\phi^{-1}\partial'^{A\dot{S}}\Phi \\
&= \phi^{-2}\nabla'_{A\dot{A}}\nabla'^{A\dot{A}}\Phi + 2\phi^{-3}(\partial'_{A\dot{A}}\phi)\partial'^{A\dot{A}}\Phi \\
&= \phi^{-3}\nabla'_{A\dot{A}}\nabla'^{A\dot{A}}(\phi\Phi) - \phi^{-3}(\nabla'_{A\dot{A}}\nabla'^{A\dot{A}}\phi)\Phi \\
&= \phi^{-3}\nabla'_{A\dot{A}}\nabla'^{A\dot{A}}(\phi\Phi) - \tfrac{1}{6}R\Phi + \tfrac{1}{6}R'\phi^{-3}\Phi,
\end{aligned}$$

that is,

$$\nabla_{A\dot{A}}\nabla^{A\dot{A}}\Phi + \tfrac{1}{6}R\Phi = \phi^{-3}\left[\nabla'_{A\dot{A}}\nabla'^{A\dot{A}}(\phi\Phi) + \tfrac{1}{6}R'(\phi\Phi)\right].$$

The zero-rest-mass field equations (3.41) and (3.42) are not satisfactory for spins greater than 1 if the conformal curvature of the space-time is different from zero (Buchdahl 1958, Plebański 1965). Making use of the Ricci identities (2.126), one obtains the integrability conditions for (3.41)

$$\begin{aligned}
0 &= \nabla_{B}{}^{\dot{A}}\nabla_{A\dot{A}}\Phi^{ABCD\dots L} \\
&= \Box_{BA}\Phi^{ABCD\dots L} \\
&= -(2s-2)C_{ABS}{}^{(C}\Phi^{D\dots L)ABS}.
\end{aligned} \tag{3.46}$$

Note that by virtue of the Bianchi identities (2.116), when the traceless part of the Ricci tensor vanishes, the Weyl spinor satisfies the zero-rest-mass field equations, but the right-hand side of (3.46) vanishes identically when $\Phi_{ABCD} = C_{ABCD}$, and therefore no algebraic constraints on the Weyl spinor arise in this way.

3.3 Einstein's Equations

With the Ricci tensor defined as in (2.63), the Einstein field equations are given by

$$R_{ab} - \tfrac{1}{2} R g_{ab} = \frac{8\pi G}{c^4} T_{ab}, \tag{3.47}$$

where G is Newton's constant of gravitation and T_{ab} denotes the energy–momentum tensor of the matter present (see, e.g., Rindler 1977, Wald 1984). Equations (3.47) yield $R = -(8\pi G/c^4) T^a{}_a$; therefore, in the case of vacuum ($T_{ab} = 0$), the Einstein equations reduce to

$$R_{ab} = 0, \tag{3.48}$$

while in the case that the source of the gravitational field is the electromagnetic field [see (3.20)],

$$R_{ab} = \frac{2G}{c^4} (F_{ac} F_b{}^c - \tfrac{1}{4} F_{cd} F^{cd} g_{ab}),$$

or equivalently, making use of (2.85) and (3.19),

$$C_{AB\dot{A}\dot{B}} = \frac{2G}{c^4} f_{AB} f_{\dot{A}\dot{B}}. \tag{3.49}$$

The tensor equivalent of $C_{ABCD} C_{\dot{A}\dot{B}\dot{C}\dot{D}}$ is a real, traceless, totally symmetric four-index tensor field T_{abcd}, known as the Bel–Robinson tensor [cf. (3.19)]. If the traceless part of the Ricci tensor vanishes, as a consequence of the Bianchi identities (2.116), we have $\nabla^{A\dot{A}} (C_{ABCD} C_{\dot{A}\dot{B}\dot{C}\dot{D}}) = 0$, that is, $\nabla^a T_{abcd} = 0$.

Since $C_{ABCD} \varepsilon_{\dot{A}\dot{B}} \varepsilon_{\dot{C}\dot{D}}$ and $C_{\dot{A}\dot{B}\dot{C}\dot{D}} \varepsilon_{AB} \varepsilon_{CD}$ are the spinor equivalents of $\frac{1}{2}(C_{abcd} + \mathrm{i}\,{}^*C_{abcd})$ and $\frac{1}{2}(C_{abcd} - \mathrm{i}\,{}^*C_{abcd})$, respectively, where

$${}^*C_{abcd} \equiv \tfrac{1}{2} \varepsilon_{abrs} C^{rs}{}_{cd}$$

and

$$C_{ABCD} C_{\dot{A}\dot{B}\dot{C}\dot{D}} = C_{ARBS} \varepsilon_{\dot{A}\dot{R}} \varepsilon_{\dot{B}\dot{S}} C_{\dot{C}}{}^{\dot{R}}{}_{\dot{D}}{}^{\dot{S}} \varepsilon_C{}^R \varepsilon_D{}^S,$$

it follows that

$$T_{abcd} = \tfrac{1}{4} (C_{arbs} + \mathrm{i}\,{}^*C_{arbs})(C_c{}^r{}_d{}^s - \mathrm{i}\,{}^*C_c{}^r{}_d{}^s).$$

Using the fact that C_{abcd}, ${}^*C_{abcd}$, and T_{abcd} are real, we conclude that

$$T_{abcd} = \tfrac{1}{4} (C_{arbs} C_c{}^r{}_d{}^s + {}^*C_{arbs}{}^*C_c{}^r{}_d{}^s) \tag{3.50}$$

[cf. (3.21)] and

$$C_{arbs}{}^*C_c{}^r{}_d{}^s - {}^*C_{arbs} C_c{}^r{}_d{}^s = 0.$$

The Bel–Robinson tensor is analogous to the energy–momentum tensor of the electromagnetic field in various senses; however, it is not possible to give a local definition of energy or momentum for the gravitational field (see, e.g., Wald 1984).

A totally symmetric, traceless, four-index tensor T_{abcd} possesses a spinor equivalent of the form $C_{ABCD}C_{\dot{A}\dot{B}\dot{C}\dot{D}}$ if and only if its spinor equivalent satisfies the relation

$$T_{ABCD\dot{A}\dot{B}\dot{C}\dot{D}}T_{EFGH\dot{E}\dot{F}\dot{G}\dot{H}} = T_{ABCD\dot{E}\dot{F}\dot{G}\dot{H}}T_{EFGH\dot{A}\dot{B}\dot{C}\dot{D}} \tag{3.51}$$

[cf. (3.26)]. However, the tensor equivalent of this condition has a complicated form (see Bergqvist and Lankinen 2004). If $T_{ABCD\dot{A}\dot{B}\dot{C}\dot{D}} = T_{(ABCD)(\dot{A}\dot{B}\dot{C}\dot{D})}$ satisfies the condition (3.51), then the corresponding spinor C_{ABCD} is uniquely defined up to a phase factor (see Exercise 3.7).

3.3.1 The Goldberg–Sachs Theorem

The following two propositions imply that the integral curves of a repeated principal null direction of a zero-rest-mass field have several special geometric properties.

Proposition 3.2. *If f_{AB} is an algebraically special solution of the source-free Maxwell equations and l_A is a (repeated) principal spinor of f_{AB}, then*

$$l^A l^B \nabla_{A\dot{A}} l_B = 0. \tag{3.52}$$

Proof. By hypothesis, $f_{AB} = \alpha l_A l_B$, for some function α, and $\nabla^{A\dot{A}} f_{AB} = 0$; hence

$$0 = l^B \nabla^{A\dot{A}}(\alpha l_A l_B) = l^B[\alpha l_A \nabla^{A\dot{A}} l_B + l_B \nabla^{A\dot{A}}(\alpha l_A)] = \alpha l^B l_A \nabla^{A\dot{A}} l_B.$$

□

It may be noticed that condition (3.52) is invariant under rescalings of the spinor field l_A. With respect to a spinor frame such that $l_A \propto \delta^2_A$, (3.52) is equivalent to $\Gamma_{111\dot{A}} = 0$ (that is, $\kappa = 0 = \sigma$, in the Newman–Penrose notation); then, making use of (2.91), one finds that $C_{1111} = 0$, which means that any spinor field l_A satisfying (3.52) is a principal spinor of C_{ABCD}.

The foregoing proposition is a special case of the following more general result.

Proposition 3.3. *If l_A is a repeated principal spinor of a totally symmetric spinor field $\phi_{AB...L}$ that satisfies the zero-rest-mass field equations*

$$\nabla^{A\dot{A}} \phi_{AB...L} = 0, \tag{3.53}$$

then l_A satisfies (3.52).

Proof. Denoting the components $\phi_{AB...L}$ by $\phi_{(j)}$, where $j = 0, 1, \ldots, 2s$ is the number of indices taking the value 2, i.e., $\phi_{(0)} = \phi_{11...1}$, $\phi_{(1)} = \phi_{21...1}$, $\ldots$, $\phi_{(2s)} = \phi_{22...2}$, equations (3.53) are explicitly given by

$$\begin{aligned} 0 = {} & [\partial_{1\dot{A}} - 2(s-1-j)\Gamma_{121\dot{A}} - (2s-j)\Gamma_{112\dot{A}}]\phi_{(j+1)} \\ & - [\partial_{2\dot{A}} - 2(s-j)\Gamma_{122\dot{A}} + (j+1)\Gamma_{221\dot{A}}]\phi_{(j)} \\ & + j\Gamma_{222\dot{A}}\phi_{(j-1)} + (2s-1-j)\Gamma_{111\dot{A}}\phi_{(j+2)}, \end{aligned} \tag{3.54}$$

$j = 0, 1, \ldots, 2s-1$. In a spin frame such that $l_A \propto \delta_A^2$, the fact that l_A is a repeated principal spinor of $\phi_{AB\ldots L}$ amounts to $\phi_{(0)} = 0 = \phi_{(1)}$, and from (3.54), assuming that $\phi_{AB\ldots L}$ is different from zero, one obtains $\Gamma_{111\dot{A}} = 0$, which is equivalent to the condition (3.52) when $l_A \propto \delta_A^2$. □

There is a converse of Proposition 3.2. Given a one-index spinor l_A that satisfies $l^A l^B \nabla_{A\dot{A}} l_B = 0$, there exist (locally) algebraically special solutions f_{AB} of the source-free Maxwell equations such that l_A is the repeated principal spinor of f_{AB}. Indeed, looking for solutions of the source-free Maxwell equations (equations (3.54) with $s = 1$) such that $\phi_{(0)} = 0 = \phi_{(1)}$ with respect to a spin frame for which $l_A \propto \delta_A^2$, we obtain

$$0 = \Gamma_{111\dot{A}}\phi_{(2)}, \qquad 0 = (\partial_{1\dot{A}} + 2\Gamma_{121\dot{A}} - \Gamma_{112\dot{A}})\phi_{(2)}.$$

The first of these equations is satisfied by virtue of the hypothesis, while the second one can be written in the equivalent form

$$\partial_{1\dot{A}} \ln \phi_{(2)} = \Gamma_{112\dot{A}} - 2\Gamma_{121\dot{A}}. \tag{3.55}$$

The integrability condition for these partial differential equations for $\phi_{(2)}$ is obtained by applying $\partial_1{}^{\dot{A}}$ to both sides of (3.55):

$$\partial_1{}^{\dot{A}} \partial_{1\dot{A}} \ln \phi_{(2)} = \partial_1{}^{\dot{A}} (\Gamma_{112\dot{A}} - 2\Gamma_{121\dot{A}}). \tag{3.56}$$

The left-hand side of this last equation contains the commutator

$$\partial_1{}^{\dot{A}} \partial_{1\dot{A}} = \partial_{1\dot{1}} \partial_{1\dot{2}} - \partial_{1\dot{2}} \partial_{1\dot{1}} = -\Gamma^S{}_{11}{}^{\dot{A}} \partial_{S\dot{A}} - \Gamma^{\dot{S}}{}_{\dot{A}}{}^{\dot{A}}{}_1 \partial_{1\dot{S}} \tag{3.57}$$

[see (2.23)], which, since $\Gamma_{111\dot{A}} = 0$, reduces to

$$\partial_1{}^{\dot{A}} \partial_{1\dot{A}} = \Gamma_{121}{}^{\dot{A}} \partial_{1\dot{A}} - \Gamma^{\dot{A}}{}_{\dot{S}}{}^{\dot{S}}{}_1 \partial_{1\dot{A}}. \tag{3.58}$$

On the other hand, from (2.91) we obtain

$$C_{1211} = (\partial_1{}^{\dot{A}} - \Gamma^{\dot{A}\dot{R}}{}_{\dot{R}1})\Gamma_{121\dot{A}} \tag{3.59}$$

and

$$2C_{1112} = (\partial_1{}^{\dot{A}} - \Gamma_{121}{}^{\dot{A}} - \Gamma^{\dot{A}\dot{R}}{}_{\dot{R}1})\Gamma_{112\dot{A}}. \tag{3.60}$$

Substituting (3.58), (3.55), (3.59), and (3.60) into (3.56), one obtains an identity, which means that the equations (3.55) are locally integrable. This result, together with Proposition 3.2, is known as the Mariot–Robinson theorem (Mariot 1954, Robinson 1961).

If the one-index spinor field l_A satisfies the condition (3.52), then the integral curves of the (real) null vector field $v = -l^A \bar{l}^{\dot{A}} \partial_{A\dot{A}}$ form a shear-free family (or *congruence*) of null geodesics. From (3.52) we obtain

$$l^A l^B \bar{l}^{\dot{A}} \nabla_{A\dot{A}} l_B = 0, \tag{3.61}$$

which means that the integral curves of v are geodesics; in fact, (3.61) is equivalent to

$$l^A\bar{l}^{\dot{A}}\nabla_{A\dot{A}}l_B = \varepsilon l_B, \tag{3.62}$$

for some complex-valued function ε, and therefore,

$$l^A\bar{l}^{\dot{A}}\nabla_{A\dot{A}}(l_B\bar{l}_{\dot{B}}) = l^A\bar{l}^{\dot{A}}(\bar{l}_{\dot{B}}\nabla_{A\dot{A}}l_B + l_B\nabla_{A\dot{A}}\bar{l}_{\dot{B}}) = (\varepsilon+\bar{\varepsilon})\,l_B\bar{l}_{\dot{B}},$$

that is, $\nabla_v v \propto v$, which is the definition of a geodesic. (Only when $\varepsilon+\bar{\varepsilon}=0$ will these geodesics be affinely parameterized.)

By replacing the spinor field l_A by $l'_A \equiv f l_A$, where f is a complex-valued function, from (3.62) we have

$$l'^A\bar{l'}^{\dot{A}}\nabla_{A\dot{A}}l'_B = |f|^2(f\varepsilon l_B + l_B l^A\bar{l}^{\dot{A}}\partial_{A\dot{A}}f).$$

Hence, choosing f as a solution of the equation $l^A\bar{l}^{\dot{A}}\partial_{A\dot{A}}\ln f = -\varepsilon$, we obtain $l'^A\bar{l'}^{\dot{A}}\nabla_{A\dot{A}}l'_B = 0$; that is, l'_A is translated parallel to itself along the congruence of geodesics and $\nabla_{v'}v' = 0$, with $v' = -l'^A\bar{l'}^{\dot{A}}\partial_{A\dot{A}}$ (the integral curves of v' are geodesics affinely parameterized). From $l'^A\bar{l'}^{\dot{A}}\nabla_{A\dot{A}}l'_B = 0$ we have

$$l'^A l'^B\bar{l'}^{\dot{A}}\nabla_{A\dot{A}}l'_B = 0,$$

which implies that

$$l'^A l'^B\nabla_{A\dot{A}}l'_B = \sigma\bar{l'}_{\dot{A}}, \tag{3.63}$$

for some function σ.

In order to exhibit the geometric meaning of σ, we consider the behavior of vectors joining each of the geodesics with its neighbors. More precisely, we determine how a vector ξ rotates and is stretched as it is dragged along the congruence of the affinely parameterized geodesics, that is, $\pounds_{v'}\xi = 0$. Since $\pounds_{v'}\xi = [v',\xi] = \nabla_{v'}\xi - \nabla_\xi v'$ and $\nabla_{v'}v' = 0$, the directional derivative of the inner product $g(v',\xi)$ along v' vanishes,

$$v'[g(v',\xi)] = g(\nabla_{v'}v',\xi) + g(v',\nabla_{v'}\xi) = g(v',\nabla_\xi v') = \tfrac{1}{2}\xi[g(v',v')] = 0,$$

where $v'[f]$ denotes the directional derivative of f along v'. Therefore, if ξ is orthogonal to v' at some point, it will remain orthogonal to v' along the geodesic passing through that point. Hence, we can assume that ξ is orthogonal to v' everywhere, which means that the spinor equivalent of ξ is of the form

$$\xi^{A\dot{A}} = zm^A\bar{l'}^{\dot{A}} + \bar{z}l'^A\bar{m}^{\dot{A}} + bl'^A\bar{l'}^{\dot{A}}, \tag{3.64}$$

where m^A is a spinor field such that $m^A l'_A = 1$ and we can impose the condition $l'^A\bar{l'}^{\dot{A}}\nabla_{A\dot{A}}m_B = 0$ (that is, l'_A and m_A are covariantly constant along v'), z is a complex-valued function, and b is a real-valued function. Substituting (3.64) into the spinor equivalent of the relation $\nabla_{v'}\xi - \nabla_\xi v' = 0$, we obtain

$$\begin{aligned}0 &= -l'^A\overline{l'}^{\dot{A}}\nabla_{A\dot{A}}\xi_{B\dot{B}} + \xi^{A\dot{A}}\nabla_{A\dot{A}}(l'_B\overline{l'}_{\dot{B}})\\&= v'[z]\,m_B\overline{l'}_{\dot{B}} + v'[\overline{z}]\,l'_B\overline{m}_{\dot{B}} + v'[b]\,l'_B\overline{l'}_{\dot{B}}\\&\quad + zm^A\overline{l'}^{\dot{A}}\nabla_{A\dot{A}}(l'_B\overline{l'}_{\dot{B}}) + \overline{z}l'^A\overline{m}^{\dot{A}}\nabla_{A\dot{A}}(l'_B\overline{l'}_{\dot{B}}).\end{aligned}$$

The contraction of this last equation with l'^B yields

$$0 = -v'[z] + \rho z + \sigma\overline{z}, \tag{3.65}$$

where

$$\rho \equiv m^A\overline{l'}^{\dot{A}}l'^B\nabla_{A\dot{A}}l'_B \tag{3.66}$$

and we have made use of (3.63). Thus, if s is an affine parameter of the geodesics and we write $z = (x+\mathrm{i}y)/\sqrt{2}$, $\rho = \theta + \mathrm{i}\omega$, and $\sigma = |\sigma|\,\mathrm{e}^{\mathrm{i}\chi}$, equation (3.65) is equivalent to a pair of real ordinary differential equations that can be expressed in the matrix form

$$\frac{\mathrm{d}}{\mathrm{d}s}\begin{pmatrix} x \\ y \end{pmatrix} = \begin{pmatrix} \theta + |\sigma|\cos\chi & -\omega + |\sigma|\sin\chi \\ \omega + |\sigma|\sin\chi & \theta - |\sigma|\cos\chi \end{pmatrix}\begin{pmatrix} x \\ y \end{pmatrix}. \tag{3.67}$$

Since

$$\xi_{A\dot{A}} = x\frac{m_A\overline{l'}_{\dot{A}} + l'_A\overline{m}_{\dot{A}}}{\sqrt{2}} + y\frac{\mathrm{i}m_A\overline{l'}_{\dot{A}} - \mathrm{i}l'_A\overline{m}_{\dot{A}}}{\sqrt{2}} + bl'_A\overline{l'}_{\dot{A}}$$

[see (3.64)], the real-valued functions x and y are the components, with respect to an orthonormal basis, of the projection of ξ on a two-dimensional plane orthogonal to v' (see Exercise 1.10). At each point of M where the real-valued function χ has some specific value χ_0, say, by means of a rotation in this plane through an angle $\chi_0/2$, which amounts to replacing z by $\mathrm{e}^{\mathrm{i}\chi_0/2}z$, one obtains an equation of the form (3.67) with $\chi = 0$, at that point only [see (3.65)], and then one can see that θ represents the rate of *expansion* of the congruence, ω is the angular velocity, or *twist*, with which the congruence rotates, and $|\sigma|$ corresponds to a distortion, or *shear*, of the congruence whose principal axes form an angle $\chi_0/2$ with respect to the basis vectors employed in (3.67).

Since $l'_A = fl_A$, from (3.52) it follows that $l'^Al'^B\nabla_{A\dot{A}}l'_B = 0$, which, according to the definition (3.63), means that $\sigma = 0$; thus, the integral curves of the vector field $v = -l^A\overline{l}^{\dot{A}}\partial_{A\dot{A}}$ form a shear-free congruence of null geodesics, as stated above.

As pointed out at the end of Section 3.2, when the traceless part of the Ricci tensor vanishes, the Bianchi identities imply that the Weyl spinor satisfies the zero-rest-mass field equations; therefore, according to the last proposition, if $C_{AB\dot{C}\dot{D}} = 0$ (as in the case in which the Einstein vacuum field equations hold), each repeated DP spinor defines a shear-free congruence of null geodesics.

There exists a converse of this result, which does not require the vanishing of the traceless part of the Ricci tensor.

Proposition 3.4. *Let l_A be a one-index spinor field such that*

$$l^Al^B\nabla_{A\dot{A}}l_B = 0 \qquad \text{and} \qquad l^Al^BC_{AB\dot{A}\dot{B}} = 0. \tag{3.68}$$

Then l_A is a repeated principal spinor of C_{ABCD}, i.e., $l^A l^B l^C C_{ABCD} = 0$.

Proof. In a spin frame such that $l_A \propto \delta_A^2$, conditions (3.68) are equivalent to $\Gamma_{11 1\dot{A}} = 0$ and $C_{11\dot{A}\dot{B}} = 0$, respectively. Then, from (2.91) we obtain $C_{1111} = 0$, together with

$$C_{1211} = (\partial_1{}^{\dot{A}} - \Gamma^{\dot{A}\dot{R}}{}_{\dot{R}1})\Gamma_{121\dot{A}} \tag{3.69}$$

and

$$2C_{1112} = (\partial_1{}^{\dot{A}} - \Gamma_{121}{}^{\dot{A}} - \Gamma^{\dot{A}\dot{R}}{}_{\dot{R}1})\Gamma_{112\dot{A}}. \tag{3.70}$$

On the other hand, with $C_{11\dot{A}\dot{B}} = 0$ and $\Gamma_{111\dot{A}} = 0$, we have $\nabla_1{}^{\dot{S}} C_{11\dot{R}\dot{S}} = \partial_1{}^{\dot{S}} C_{11\dot{R}\dot{S}} + 2\Gamma^A{}_{11}{}^{\dot{S}} C_{A1\dot{R}\dot{S}} + \Gamma^{\dot{A}}{}_{\dot{R}}{}^{\dot{S}}{}_1 C_{11\dot{A}\dot{S}} + \Gamma^{\dot{A}}{}_{\dot{S}}{}^{\dot{S}}{}_1 C_{11\dot{R}\dot{A}} = 0$, and taking into account that $C_{1111} = 0$, the Bianchi identities (2.116) yield

$$\begin{aligned} 0 &= \nabla^A{}_{\dot{R}} C_{111A} \\ &= \partial^A{}_{\dot{R}} C_{111A} + 3\Gamma^S{}_1{}^A{}_{\dot{R}} C_{11SA} + \Gamma^S{}_A{}^A{}_{\dot{R}} C_{111S} \\ &= (\partial_{1\dot{R}} - 2\Gamma_{121\dot{R}} - 4\Gamma_{112\dot{R}}) C_{1112}, \end{aligned}$$

i.e.,

$$\partial_{1\dot{R}} C_{1112} = (2\Gamma_{121\dot{R}} + 4\Gamma_{112\dot{R}}) C_{1112}. \tag{3.71}$$

Then, making use of (3.71), (3.70), and (3.69), expressed in the equivalent form

$$C_{1211} = (\partial_1{}^{\dot{A}} - \Gamma_{121}{}^{\dot{A}} - \Gamma^{\dot{A}\dot{R}}{}_{\dot{R}1})\Gamma_{121\dot{A}}, \tag{3.72}$$

we obtain

$$\begin{aligned} \partial_1{}^{\dot{R}} \partial_{1\dot{R}} C_{1112} &= (2\Gamma_{121\dot{R}} + 4\Gamma_{112\dot{R}})\partial_1{}^{\dot{R}} C_{1112} + C_{1112}\partial_1{}^{\dot{R}}(2\Gamma_{121\dot{R}} + 4\Gamma_{112\dot{R}}) \\ &= C_{1112}\partial_1{}^{\dot{R}}(2\Gamma_{121\dot{R}} + 4\Gamma_{112\dot{R}}) \\ &= \{2[C_{1112} + (\Gamma_{121}{}^{\dot{A}} + \Gamma^{\dot{A}\dot{R}}{}_{\dot{R}1})\Gamma_{121\dot{A}}] \\ &\quad + 4[2C_{1112} + (\Gamma_{121}{}^{\dot{A}} + \Gamma^{\dot{A}\dot{R}}{}_{\dot{R}1})\Gamma_{112\dot{A}}]\} C_{1112} \\ &= \left[10C_{1112} + (\Gamma_{121}{}^{\dot{A}} + \Gamma^{\dot{A}\dot{R}}{}_{\dot{R}1})(2\Gamma_{121\dot{A}} + 4\Gamma_{112\dot{A}})\right] C_{1112}. \end{aligned}$$

According to (3.58) and (3.71), the left-hand side of this last equation is also equal to

$$(\Gamma_{121}{}^{\dot{A}} + \Gamma^{\dot{A}\dot{R}}{}_{\dot{R}1})\partial_{1\dot{A}} C_{1112} = (\Gamma_{121}{}^{\dot{A}} + \Gamma^{\dot{A}\dot{R}}{}_{\dot{R}1})(2\Gamma_{121\dot{A}} + 4\Gamma_{112\dot{A}}) C_{1112}.$$

Thus, we conclude that $C_{1112} = 0$, which, together with $C_{1111} = 0$, is equivalent to $C_{111A} = 0$ and to the covariant expression $l^A l^B l^C C_{ABCD} = 0$ in a frame such that $l_A \propto \delta_A^2$. □

Thus, if the Einstein vacuum field equations hold, the conformal curvature is algebraically special if and only if the space-time admits a shear-free congruence of null geodesics. This result is known as the Goldberg–Sachs theorem (Goldberg and Sachs 1962).

In the case of a solution of the Einstein–Maxwell equations with an algebraically special electromagnetic field, from (3.49) we have $l^A l^B C_{AB\dot{A}\dot{B}} = 0$ if l_A is the repeated principal spinor of f_{AB}, and from (3.52) and (3.68) it follows that l_A is also a repeated principal spinor of C_{ABCD}, which is therefore algebraically special.

The space-times that admit a shear-free congruence of null geodesics defined by a multiple DP spinor have several remarkable properties.

Proposition 3.5. *If l_A is a repeated principal spinor of the Weyl spinor C_{ABCD} ($C_{ABCD}l^A l^B l^C = 0$) such that $l^A l^B \nabla_{A\dot{A}} l_B = 0$, then there exists locally a complex-valued function ϕ such that*

$$l^B \nabla_{A\dot{A}} l_B = l_A l^B \partial_{B\dot{A}} \ln \phi. \tag{3.73}$$

Proof. In a spinor frame such that $l_A = \delta_A^2$, equation (3.73) is equivalent to

$$\Gamma_{11 1\dot{A}} = 0, \qquad \Gamma_{112\dot{A}} = \partial_{1\dot{A}} \ln \phi. \tag{3.74}$$

The integrability condition for the function ϕ is obtained by applying the operator $\partial_1{}^{\dot{A}}$ to both sides of the second equation in (3.74), which, making use of (3.58) and (3.74), is equivalent to the condition

$$\partial_1{}^{\dot{A}} \Gamma_{112\dot{A}} = \Gamma_{211}{}^{\dot{A}} \Gamma_{112\dot{A}} - \Gamma^{\dot{S}}{}_{\dot{A}}{}^{\dot{A}}{}_1 \Gamma_{112\dot{S}},$$

which is satisfied, since by hypothesis, $C_{1112} = 0$ [see (3.70)]. □

In a similar manner one finds that (3.72) with $C_{1112} = 0$ implies the local existence of a function ζ such that

$$\Gamma_{121\dot{A}} = \partial_{1\dot{A}} \ln \zeta. \tag{3.75}$$

As a first application of the existence of the functions ϕ and ζ we can now establish the following proposition.

Proposition 3.6. *In a space-time that admits a shear-free congruence of null geodesics defined by a multiple DP spinor l_A, up to gauge transformations the most general solution of the self-duality condition (3.10) can be expressed locally as*

$$\Phi_{1\dot{B}} = 0, \qquad \Phi_{2\dot{B}} = (\partial_{1\dot{B}} + 2\Gamma_{121\dot{B}} + \Gamma_{112\dot{B}})\chi, \tag{3.76}$$

with respect to a spin frame such that $l_A \propto \delta_A^2$, where χ is a complex potential satisfying the equation

$$(\partial_2{}^{\dot{B}} + \Gamma_{122}{}^{\dot{B}} - \Gamma^{\dot{B}\dot{S}}{}_{\dot{S}2})(\partial_{1\dot{B}} + 2\Gamma_{121\dot{B}} + \Gamma_{112\dot{B}})\chi = 0.$$

Proof. The self-duality condition $\nabla_{(B}{}^{\dot{B}} \Phi_{C)\dot{B}} = 0$ is explicitly given by the equations

$$0 = \partial_{(B}{}^{\dot{B}} \Phi_{C)\dot{B}} + \Gamma^S{}_{(BC)}{}^{\dot{B}} \Phi_{S\dot{B}} + \Gamma^{\dot{S}}{}_{\dot{B}}{}^{\dot{B}}{}_{(B} \Phi_{C)\dot{S}}, \tag{3.77}$$

which are equivalent to (3.12). Taking $B = 1 = C$ in (3.77), we have

$$0 = (\partial_1{}^{\dot{B}} - \Gamma_{121}{}^{\dot{B}} - \Gamma^{\dot{B}\dot{S}}{}_{\dot{S}1})\Phi_{1\dot{B}} + \Gamma_{111}{}^{\dot{B}}\Phi_{2\dot{B}}. \qquad (3.78)$$

Making use of the fact that we have $\Gamma_{111}{}^{\dot{B}} = 0$ with respect to a spin frame such that the multiple DP spinor that defines a shear-free geodesic congruence has the form $l_A \propto \delta_A^2$, from (3.78) it follows that there exists locally a function ξ such that $\Phi_{1\dot{B}} = -\partial_{1\dot{B}}\xi$. Indeed, the integrability condition for ξ is given by

$$\partial_1{}^{\dot{A}}\partial_{1\dot{A}}\xi = \Gamma_{121}{}^{\dot{A}}\partial_{1\dot{A}}\xi - \Gamma^{\dot{A}}{}_{\dot{S}}{}^{\dot{S}}{}_1\partial_{1\dot{A}}\xi$$

[see (3.58)], which amounts to

$$-\partial_1{}^{\dot{A}}\Phi_{1\dot{A}} = -\Gamma_{121}{}^{\dot{A}}\Phi_{1\dot{A}} + \Gamma^{\dot{A}}{}_{\dot{S}}{}^{\dot{S}}{}_1\Phi_{1\dot{A}}$$

and coincides with (3.78). Then, the gauge transformation $\Phi_{B\dot{B}} \mapsto \Phi_{B\dot{B}} + \partial_{B\dot{B}}\xi$ yields $\Phi_{1\dot{B}} = 0$.

Taking $B = 1, C = 2$ in (3.77) with $\Phi_{1\dot{B}} = 0$, we obtain the condition

$$0 = (\partial_1{}^{\dot{B}} + \Gamma_{121}{}^{\dot{B}} + \Gamma_{112}{}^{\dot{B}} - \Gamma^{\dot{B}\dot{S}}{}_{\dot{S}1})\Phi_{2\dot{B}}. \qquad (3.79)$$

Making use of (3.74) and (3.75), this last condition can also be written as

$$0 = (\partial_1{}^{\dot{B}} - \Gamma_{121}{}^{\dot{B}} - \Gamma^{\dot{B}\dot{S}}{}_{\dot{S}1})(\zeta^2\phi\Phi_{2\dot{B}})$$

[cf. (3.78)], which implies the local existence of a function η such that $\zeta^2\phi\Phi_{2\dot{B}} = \partial_{1\dot{B}}\eta$. Hence, letting $\chi \equiv \zeta^{-2}\phi^{-1}\eta$, we have

$$\Phi_{2\dot{B}} = \zeta^{-2}\phi^{-1}\partial_{1\dot{B}}(\zeta^2\phi\chi) = (\partial_{1\dot{B}} + 2\Gamma_{121\dot{B}} + \Gamma_{112\dot{B}})\chi, \qquad (3.80)$$

where we have made use again of (3.74) and (3.75).

Taking $A = 2 = C$ in (3.77) we obtain

$$0 = (\partial_2{}^{\dot{B}} + \Gamma_{122}{}^{\dot{B}} - \Gamma^{\dot{B}\dot{S}}{}_{\dot{S}2})\Phi_{2\dot{B}},$$

which, by virtue of (3.80), implies that the potential χ must satisfy the equation

$$0 = (\partial_2{}^{\dot{B}} + \Gamma_{122}{}^{\dot{B}} - \Gamma^{\dot{B}\dot{S}}{}_{\dot{S}2})(\partial_{1\dot{B}} + 2\Gamma_{121\dot{B}} + \Gamma_{112\dot{B}})\chi \qquad (3.81)$$

(Torres del Castillo 1999). □

It can be readily verified that the covariant expression

$$\Phi_{B\dot{B}} = \phi^{-2}\nabla^S{}_{\dot{B}}(\phi^2 l_S l_B \psi) \qquad (3.82)$$

reduces to (3.76), with $\chi = (l_2)^2 \psi$, in a spinor frame such that $l_A \propto \delta_A^2$. If we assume that ϕ and ψ are scalar functions, then both sides of (3.82) are spinor fields of the same type, and therefore the equality (3.82) is valid in any spinor frame.

Hence, according to the results of Section 3.1, the most general real source-free electromagnetic field is given locally by

$$f_{\dot{A}\dot{B}} = \nabla^B{}_{(\dot{A}}[\phi^{-2}\nabla^S{}_{\dot{B})}(\phi^2 l_S l_B \psi)] \tag{3.83}$$

(see also Cohen and Kegeles 1974, Wald 1978, Mustafa and Cohen 1987). The second-order linear partial differential equation (3.81) can be solved by separation of variables in all the type-D solutions of the Einstein vacuum field equations (Kamran 1987, Torres del Castillo 1988).

In the case of flat space-time there exist covariantly constant one-index spinors (which form a complex two-dimensional vector space), and any of them satisfies the conditions of Proposition 3.5. Then the function ϕ defined by (3.73) can be taken equal to 1, and (3.83) reduces to

$$f_{\dot{A}\dot{B}} = \nabla^A{}_{\dot{A}}\nabla^B{}_{\dot{B}}(l_A l_B \psi),$$

where l_A is a covariantly constant spinor and ψ is a complex-valued function that has to satisfy the wave equation (see also Penrose 1965).

The solution of the Weyl equation can also be expressed locally in terms of a complex potential when the space-time admits a shear-free congruence of null geodesics defined by a multiple DP spinor. The Weyl equation $0 = \nabla_A{}^{\dot{A}}\eta_{\dot{A}}$ is given explicitly by

$$0 = (\partial_1{}^{\dot{A}} - \Gamma^{\dot{A}\dot{S}}{}_{\dot{S}1})\eta_{\dot{A}}, \tag{3.84}$$

$$0 = (\partial_2{}^{\dot{A}} - \Gamma^{\dot{A}\dot{S}}{}_{\dot{S}2})\eta_{\dot{A}}. \tag{3.85}$$

Making use of (3.75), equation (3.84) can be rewritten as

$$0 = (\partial_1{}^{\dot{A}} - \Gamma_{121}{}^{\dot{A}} - \Gamma^{\dot{A}\dot{S}}{}_{\dot{S}1})(\zeta\eta_{\dot{A}}),$$

which implies the local existence of a complex-valued function η such that $\zeta\eta_{\dot{A}} = \partial_{1\dot{A}}\eta$, and writing $\eta = \zeta\chi$, we have

$$\eta_{\dot{A}} = \zeta^{-1}\partial_{1\dot{A}}(\zeta\chi) = (\partial_{1\dot{A}} + \Gamma_{121\dot{A}})\chi. \tag{3.86}$$

Substituting this expression into (3.85), we obtain

$$0 = (\partial_2{}^{\dot{A}} - \Gamma^{\dot{A}\dot{S}}{}_{\dot{S}2})(\partial_{1\dot{A}} + \Gamma_{121\dot{A}})\chi. \tag{3.87}$$

Then, as in the case of (3.82), one finds that with respect to an arbitrary frame, the solution of the Weyl equation is given by

$$\eta_{\dot{B}} = \phi^{-1}\nabla^S{}_{\dot{B}}(\phi l_S \psi), \tag{3.88}$$

where ψ is a complex potential that has to satisfy the condition

$$\nabla^{B\dot{B}}\left[\phi^{-1}\nabla^{S}{}_{\dot{B}}(\phi l_S \psi)\right] = 0$$

(the left-hand side of this last equation is proportional to l^B, and therefore this equation imposes a single condition on ψ).

In any type-D solution of the Einstein vacuum field equations, equation (3.87) can be solved by separation of variables.

3.3.2 Space-Times with Symmetries. Ernst Potentials

With each nonnull Killing vector of a solution of the Einstein vacuum field equations, there is associated a complex function, called the Ernst potential, which can be used to construct the metric. In the case of a solution of the Einstein–Maxwell equations, there is a pair of Ernst potentials associated with each nonnull Killing vector, provided that the Lie derivative of the electromagnetic field with respect to the Killing vector vanishes. In this subsection the equations for the Ernst potentials are obtained by means of the spinor formalism (alternative derivations, making use of the tensor formalism or of differential forms, can be found, for instance, in Chandrasekhar 1983, Heusler 1996, Stephani et al. 2003). Owing to their nonlinearity, it is difficult to find explicit solutions of the Ernst equations; fortunately, the symmetries of the Ernst equations make it possible to generate new solutions from a given one (see, e.g., Stephani et al. 2003).

An arbitrary vector field K is (locally) hypersurface orthogonal if and only if

$$\omega_a \equiv \varepsilon_{abcd} K^b \nabla^c K^d$$

vanishes (see below). Making use of (1.64), one finds that the spinor equivalent of ω_a is given by

$$\omega_{A\dot{A}} = -\mathrm{i}(K^{B\dot{B}}\nabla_{A\dot{B}}K_{B\dot{A}} - K^{B\dot{B}}\nabla_{B\dot{A}}K_{A\dot{B}}).$$

According to (2.142) and (2.71), the components $\omega_{A\dot{A}}$ can be obtained by means of the relation

$$K_a\theta^a \wedge \mathrm{d}(K_b\theta^b) = \mathrm{i}\omega_{A\dot{A}}\breve{\theta}^{A\dot{A}}. \tag{3.89}$$

In particular, if K is a Killing vector, from (2.140) it follows that

$$\omega_{A\dot{A}} = 2\mathrm{i}(K^B{}_{\dot{A}}L_{AB} - K_A{}^{\dot{B}}L_{\dot{A}\dot{B}}). \tag{3.90}$$

Similarly, if we let

$$f \equiv K^{B\dot{B}}K_{B\dot{B}} \tag{3.91}$$

(i.e., $f = -K^a K_a$), then using again (2.140), we have

$$\nabla_{A\dot{A}}f = 2K^{B\dot{B}}\nabla_{A\dot{A}}K_{B\dot{B}} = 2(K^B{}_{\dot{A}}L_{AB} + K_A{}^{\dot{B}}L_{\dot{A}\dot{B}}), \tag{3.92}$$

[cf. (3.90)]. Hence

$$\omega_{A\dot{A}} = 4\mathrm{i}K^{B}{}_{\dot{A}}L_{AB} - \mathrm{i}\nabla_{A\dot{A}}f. \tag{3.93}$$

We can now prove the assertion made at the beginning of this subsection. When $\omega_a = 0$, from (3.93) we see that $\nabla_{A\dot{A}}f = 4K^{B}{}_{\dot{A}}L_{AB}$; therefore, if K is nonnull, $K_{(C}{}^{\dot{A}}\nabla_{A)\dot{A}}f = -2fL_{AC}$, or equivalently, $K_{(C}{}^{\dot{A}}\nabla_{A)\dot{A}}f = f\nabla_{(A|\dot{A}|}K_{C)}{}^{\dot{A}}$ [see (2.141)]. This last equation can also be written in the form $\nabla_{(A|\dot{A}|}[f^{-1}K_{C)}{}^{\dot{A}}] = 0$, and taking into account that the spinor equivalent of the curl of ω_a is given by

$$\nabla_{A\dot{A}}\omega_{B\dot{B}} - \nabla_{B\dot{B}}\omega_{A\dot{A}} = \varepsilon_{\dot{A}\dot{B}}\nabla_{(A}{}^{\dot{R}}\omega_{B)\dot{R}} + \varepsilon_{AB}\nabla^{R}{}_{(\dot{A}}\omega_{|R|\dot{B})},$$

it follows that there exists (locally) a real-valued function u such that

$$K_{A\dot{A}} = f\nabla_{A\dot{A}}u, \tag{3.94}$$

thus showing that K is orthogonal to the hypersurfaces $u =$ constant. Alternatively, if $\omega_a = 0$, from (3.89) and Frobenius's theorem it follows that the one-form $K_a\theta^a$ is locally integrable; that is, locally there exist functions F and G such that $K_a\theta^a = F\mathrm{d}G$, which amounts to

$$K_{A\dot{A}} = F\nabla_{A\dot{A}}G.$$

(Note that this last proof is valid also if K is null and even if K is not a Killing vector; on the other hand, (3.94) explicitly gives an integrating factor for the one-form $K_a\theta^a$.)

By virtue of (2.140) and (2.144), the anti-self-dual part of the curl of ω_a is determined by

$$\nabla_{(C}{}^{\dot{A}}\omega_{A)\dot{A}} = 4\mathrm{i}K^{B}{}_{\dot{A}}\nabla_{(C}{}^{\dot{A}}L_{A)B} = 4\mathrm{i}K^{B\dot{B}}C_{B(A|\dot{B}|}{}^{\dot{S}}K_{C)\dot{S}}. \tag{3.95}$$

The right-hand side of (3.95) vanishes if K is an eigenvector of the Ricci tensor or if the traceless part of the Ricci tensor is equal to zero. Thus, if we assume that the Einstein vacuum field equations are satisfied, from (3.95) it follows that there exists (at least locally) a real-valued function ω, the *twist potential*, such that

$$\omega_{A\dot{A}} = -\nabla_{A\dot{A}}\omega, \tag{3.96}$$

and (3.93) yields

$$4K^{B}{}_{\dot{A}}L_{AB} = \nabla_{A\dot{A}}\chi, \tag{3.97}$$

where

$$\chi \equiv f + \mathrm{i}\omega. \tag{3.98}$$

Therefore, if K is nonnull (i.e., $f \neq 0$), we have

$$L_{AB} = -\frac{1}{2f}K_{B}{}^{\dot{A}}\nabla_{A\dot{A}}\chi. \tag{3.99}$$

Then (3.97), (2.144), with $R_{ab} = 0$, (2.140), and (3.99), give

$$\begin{aligned}\nabla^{A\dot{A}}\nabla_{A\dot{A}}\chi &= 4K^{B}{}_{\dot{A}}\nabla^{A\dot{A}}L_{AB} + 4L_{AB}\nabla^{A\dot{A}}K^{B}{}_{\dot{A}} \\ &= 8L_{AB}L^{AB} \\ &= \frac{1}{f}(\nabla^{A\dot{A}}\chi)(\nabla_{A\dot{A}}\chi).\end{aligned}$$

Thus, χ obeys the nonlinear equation

$$(\mathrm{Re}\,\chi)\nabla^a\nabla_a\chi = (\nabla^a\chi)(\nabla_a\chi). \tag{3.100}$$

This equation is known as the Ernst equation and χ is the *Ernst potential* (Ernst 1968a). (Note that from (3.97) it follows that $K^{A\dot{A}}\nabla_{A\dot{A}}\chi = 0$.)

The Ernst potential is useful in finding exact solutions of the Einstein vacuum field equations. For instance, the metric of a stationary axisymmetric space-time can be locally expressed in the form

$$g = -\mathrm{e}^{2U}(\mathrm{d}t + A\mathrm{d}\varphi)^2 + \mathrm{e}^{-2U}\left[\mathrm{e}^{2\gamma}(\mathrm{d}\rho^2 + \mathrm{d}z^2) + \rho^2\mathrm{d}\varphi^2\right],$$

where t, ρ, φ, z form a coordinate system and the functions U, γ, and A depend on ρ and z only (Papapetrou 1953; see also Chandrasekhar 1983, Stephani et al. 2003). This metric possesses the timelike Killing vector $K = \partial/\partial t$, for which $f = K^{A\dot{A}}K_{A\dot{A}} = -K^aK_a = \mathrm{e}^{2U}$; hence, we can also write

$$g = -f(\mathrm{d}t + A\mathrm{d}\varphi)^2 + f^{-1}\left[\mathrm{e}^{2\gamma}(\mathrm{d}\rho^2 + \mathrm{d}z^2) + \rho^2\mathrm{d}\varphi^2\right].$$

Then, the Ernst equation (3.100) takes the explicit form

$$(\mathrm{Re}\,\chi)\left[\frac{1}{\rho}\frac{\partial}{\partial\rho}\left(\rho\frac{\partial\chi}{\partial\rho}\right) + \frac{\partial^2\chi}{\partial z^2}\right] = \left(\frac{\partial\chi}{\partial\rho}\right)^2 + \left(\frac{\partial\chi}{\partial z}\right)^2,$$

which does not involve the unknown functions f, γ, and A. The real part of a given solution of the Ernst equation gives f [see (3.98)], while the imaginary part of χ determines A by means of

$$\frac{f^2}{\rho}\left(\frac{\partial A}{\partial\rho} + \mathrm{i}\frac{\partial A}{\partial z}\right) = \frac{\partial\omega}{\partial z} - \mathrm{i}\frac{\partial\omega}{\partial\rho}. \tag{3.101}$$

In fact, $K_a\theta^a = -f(\mathrm{d}t + A\mathrm{d}\varphi)$, and therefore $K_a\theta^a \wedge \mathrm{d}(K_b\theta^b) = f^2\mathrm{d}t \wedge \mathrm{d}A \wedge \mathrm{d}\varphi$. Then (3.89) and (3.96) lead to (3.101).

The function γ is also determined by the Ernst potential, as a consequence of the Einstein equations (see, e.g., Heusler 1996, Sec. 4.1, Stephani et al. 2003, §19.5).

Solutions of the Einstein–Maxwell Equations

When there is an electromagnetic field present, expressions analogous to (3.99) and (3.100) can be given, provided that the electromagnetic field is invariant under a one-parameter group of isometries. The electromagnetic field is invariant under the one-parameter group of isometries generated by a Killing vector $K = -K^{A\dot{A}}\partial_{A\dot{A}}$ if the Lie derivative of the electromagnetic spinor f_{AB} with respect to K vanishes, i.e.,

$$-K^{C\dot{C}}\nabla_{C\dot{C}}f_{AB} + 2L^{C}{}_{(A}f_{B)C} = 0$$

[see (2.143)], or equivalently, making use of (1.27) and (2.140),

$$\begin{aligned}
0 &= -K^{C\dot{C}}\nabla_{(A|\dot{C}}f_{C|B)} + K_{(A}{}^{\dot{C}}\nabla^{S}{}_{|\dot{C}}f_{S|B)} + 2L^{C}{}_{(A}f_{B)C} \\
&= -\nabla_{(A|\dot{C}}[K^{C\dot{C}}f_{C|B)}] + f_{C(A}\nabla_{B)\dot{C}}K^{C\dot{C}} + K_{(A}{}^{\dot{C}}\nabla^{S}{}_{|\dot{C}}f_{S|B)} + 2L^{C}{}_{(A}f_{B)C} \\
&= -\nabla_{(A|\dot{C}}[K^{C\dot{C}}f_{C|B)}] + K_{(A}{}^{\dot{C}}\nabla^{S}{}_{|\dot{C}}f_{S|B)}.
\end{aligned}$$

Thus, if the source-free Maxwell equations are satisfied, letting $P_{B\dot{C}} \equiv K^{C}{}_{\dot{C}}f_{BC}$, we have $\nabla_{(A}{}^{\dot{C}}P_{B)\dot{C}} = 0$. Furthermore,

$$\nabla^{B}{}_{(\dot{A}}P_{|B|\dot{C})} = \nabla^{B}{}_{(\dot{A}}[K^{C}{}_{\dot{C})}f_{BC}] = f_{BC}\nabla^{B}{}_{(\dot{A}}K^{C}{}_{\dot{C})} = 0$$

[see (2.140)]. Hence the curl of the tensor equivalent of $P_{A\dot{A}}$ is equal to zero, and therefore, there exists, locally, a complex-valued function Φ such that

$$K^{C}{}_{\dot{C}}f_{BC} = \frac{c^2}{2\sqrt{G}}\nabla_{B\dot{C}}\Phi, \tag{3.102}$$

where the constant factor is introduced for later convenience. The symmetry of f_{BC} implies that $K^{B\dot{C}}\nabla_{B\dot{C}}\Phi = 0$. Thus, if K_a is nonnull,

$$f_{BC} = -\frac{c^2}{f\sqrt{G}}K_{C}{}^{\dot{C}}\nabla_{B\dot{C}}\Phi. \tag{3.103}$$

The source-free Maxwell equations imply $K^{B\dot{B}}\nabla^{A}{}_{\dot{B}}f_{AB} = 0$, which, making use of (3.103) and (2.140), reduces to

$$f\nabla^{A\dot{A}}\nabla_{A\dot{A}}\Phi = 4K^{B\dot{A}}L^{A}{}_{B}\nabla_{A\dot{A}}\Phi. \tag{3.104}$$

(See Exercise 3.9.)

Substituting the Einstein equations (3.49) and (3.102) into (3.95), one finds that

$$\begin{aligned}
\nabla_{(C}{}^{\dot{A}}\omega_{A)\dot{A}} &= -2\mathrm{i}(\nabla_{(C}{}^{\dot{A}}\Phi)(\nabla_{A)\dot{A}}\overline{\Phi}) \\
&= \mathrm{i}\nabla_{(C}{}^{\dot{A}}\left[\overline{\Phi}\nabla_{A)\dot{A}}\Phi - \Phi\nabla_{A)\dot{A}}\overline{\Phi}\right],
\end{aligned}$$

which implies (locally) the existence of a real-valued function ω such that $\omega_{A\dot{A}} = \mathrm{i}(\overline{\Phi}\nabla_{A\dot{A}}\Phi - \Phi\nabla_{A\dot{A}}\overline{\Phi}) - \nabla_{A\dot{A}}\omega$, and therefore (3.93) gives

$$4K^{B}{}_{\dot{A}}L_{AB} = \nabla_{A\dot{A}}\chi + 2\overline{\Phi}\nabla_{A\dot{A}}\Phi, \tag{3.105}$$

where

$$\chi \equiv f - \overline{\Phi}\Phi + \mathrm{i}\omega. \tag{3.106}$$

From (3.105), making use of (2.140), (2.144), (3.104), the Einstein equations, and (3.102), we obtain

$$\begin{aligned}
\nabla^{A\dot{A}}\nabla_{A\dot{A}}\chi &= 4L_{AB}\nabla^{A\dot{A}}K^{B}{}_{\dot{A}} + 4K^{B}{}_{\dot{A}}\nabla^{A\dot{A}}L_{AB} - 2(\nabla^{A\dot{A}}\overline{\Phi})(\nabla_{A\dot{A}}\Phi) \\
&\quad - 2\overline{\Phi}\nabla^{A\dot{A}}\nabla_{A\dot{A}}\Phi \\
&= 8L_{AB}L^{AB} + \frac{8G}{c^4}K^{B}{}_{\dot{A}}f_{BR}K^{R\dot{S}}f_{\dot{S}}{}^{\dot{A}} - 2(\nabla^{A\dot{A}}\overline{\Phi})(\nabla_{A\dot{A}}\Phi) \\
&\quad - 8f^{-1}K^{B\dot{A}}L^{A}{}_{B}\overline{\Phi}\nabla_{A\dot{A}}\Phi \\
&= f^{-1}(\nabla^{A\dot{A}}\chi + 2\overline{\Phi}\nabla^{A\dot{A}}\Phi)\nabla_{A\dot{A}}\chi.
\end{aligned}$$

Thus, from (3.104)–(3.106) we finally obtain the system of nonlinear equations (Ernst 1968b)

$$\begin{aligned}
(\mathrm{Re}\,\chi + \overline{\Phi}\Phi)\nabla^a\nabla_a\chi &= (\nabla^a\chi + 2\overline{\Phi}\nabla^a\Phi)\nabla_a\chi, \\
(\mathrm{Re}\,\chi + \overline{\Phi}\Phi)\nabla^a\nabla_a\Phi &= (\nabla^a\chi + 2\overline{\Phi}\nabla^a\Phi)\nabla_a\Phi.
\end{aligned} \tag{3.107}$$

The Ernst equations (3.107) can be used to find stationary axisymmetric solutions of the Einstein–Maxwell equations, and as in the case of the Ernst equation (3.100), the symmetries of these equations make it possible to generate new solutions from a given one.

Relations Between Killing Vectors and the Conformal Curvature

The equations derived above are also useful in the derivation of some relations between the properties of a Killing vector and the conformal curvature (equivalent results using other formalisms have been given in Debney 1971a,b and Catenacci et al. 1980). In the rest of this section we shall restrict ourselves to solutions of the Einstein vacuum field equations.

If the space-time admits a null Killing vector K, we can write

$$K_{A\dot{B}} = \pm\alpha_A\overline{\alpha}_{\dot{B}}, \tag{3.108}$$

where α_A is some one-index spinor field, and from the Killing equations (2.140) we obtain

$$0 = \alpha^A\alpha^C(\nabla_{A\dot{B}}K_{C\dot{D}} + \nabla_{C\dot{D}}K_{A\dot{B}}) = \pm(\alpha^A\overline{\alpha}_{\dot{D}}\alpha^C\nabla_{A\dot{B}}\alpha_C + \alpha^A\overline{\alpha}_{\dot{B}}\alpha^C\nabla_{C\dot{D}}\alpha_A),$$

which implies

$$\alpha^A\alpha^C\nabla_{A\dot{B}}\alpha_C = 0,$$

that is, K is tangent to a shear-free congruence of null geodesics [see (3.52)]. Then, by virtue of the Goldberg–Sachs theorem (Proposition 3.4), since the Einstein vacuum field equations are assumed to hold, α_A is a repeated DP spinor and the conformal curvature is algebraically special.

In the present case, in which the function $f = K^{B\dot{B}}K_{B\dot{B}}$ vanishes, from (3.92) and (3.108) we have

$$0 = K^B{}_{\dot{A}} L_{AB} + K_A{}^{\dot{B}} L_{\dot{A}\dot{B}} = \pm(\alpha^B \overline{\alpha}_{\dot{A}} L_{AB} + \alpha_A \overline{\alpha}^{\dot{B}} L_{\dot{A}\dot{B}}),$$

which implies that $L_{AB}\alpha^B = \mathrm{i}B\alpha_A$, where B is a real-valued function; then (3.90) and (3.108) give $\omega_{A\dot{A}} = -4BK_{A\dot{A}}$.

The condition $\nabla_{(C}{}^{\dot{A}}\omega_{A)\dot{A}} = 0$ [see (3.95)] yields

$$0 = B\nabla_{(C}{}^{\dot{A}}K_{A)\dot{A}} + K_{(A|\dot{A}|}\nabla_{C)}{}^{\dot{A}}B = 2BL_{CA} \pm \alpha_{(A}\overline{\alpha}_{|\dot{A}|}\nabla_{C)}{}^{\dot{A}}B.$$

Contracting this relation with α^C, we obtain $2\mathrm{i}B^2\alpha_A \pm \frac{1}{2}\alpha_A\overline{\alpha}_{\dot{A}}\alpha^C\nabla_C{}^{\dot{A}}B = 0$, which implies $B = 0$.

Thus, $L_{AB}\alpha^B = 0$; that is, α_A is a double principal spinor of L_{AB}, and L_{AB} is algebraically special. Hence, we can write

$$L_{AB} = \psi\alpha_A\alpha_B \tag{3.109}$$

for some complex-valued function ψ. Furthermore, $\omega_{A\dot{A}} = 0$; that is, K is hypersurface orthogonal.

Substituting (3.108) and (3.109) into (2.144), we see that

$$\nabla_{A\dot{A}}(\psi\alpha_B\alpha_C) = \mp C^D{}_{ABC}\alpha_D\overline{\alpha}_{\dot{A}}.$$

Hence, contracting with $\alpha^B\overline{\alpha}^{\dot{A}}$ we obtain (assuming $\psi \neq 0$)

$$\alpha^B\overline{\alpha}^{\dot{A}}\nabla_{A\dot{A}}\alpha_B = 0,$$

which shows that the twist and the expansion of the congruence of null geodesics defined by α_A are equal to zero [see (3.66)].

On the other hand, if we assume that L_{AB} is algebraically special [as in (3.109)] but that the corresponding Killing vector K is nonnull, writing $L_{AB} = \alpha_A\alpha_B$, from (2.144), assuming again that the Einstein vacuum field equations hold, we arrive at

$$0 = \alpha^B\alpha^C\nabla_{A\dot{A}}(\alpha_B\alpha_C) = -C^D{}_{ABC}\alpha^B\alpha^C K_{D\dot{A}}.$$

Since K is nonnull, it follows that $C_{ABCD}\alpha^B\alpha^C = 0$, thus showing that the conformal curvature is of type III or N, or equal to zero.

3.4 Killing Spinors

Walker and Penrose (1970) showed that any type-D solution of the Einstein vacuum field equations admits a quadratic integral of the null geodesic equations (see also Hughston et al. 1972, Hughston and Sommers 1973). The existence of this first integral follows from that of a certain two-index spinor field, which has been called a Killing spinor.

Apart from their relationship with first integrals of the geodesic equations, the Killing spinors also appear in "symmetry operators" for the Weyl, Dirac, and Maxwell equations (Carter and McLenaghan 1979, Kamran and McLenaghan 1983, 1984a, Torres del Castillo 1985, 1986, McLenaghan et al. 2000).

A $D(k,0)$ Killing spinor is a totally symmetric $2k$-index spinor field $M_{AB...D}$ such that

$$\nabla_{(A}{}^{\dot{A}} M_{BC...L)} = 0. \tag{3.110}$$

(Note that the equations for a conformal Killing vector can be expressed in the form $\nabla_{(A}{}^{(\dot{A}} K_{B)}{}^{\dot{B})} = 0$.) It can readily be seen that the symmetrized tensor product of Killing spinors is also a Killing spinor. Equation (3.110) is also known as the twistor equation.

By virtue of the Ricci identities (2.126),

$$\nabla_{(A|\dot{A}|} \nabla_B{}^{\dot{A}} M_{CD...L)} = -\Box_{(AB} M_{CD...L)} = -2k C^S{}_{(ABC} M_{D...L)S};$$

hence, an integrability condition for the set of equations (3.110) is

$$C^S{}_{(ABC} M_{D...L)S} = 0. \tag{3.111}$$

As we shall show below, if $k \neq 2,4,6,\ldots$, or if the traceless part of the Ricci tensor vanishes, these equations imply that the conformal curvature must be of type D or N, or equal to zero [see (2.130)]. Denoting the components of $M_{AB...D}$ by $M_{(j)}$, where $j = 0,1,\ldots,2k$ is the number of indices taking the value 2, i.e., $M_{(0)} = M_{11...1}$, $M_{(1)} = M_{21...1}$, $\ldots$, $M_{(2k)} = M_{22...2}$, the integrability conditions (3.111) are given explicitly by the set of $2k+3$ equations

$$\begin{aligned} 0 = {} & (2k+2-j)(2k+1-j)(2k-j)\,C^{(5)} M_{(j+1)} \\ & + 2(2j-k)(2k+2-j)(2k+1-j)\,C^{(4)} M_{(j)} \\ & + 6j(2k+2-j)(j-k-1)\,C^{(3)} M_{(j-1)} \\ & + 2j(j-1)(2j-3k-4)\,C^{(2)} M_{(j-2)} \\ & - j(j-1)(j-2)\,C^{(1)} M_{(j-3)} = 0, \end{aligned} \tag{3.112}$$

where j can take the values $0,1,\ldots,2k+2$, $C^{(5)} \equiv C_{1111}$, $C^{(4)} \equiv C_{1112}$, $C^{(3)} \equiv C_{1122}$, $C^{(2)} \equiv C_{1222}$, and $C^{(1)} \equiv C_{2222}$.

If l_A is a principal spinor of $M_{AB...D}$, then in a spinor frame such that $l_A \propto \delta^2_A$, we have $M_{(0)} = 0$, and from (3.112) it follows that $C^{(5)} = 0$, which means that a principal spinor of $M_{AB...D}$ is also a principal spinor of C_{ABCD}. Substituting the values $M_{(0)} = 0$ and $C^{(5)} = 0$ into (3.112), one finds that if $k \neq 2,4,6,\ldots$, then $C^{(4)} = 0$; hence, when k is not an even integer, each principal spinor of $M_{AB...D}$ is at least a double principal spinor of C_{ABCD}. Therefore (if $C_{ABCD} \neq 0$), there are at most two principal spinors of $M_{AB...D}$ that are not proportional to each other. When $M_{AB...D}$ has a $2k$-fold repeated principal spinor, from (3.111) or (3.112) one concludes that C_{ABCD} is of type N; otherwise, C_{ABCD} must be of type D (Torres del Castillo 1984a).

The fact that a principal spinor of a $D(k,0)$ Killing spinor, $M_{AB...D}$, is a principal spinor of C_{ABCD} also follows from the fact that each principal spinor of $M_{AB...D}$ satisfies condition (3.52) (Hughston et al. 1972). Indeed, equations (3.110) amount to

$$\partial_{(A}{}^{\dot{A}} M_{BC...L)} + 2k\Gamma^S{}_{(AB}{}^{\dot{A}} M_{C...L)S} = 0,$$

or more explicitly,

$$\begin{aligned} 0 = {} & j\,[\partial_{2\dot{A}} - 2(k+1-j)\Gamma_{122\dot{A}} - (2k+1-j)\Gamma_{221\dot{A}}]\,M_{(j-1)} \\ & + (2k+1-j)[\partial_{1\dot{A}} - 2(k-j)\Gamma_{121\dot{A}} + j\Gamma_{112\dot{A}}]\,M_{(j)} \\ & - j(j-1)\Gamma_{222\dot{A}} M_{(j-2)} + (2k+1-j)(2k-j)\Gamma_{111\dot{A}} M_{(j+1)}, \end{aligned} \tag{3.113}$$

where j can take the values $0,1,\ldots,2k+1$. Substituting $M_{(0)} = 0$ into (3.113), one finds that $\Gamma_{111\dot{A}} = 0$, which implies $C_{1111} = 0$ [see (2.91)]. Furthermore, as we have seen, $\Gamma_{111\dot{A}} = 0$ means that any one-index spinor l_A such that $l_A \propto \delta^2_A$ satisfies condition (3.52); hence, if the traceless part of the Ricci tensor vanishes, Proposition 3.4 implies that each principal spinor of $M_{AB...D}$ is at least a double principal spinor of C_{ABCD} (without restriction on the value of k). Therefore, when the traceless part of the Ricci tensor vanishes, C_{ABCD} must be of type N or D, or equal to zero.

If we do not assume that the traceless part of the Ricci tensor vanishes, when $k = 2$, for instance, the integrability conditions (3.111) imply that M_{ABCD} is proportional to C_{ABCD}, without restriction on the algebraic type of the conformal curvature.

When C_{ABCD} is of type D one can find a spin frame such that only $C^{(3)}$ is different from zero; then (3.112) implies that $M_{(k)}$ is the only nonvanishing component of $M_{AB...L}$, and necessarily k must be an integer. In the case that C_{ABCD} is of type N, one can find a spin frame such that only $C^{(1)}$ is different from zero, and (3.112) implies that $M_{(2k)}$ is the only nonvanishing component of $M_{AB...L}$.

In the case that C_{ABCD} is of type D, assuming that $M_{(k)}$ is the only nonvanishing component of $M_{AB...D}$, equations (3.113) reduce to

$$\begin{aligned} 0 &= \Gamma_{111\dot{A}}, \\ 0 &= \Gamma_{222\dot{A}}, \\ 0 &= (\partial_{1\dot{A}} + k\Gamma_{112\dot{A}})M_{(k)}, \\ 0 &= (\partial_{2\dot{A}} - k\Gamma_{221\dot{A}})M_{(k)}. \end{aligned} \tag{3.114}$$

These equations are integrable if the traceless part of the Ricci tensor vanishes (Walker and Penrose 1970); in fact, the Bianchi identities (2.116) with $C_{AB\dot{A}\dot{B}} = 0$ imply that the Weyl spinor satisfies the zero-rest-mass equations (3.53), and assuming that $C^{(3)}$ is the only nonvanishing component of C_{ABCD}, equations (3.54) reduce to

$$
\begin{aligned}
0 &= \Gamma_{111\dot{A}},\\
0 &= \Gamma_{222\dot{A}},\\
0 &= (\partial_{1\dot{A}} - 3\Gamma_{112\dot{A}})C^{(3)},\\
0 &= (\partial_{2\dot{A}} + 3\Gamma_{221\dot{A}})C^{(3)}.
\end{aligned}
$$

Therefore, the solution of (3.114) is given by

$$
M_{(k)} = \text{const. } [C^{(3)}]^{-k/3}. \tag{3.115}
$$

Thus, a type-D solution of the Einstein vacuum field equations (such as the Schwarzschild or the Kerr metric) admits a $D(k,0)$ Killing spinor, for $k = 1, 2, \ldots$. (According to (3.115), up to a constant factor, all these Killing spinors are symmetrized products of the $D(1,0)$ Killing spinor with itself.)

Similarly, one can give the solution of (3.114) if the space-time metric satisfies the Einstein–Maxwell equations with an algebraically general electromagnetic field such that each principal spinor of f_{AB} satisfies (3.52) (as in the case of the Reissner–Nordström and the Kerr–Newman solutions). With these assumptions, the Einstein equations (3.49) imply that if l^A is a principal spinor of f_{AB}, then $l^A l^B C_{AB\dot{C}\dot{D}} = 0$, and by Proposition 3.4, l^A is a repeated DP spinor, which implies that the conformal curvature is of type D. With respect to a spinor frame such that f_{12} is the only nonvanishing component of f_{AB}, we have $\Gamma_{111\dot{A}} = 0 = \Gamma_{222\dot{A}}$, and the source-free Maxwell equations [(3.54) with $s = 1$] give

$$
\begin{aligned}
0 &= (\partial_{1\dot{A}} - 2\Gamma_{112\dot{A}})\phi_{(1)},\\
0 &= (\partial_{2\dot{A}} + 2\Gamma_{221\dot{A}})\phi_{(1)}.
\end{aligned}
$$

Thus, the solution of (3.114) is

$$
M_{(k)} = \text{const. } (\phi_{(1)})^{-k/2} = \text{const. } (f_{12})^{-k/2}.
$$

When C_{ABCD} is of type N, assuming that $M_{(2k)}$ is the only nonvanishing component of $M_{AB\ldots D}$, equations (3.113) give

$$
\begin{aligned}
0 &= \Gamma_{111\dot{A}},\\
0 &= (\partial_{1\dot{A}} + 2k\Gamma_{121\dot{A}} + 2k\Gamma_{112\dot{A}})M_{(2k)},\\
0 &= (\partial_{2\dot{A}} + 2k\Gamma_{122\dot{A}})M_{(2k)}.
\end{aligned} \tag{3.116}
$$

The system of equations (3.116) is integrable for $k = 1/2, 1, 3/2, 2, \ldots$ if and only if it is integrable for $k = 1/2$; in fact, if we denote by Q the solution of equations

(3.116) for $k = 1/2$, that is,

$$(\partial_{1\dot{A}} + \Gamma_{121\dot{A}} + \Gamma_{112\dot{A}})Q = 0, \qquad (\partial_{2\dot{A}} + \Gamma_{122\dot{A}})Q = 0,$$

then the solution of equations (3.116) for any other value of k is $M_{(2k)} = Q^{2k}$, which means that in the present case, any $D(k,0)$ Killing spinor $M_{AB...L}$ is of the form $M_{AB...L} = M_A M_B \cdots M_L$, where M_A is a $D(1/2,0)$ Killing spinor.

As pointed out at the beginning of this section, the existence of a $D(1,0)$ Killing spinor allows us to construct a quadratic integral of the *null* geodesic equations. Indeed, if $\alpha_A \overline{\alpha}_{\dot{A}}$ is the spinor equivalent of the tangent vector of an affinely parameterized null geodesic, that is, $\alpha^A \overline{\alpha}^{\dot{A}} \nabla_{A\dot{A}} (\alpha_B \overline{\alpha}_{\dot{B}}) = 0$, then from (3.110) it follows that if M_{AB} is a $D(1,0)$ Killing spinor, the directional derivative of the real-valued function $M_{AB} \overline{M}_{\dot{A}\dot{B}} \alpha^A \overline{\alpha}^{\dot{A}} \alpha^B \overline{\alpha}^{\dot{B}}$ along the geodesic is equal to zero. If M_{AB} satisfies the additional condition $\nabla^B{}_{\dot{C}} M_{AB} + \nabla_A{}^{\dot{B}} \overline{M}_{\dot{C}\dot{B}} = 0$, one can find a quadratic constant of motion for any geodesic (see Exercise 3.14).

According to the definition (3.110), $M_{AB...D}$ is a $D(k,0)$ Killing spinor if and only if there exists $K_{AB...D\dot{L}} = K_{(AB...D)\dot{L}}$ such that

$$\nabla_A{}^{\dot{A}} M_{BC...L} = \frac{2k}{2k+1} \varepsilon_{A(B} K_{C...L)}{}^{\dot{A}}. \tag{3.117}$$

Then

$$K_{AB...L}{}^{\dot{A}} = \nabla^{S\dot{A}} M_{SAB...L}, \tag{3.118}$$

and making use of (2.117) and (3.117), one finds that

$$\begin{aligned}
\nabla_{(A}{}^{(\dot{A}} K_{B...L)}{}^{\dot{B})} &= \nabla_{(A}{}^{(\dot{A}} \nabla^{|S|\dot{B})} M_{B...L)S} \\
&= \nabla^{S(\dot{B}} \nabla_{(A}{}^{\dot{A})} M_{B...L)S} - \Box^{\dot{A}\dot{B}} M_{AB...L} \\
&= \frac{1}{2k+1} \nabla_{(A}{}^{(\dot{B}} K_{B...L)}{}^{\dot{A})} - \Box^{\dot{A}\dot{B}} M_{AB...L};
\end{aligned}$$

hence,

$$\nabla_{(A}{}^{(\dot{A}} K_{B...L)}{}^{\dot{B})} = -(2k+1) C_{(A}{}^{S\dot{A}\dot{B}} M_{B...L)S}, \tag{3.119}$$

where we have made use of the Ricci identities (2.127).

D(1, 0) Killing Spinors. Symmetry Operators

According to the definitions given above, a $D(1,0)$ Killing spinor M_{AB} is a symmetric two-index spinor field such that

$$\nabla_A{}^{\dot{A}} M_{BC} = \tfrac{1}{3}(\varepsilon_{AB} K_C{}^{\dot{A}} + \varepsilon_{AC} K_B{}^{\dot{A}}), \tag{3.120}$$

with $K_A{}^{\dot{A}} = \nabla^{S\dot{A}} M_{SA}$ [see (3.117) and (3.118)]. Then, making use of (2.117) and (3.120), we obtain

$$
\begin{aligned}
\nabla^{B\dot{A}} K^{A}{}_{\dot{A}} &= \nabla^{B\dot{A}} \nabla^{S}{}_{\dot{A}} M_{S}{}^{A} \\
&= \nabla^{S}{}_{\dot{A}} \nabla^{B\dot{A}} M_{S}{}^{A} + 2\square^{BS} M_{S}{}^{A} \\
&= \tfrac{1}{3} \nabla^{S}{}_{\dot{A}} (\varepsilon^{B}{}_{S} K^{A\dot{A}} + \varepsilon^{BA} K_{S}{}^{\dot{A}}) + 2\square^{BS} M_{S}{}^{A} \\
&= \tfrac{1}{3} \nabla^{B}{}_{\dot{A}} K^{A\dot{A}} - \tfrac{1}{3} \varepsilon^{BA} \nabla^{S\dot{A}} K_{S\dot{A}} + 2\square^{BS} M_{S}{}^{A}.
\end{aligned}
$$

Thus, by virtue of (2.126),

$$
\tfrac{4}{3} \nabla^{B\dot{A}} K^{A}{}_{\dot{A}} = -\tfrac{1}{3} \varepsilon^{BA} \nabla^{S\dot{A}} K_{S\dot{A}} + 2C^{BASC} M_{SC} - \tfrac{1}{3} R M^{BA}.
$$

Contracting the last equation with ε_{BA}, one finds that

$$
\nabla^{A\dot{A}} K_{A\dot{A}} = 0, \tag{3.121}
$$

and therefore

$$
\tfrac{4}{3} \nabla^{B\dot{A}} K^{A}{}_{\dot{A}} = 2C^{BASC} M_{SC} - \tfrac{1}{3} R M^{BA}. \tag{3.122}
$$

On the other hand, from (3.119) and (3.121) it follows that $K_{A\dot{A}}$ is the spinor equivalent of a (possibly complex) Killing vector if and only if $C_{(A}{}^{S\dot{A}\dot{B}} M_{B)S} = 0$. Thus, in particular, $K_{A\dot{A}}$ corresponds to a possibly complex Killing vector if the space-time metric is a type-D solution of the Einstein vacuum field equations or of the Einstein–Maxwell equations with an algebraically general electromagnetic field such that each principal spinor of f_{AB} satisfies (3.52), since, as shown above, in this case M_{AB} is proportional to f_{AB}.

Proposition 3.7. *Let M_{AB} be a $D(1,0)$ Killing spinor and $K_{A\dot{A}} \equiv \nabla^{B}{}_{\dot{A}} M_{BA}$. Then for an arbitrary spinor field Φ_A,*

$$
\nabla^{C\dot{A}} (M^{AB} \nabla_{B\dot{A}} - \tfrac{2}{3} K^{A}{}_{\dot{A}}) \Phi_{A} = (M^{CB} \nabla_{B\dot{A}} - \tfrac{1}{3} K^{C}{}_{\dot{A}}) \nabla^{A\dot{A}} \Phi_{A}. \tag{3.123}
$$

Proof. We begin by noticing that, making use of (1.27), (2.117), and (2.122),

$$
\begin{aligned}
M^{AB} \nabla^{C\dot{A}} \nabla_{B\dot{A}} \Phi_{A} &= M^{CB} \nabla^{A\dot{A}} \nabla_{B\dot{A}} \Phi_{A} + \varepsilon^{AC} M^{SB} \nabla_{S}{}^{\dot{A}} \nabla_{B\dot{A}} \Phi_{A} \\
&= M^{CB} \nabla_{B\dot{A}} \nabla^{A\dot{A}} \Phi_{A} + 2M^{CB} \square^{A}{}_{B} \Phi_{A} + M^{SB} \square_{SB} \Phi^{C} \\
&= M^{CB} \nabla_{B\dot{A}} \nabla^{A\dot{A}} \Phi_{A} - \tfrac{1}{6} R M^{CB} \Phi_{B} + M^{SB} C^{RC}{}_{SB} \Phi_{R}.
\end{aligned}
$$

Hence, with the aid of (3.120), (3.122), and (1.27), we obtain

$$
\begin{aligned}
\nabla^{C\dot{A}} (M^{AB} \nabla_{B\dot{A}} - \tfrac{2}{3} K^{A}{}_{\dot{A}}) \Phi_{A} &= M^{AB} \nabla^{C\dot{A}} \nabla_{B\dot{A}} \Phi_{A} + (\nabla^{C\dot{A}} M^{AB}) \nabla_{B\dot{A}} \Phi_{A} - \tfrac{2}{3} K^{A}{}_{\dot{A}} \nabla^{C\dot{A}} \Phi_{A} \\
&\quad - \tfrac{2}{3} (\nabla^{C\dot{A}} K^{A}{}_{\dot{A}}) \Phi_{A} \\
&= M^{CB} \nabla_{B\dot{A}} \nabla^{A\dot{A}} \Phi_{A} - \tfrac{1}{6} R M^{CB} \Phi_{B} + M^{SB} C^{RC}{}_{SB} \Phi_{R} \\
&\quad + \tfrac{1}{3} (\varepsilon^{CA} K^{B\dot{A}} + \varepsilon^{CB} K^{A\dot{A}}) \nabla_{B\dot{A}} \Phi_{A} - \tfrac{2}{3} K^{A}{}_{\dot{A}} \nabla^{C\dot{A}} \Phi_{A} \\
&\quad - C^{CASB} M_{SB} \Phi_{A} + \tfrac{1}{6} R M^{CA} \Phi_{A} \\
&= (M^{CB} \nabla_{B\dot{A}} - \tfrac{1}{3} K^{C}{}_{\dot{A}}) \nabla^{A\dot{A}} \Phi_{A}.
\end{aligned}
$$

□

Thus, if the spinor field Φ_A satisfies the Weyl equation $\nabla^{A\dot{A}}\Phi_A = 0$ and M_{AB} is a $D(1,0)$ Killing spinor, then

$$\chi_{\dot{A}} \equiv (M^{AB}\nabla_{B\dot{A}} - \tfrac{2}{3}K^A{}_{\dot{A}})\Phi_A$$

satisfies the Weyl equation $\nabla^{C\dot{A}}\chi_{\dot{A}} = 0$ (see also Kamran and McLenaghan 1984a,b). A similar result holds in the case of the source-free Maxwell equations; if f_{AB} is a solution of the source-free Maxwell equations $\nabla^{A\dot{A}} f_{AB} = 0$, then

$$f_{\dot{R}\dot{S}} = \nabla^A{}_{(\dot{R}}\left[(M_{|AB}M_{CD|}\nabla^C{}_{\dot{S})} + \tfrac{2}{3}M_{|AB}K_{D|\dot{S})} + \tfrac{1}{3}M_{|BD}K_{A|\dot{S})})f^{BD}\right]$$

satisfies the source-free Maxwell equations $\nabla^{R\dot{R}} f_{\dot{R}\dot{S}} = 0$ (Torres del Castillo 1985). In fact, $f_{\dot{R}\dot{S}} = \nabla^A{}_{(\dot{R}}\Phi_{|A|\dot{S})}$, with

$$\Phi_{A\dot{S}} = \left(M_{AB}M_{CD}\nabla^C{}_{\dot{S}} + \tfrac{2}{3}M_{AB}K_{D\dot{S}} + \tfrac{1}{3}M_{BD}K_{A\dot{S}}\right) f^{BD},$$

and making use of (3.120), (3.122), (2.64), and the result of Exercise 3.2, one finds that $\Phi_{A\dot{S}}$ satisfies the self-duality condition (3.10).

Equation (3.123) can be written in a more symmetric form involving M_{AB} and its complex conjugate; indeed, making use of (3.123), we obtain

$$\begin{aligned}
&\nabla^{C\dot{A}}(M^{AB}\nabla_{B\dot{A}} - \tfrac{2}{3}K^A{}_{\dot{A}} + \overline{M}_{\dot{A}}{}^{\dot{B}}\nabla^A{}_{\dot{B}})\Phi_A \\
&\qquad = (M^{CB}\nabla_{B\dot{A}} - \tfrac{1}{3}K^C{}_{\dot{A}} + \overline{M}_{\dot{A}}{}^{\dot{B}}\nabla^C{}_{\dot{B}})\nabla^{A\dot{A}}\Phi_A + (\nabla^{C\dot{A}}\overline{M}_{\dot{A}}{}^{\dot{B}})\nabla^A{}_{\dot{B}}\Phi_A \\
&\qquad = (M^{CB}\nabla_{B\dot{A}} - \tfrac{1}{3}K^C{}_{\dot{A}} - \overline{K}^C{}_{\dot{A}} + \overline{M}_{\dot{A}}{}^{\dot{B}}\nabla^C{}_{\dot{B}})\nabla^{A\dot{A}}\Phi_A,
\end{aligned}$$

with $\overline{M}_{\dot{A}\dot{B}} \equiv \overline{M_{AB}}$ and $\overline{K}_{A\dot{B}} = \nabla_A{}^{\dot{S}}\overline{M}_{\dot{B}\dot{S}}$. Hence, if

$$\overline{K}_{A\dot{B}} = -K_{A\dot{B}}, \tag{3.124}$$

we have

$$\begin{aligned}
&\nabla^{C\dot{A}}(M^{AB}\nabla_{B\dot{A}} - \tfrac{2}{3}K^A{}_{\dot{A}} + \overline{M}_{\dot{A}}{}^{\dot{B}}\nabla^A{}_{\dot{B}})\Phi_A \\
&\qquad = (M^{CB}\nabla_{B\dot{A}} - \tfrac{2}{3}\overline{K}^C{}_{\dot{A}} + \overline{M}_{\dot{A}}{}^{\dot{B}}\nabla^C{}_{\dot{B}})\nabla^{A\dot{A}}\Phi_A.
\end{aligned} \tag{3.125}$$

Making use of (3.120) and its complex conjugate, one can verify that condition (3.124) holds if and only if $M_{AB}\varepsilon_{\dot{A}\dot{B}} + \overline{M}_{\dot{A}\dot{B}}\varepsilon_{AB}$ is the spinor equivalent of a two-index Killing–Yano tensor Y_{ab}, which is an antisymmetric two-index tensor satisfying

$$\nabla_a Y_{bc} + \nabla_b Y_{ac} = 0 \tag{3.126}$$

(see (3.127) below). Note that if M_{AB} is a $D(1,0)$ Killing spinor, then for any complex constant λ, λM_{AB} is also a $D(1,0)$ Killing spinor, but when $K_{A\dot{A}}$ is different from zero, M_{AB} and λM_{AB} satisfy condition (3.124) if and only if λ is real. On the other hand, if M_{AB} is a $D(1,0)$ Killing spinor such that $K_{A\dot{A}}$ is equal to zero, then for any complex constant λ, $\lambda M_{AB}\varepsilon_{\dot{A}\dot{B}} + \overline{\lambda}\overline{M}_{\dot{A}\dot{B}}\varepsilon_{AB}$ will be the spinor equivalent of a two-index Killing–Yano tensor, which implies (taking λ real or pure imaginary) that the tensor equivalent of $M_{AB}\varepsilon_{\dot{A}\dot{B}} + \overline{M}_{\dot{A}\dot{B}}\varepsilon_{AB}$ and its dual are two linearly independent Killing–Yano tensors. (Cf. Taxiarchis 1985 and the references cited therein.)

By virtue of (3.120), (3.124), (1.27), and (1.64), the spinor equivalent of $\nabla_a Y_{bc}$ is

$$\begin{aligned}
&\nabla_{A\dot{A}}(M_{BC}\varepsilon_{\dot{B}\dot{C}} + \overline{M}_{\dot{B}\dot{C}}\varepsilon_{BC}) \\
&\quad = \tfrac{1}{3}\left[(\varepsilon_{AB}K_{C\dot{A}} + \varepsilon_{AC}K_{B\dot{A}})\varepsilon_{\dot{B}\dot{C}} - \varepsilon_{BC}(\varepsilon_{\dot{A}\dot{B}}K_{A\dot{C}} + \varepsilon_{\dot{A}\dot{C}}K_{A\dot{B}})\right] \\
&\quad = \tfrac{1}{3}\left[(2\varepsilon_{AC}K_{B\dot{A}} + \varepsilon_{BC}\varepsilon_{A}{}^{R}K_{R\dot{A}})\varepsilon_{\dot{B}\dot{C}} - \varepsilon_{BC}(2\varepsilon_{\dot{A}\dot{C}}K_{A\dot{B}} + \varepsilon_{\dot{B}\dot{C}}\varepsilon_{\dot{A}}{}^{\dot{R}}K_{A\dot{R}})\right] \\
&\quad = \tfrac{2}{3}(\varepsilon_{AC}\varepsilon_{BD}\varepsilon_{\dot{A}\dot{D}}\varepsilon_{\dot{B}\dot{C}} - \varepsilon_{AD}\varepsilon_{BC}\varepsilon_{\dot{A}\dot{C}}\varepsilon_{\dot{B}\dot{D}})K^{D\dot{D}} \\
&\quad = -\tfrac{2}{3}\mathrm{i}\,\varepsilon_{A\dot{A}B\dot{B}C\dot{C}D\dot{D}}K^{D\dot{D}},
\end{aligned}$$

that is,

$$\nabla_a Y_{bc} = \tfrac{2}{3}\mathrm{i}\,\varepsilon_{abcd}K^{d}, \tag{3.127}$$

where K_a is the vector equivalent of $K_{A\dot{A}}$, which implies the validity of (3.126). (Note that condition (3.124) means that K_a is pure imaginary, and therefore the right-hand side of (3.127) is real.)

If we now consider a pair of spinor fields ψ_A, $\phi_{\dot{A}}$ satisfying the Dirac equation $\sqrt{2}\hbar\nabla_{A\dot{A}}\psi^{A} = m_0 c\,\phi_{\dot{A}}$, $\sqrt{2}\hbar\nabla_{A\dot{A}}\phi^{\dot{A}} = m_0 c\,\psi_A$, the identity (3.125) and its complex conjugate imply that the spinor fields

$$\begin{aligned}
\widetilde{\psi}_A &\equiv -(\overline{M}^{\dot{A}\dot{B}}\nabla_{A\dot{B}} - \tfrac{2}{3}\overline{K}_{A}{}^{\dot{A}} + M_{A}{}^{B}\nabla_{B}{}^{\dot{A}})\phi_{\dot{A}}, \\
\widetilde{\phi}_{\dot{A}} &\equiv (M^{AB}\nabla_{B\dot{A}} - \tfrac{2}{3}K^{A}{}_{\dot{A}} + \overline{M}_{\dot{A}}{}^{\dot{B}}\nabla^{A}{}_{\dot{B}})\psi_A,
\end{aligned} \tag{3.128}$$

also satisfy the Dirac equation (see also Exercise 3.13).

In the case of a type-D space-time that admits a $D(1,0)$ Killing spinor (e.g., any type-D solution of the Einstein vacuum field equations), according to (3.114), with respect to a frame such that only C_{1122} is different from zero, the components of the vector field $K_{A\dot{A}}$, defined by (3.120), can be explicitly expressed in the form

$$\begin{aligned}
K_{1\dot{1}} &= -3\Gamma_{112\dot{1}}M_{12}, & K_{1\dot{2}} &= -3\Gamma_{112\dot{2}}M_{12}, \\
K_{2\dot{1}} &= -3\Gamma_{221\dot{1}}M_{12}, & K_{2\dot{2}} &= -3\Gamma_{221\dot{2}}M_{12}.
\end{aligned} \tag{3.129}$$

Hence, with respect to such a frame, expressions (3.128) can be written explicitly in the form

$$\begin{aligned}
\widetilde{\psi}_1 &= 2\overline{M}_{\dot{1}\dot{2}}(\partial_{1\dot{2}} - \Gamma_{\dot{1}\dot{2}\dot{2}1} + \Gamma_{\dot{2}\dot{2}\dot{1}1})\phi_{\dot{1}} + (M_{12} + \overline{M}_{\dot{1}\dot{2}})\nabla_{1}{}^{\dot{A}}\phi_{\dot{A}}, \\
\widetilde{\psi}_2 &= 2\overline{M}_{\dot{1}\dot{2}}(\partial_{2\dot{1}} + \Gamma_{\dot{1}\dot{2}\dot{1}2} - \Gamma_{\dot{1}\dot{1}\dot{2}2})\phi_{\dot{2}} - (M_{12} + \overline{M}_{\dot{1}\dot{2}})\nabla_{2}{}^{\dot{A}}\phi_{\dot{A}}, \\
\widetilde{\phi}_{\dot{1}} &= -2M_{12}(\partial_{2\dot{1}} - \Gamma_{122\dot{1}} + \Gamma_{221\dot{1}})\psi_1 - (M_{12} + \overline{M}_{\dot{1}\dot{2}})\nabla^{A}{}_{\dot{1}}\psi_A, \\
\widetilde{\phi}_{\dot{2}} &= -2M_{12}(\partial_{1\dot{2}} + \Gamma_{121\dot{2}} - \Gamma_{112\dot{2}})\psi_2 + (M_{12} + \overline{M}_{\dot{1}\dot{2}})\nabla^{A}{}_{\dot{2}}\psi_A.
\end{aligned} \tag{3.130}$$

In the case of the Schwarzschild metric, (3.115) and (2.112) imply that M_{12} is proportional to r, and making use of (2.109) and (3.129) one finds that (3.124) is satisfied if we take, for instance,

$$M_{12} = \frac{\mathrm{i}r}{\sqrt{2}}.$$

Then, making use of (2.108), equations (3.130) reduce to

$$\begin{aligned}
\widetilde{\psi}_1 &= -\mathrm{i}\left(\frac{\partial}{\partial\theta} + \frac{\mathrm{i}}{\sin\theta}\frac{\partial}{\partial\varphi} + \frac{1}{2}\cot\theta\right)\phi_{\dot{1}}, \\
\widetilde{\psi}_2 &= -\mathrm{i}\left(\frac{\partial}{\partial\theta} - \frac{\mathrm{i}}{\sin\theta}\frac{\partial}{\partial\varphi} + \frac{1}{2}\cot\theta\right)\phi_{\dot{2}}, \\
\widetilde{\phi}_{\dot{1}} &= -\mathrm{i}\left(\frac{\partial}{\partial\theta} - \frac{\mathrm{i}}{\sin\theta}\frac{\partial}{\partial\varphi} + \frac{1}{2}\cot\theta\right)\psi_1, \\
\widetilde{\phi}_{\dot{2}} &= -\mathrm{i}\left(\frac{\partial}{\partial\theta} + \frac{\mathrm{i}}{\sin\theta}\frac{\partial}{\partial\varphi} + \frac{1}{2}\cot\theta\right)\psi_2.
\end{aligned} \tag{3.131}$$

The eigenvalues of the operator that sends ψ_A, $\phi_{\dot{A}}$ into $\widetilde{\psi}_A$, $\widetilde{\phi}_{\dot{A}}$, given by (3.131), are of the form $\pm(j+\frac{1}{2})$, where j is a half-integer [see (3.37)]. It may be noticed that this operator does not involve the parameter r_g, contained in the Schwarzschild metric; therefore, also in the Minkowski space-time this operator maps any solution of the Dirac equation into another solution. In the flat space-time limit, this operator is essentially that given in the standard notation as $\beta(\sigma\cdot\mathbf{L}+\hbar)$ (see, e.g., Messiah 1962, Davydov 1988).

Exercises

3.1. Show that if $t_{A\dot{A}}$ is the spinor equivalent of a future-pointing timelike vector, then $\langle\alpha,\beta\rangle \equiv t^{A\dot{A}}\beta_A\overline{\alpha}_{\dot{A}}$ is a positive definite Hermitian inner product for the undotted one-index spinors.

3.2. Show that if f_{AB} satisfies the source-free Maxwell equations, then $\nabla^{C\dot{C}}\nabla_{C\dot{C}}f_{AB} = 2C_{ABCD}f^{CD} - \frac{1}{3}Rf_{AB}$.

3.3. Show that at each point where the electromagnetic field is algebraically general, the minimum value of $T_{ab}t^a t^b$, with the constraint $t^a t_a = -1$, is

$$\frac{1}{16\pi}[(F^{ab}F_{ab})^2 + ({}^*F^{ab}F_{ab})^2]^{1/2}.$$

(Physically, $T_{ab}t^a t^b$ is the energy density of the electromagnetic field measured by an observer with four-velocity t^a.)

3.4. Let F_{ab} be a bivector. Prove that there exists a nonzero vector l^a such that $F_{ab}l^b = 0 = {}^*F_{ab}l^b$ if and only if F_{ab} is algebraically special.

3.5. Let T_{ab} be the energy–momentum tensor of the electromagnetic field. Prove that there exists a nonzero vector l^a such that $T_{ab}l^a = 0$ if and only if T_{ab} corresponds to an algebraically special electromagnetic field F_{ab} and l^a points along the repeated principal null direction of F_{ab}.

3.6. A set of equations for a zero-rest-mass spin-3/2 field is given by

$$\nabla_{A\dot{A}}\psi^A{}_{B\dot{B}} = \nabla_{B\dot{B}}\psi^A{}_{A\dot{A}}.$$

Show that these equations can be also expressed in the form $H_{A\dot{B}\dot{C}} = 0$, $H_{ABC} = H_{(ABC)}$, where

$$H_{A\dot{B}\dot{C}} \equiv \nabla^D{}_{(\dot{B}}\psi_{|AD|\dot{C})}, \qquad H_{ABC} \equiv \nabla_{(B}{}^{\dot{D}}\psi_{|A|C)\dot{D}},$$

and that if the Ricci tensor vanishes, H_{ABC} satisfies

$$\nabla^A{}_{\dot{A}}H_{ABC} = C_{ABC}{}^D\psi^A{}_{D\dot{A}}$$

[cf. (3.41)]. Show that if the Ricci tensor is equal to zero, there are no algebraic consistency conditions of the Buchdahl–Plebański type for these equations.

3.7. Using the spinor formalism, show that if the spinor equivalent of a real, traceless, totally symmetric four-index tensor T_{abcd} satisfies the condition (3.51), then T_{abcd} can be expressed as $T_{abcd} = \frac{1}{4}(C_{arbs}C_c{}^r{}_d{}^s + {}^*C_{arbs}{}^*C_c{}^r{}_d{}^s)$ with

$$C_{abcd} = 2\frac{T_{acgh}\phi_b{}^g{}_d{}^h - T_{bcgh}\phi_a{}^g{}_d{}^h + T_{bdgh}\phi_a{}^g{}_c{}^h - T_{adgh}\phi_b{}^g{}_c{}^h}{[T^{abcd}(\phi_{arbs}\phi_c{}^r{}_d{}^s + {}^*\phi_{arbs}{}^*\phi_c{}^r{}_d{}^s)]^{1/2}},$$

where ϕ_{abcd} is any real four-index tensor with the symmetries of the Weyl tensor such that $T^{abcd}(\phi_{arbs}\phi_c{}^r{}_d{}^s + {}^*\phi_{arbs}{}^*\phi_c{}^r{}_d{}^s)$ is different from zero [cf. (3.28)].

3.8. Assuming that the Einstein vacuum field equations are satisfied (i.e., $R_{ab} = 0$) and that K is a Killing vector, show that

$$\nabla_{[a}\left(2f\nabla_b K_{c]} + \omega\,\varepsilon_{bc]de}\nabla^d K^e\right) = 0,$$

where $f = -K^a K_a$ and ω is the twist potential of K; that is, show that

$$\nabla^{A\dot{A}}\left(-\mathrm{i}\bar{\chi}L_{AB}\varepsilon_{\dot{A}\dot{B}} + \mathrm{i}\chi L_{\dot{A}\dot{B}}\varepsilon_{AB}\right) = 0,$$

where χ is the Ernst potential (3.98). This implies the local existence of a vector field β_a such that

$$2f\nabla_{[a}K_{b]} + \omega\,\varepsilon_{abcd}\nabla^c K^d = -\nabla_{[a}\beta_{b]}.$$

Similarly, show that

$$\nabla_{[a}\left(\varepsilon_{bc]de}\nabla^{d}K^{e}\right)=0,$$

that is, $\nabla^{A\dot{A}}(L_{AB}\varepsilon_{\dot{A}\dot{B}}+L_{\dot{A}\dot{B}}\varepsilon_{AB})=0$, which implies the local existence of a vector field α_a such that

$$\varepsilon_{abcd}\nabla^{c}K^{d}=2\nabla_{[a}\alpha_{b]}.$$

The vector fields α_a and β_a can be used to construct a one-parameter family of solutions of the Einstein vacuum field equations (for details see Geroch 1971).

3.9. Show that the spinor field (3.103) satisfies $K_{(B}{}^{\dot{C}}\nabla^{A}{}_{|\dot{C}}f_{A|C)}=0$, for any differentiable function Φ.

3.10. Show that if M_{AB} is a $D(1,0)$ Killing spinor and Q_{ab} is the tensor equivalent of $M_{AB}\varepsilon_{\dot{A}\dot{B}}+\overline{M}_{\dot{A}\dot{B}}\varepsilon_{AB}$, then

$$\nabla_{a}Q_{bc}=-\tfrac{1}{3}\varepsilon_{abcd}\nabla_{s}{}^{*}Q^{sd}+\tfrac{1}{3}g_{ab}\nabla^{s}Q_{sc}-\tfrac{1}{3}g_{ac}\nabla^{s}Q_{sb},$$

and therefore $\nabla_{(a}Q_{b)c}=\frac{1}{3}g_{ab}\nabla^{s}Q_{sc}-\frac{1}{3}g_{c(a}\nabla^{s}Q_{|s|b)}$. Show that Q_{ab} is a Killing–Yano tensor if $\nabla^{a}Q_{ab}=0$.

3.11. Show that if $\Phi_{AB\ldots L}$ is a totally symmetric spinor field that satisfies the spin-s zero-rest-mass field equations and $M_{AB\ldots D}$ is a $D(k,0)$ Killing spinor, with $s>k$, then $\Phi_{AB\ldots D\ldots L}M^{AB\ldots D}$ also satisfies the zero-rest-mass field equations.

3.12. Show that if M_A is a $D(1/2,0)$ Killing spinor, then

$$\nabla_{A\dot{A}}K_{\dot{B}}=-\tfrac{1}{12}R\,\varepsilon_{\dot{A}\dot{B}}M_{A}-2C^{D}{}_{A\dot{A}\dot{B}}M_{D},$$

with $K_{\dot{B}}\equiv\nabla^{A}{}_{\dot{B}}M_{A}$. Show that if $\Phi_{\dot{A}}$ satisfies the Weyl equation, then

$$f_{\dot{A}\dot{B}}\equiv M^{C}\nabla_{C\dot{A}}\Phi_{\dot{B}}-\Phi_{(\dot{A}}\nabla^{C}{}_{\dot{B})}M_{C}$$

satisfies the source-free Maxwell equations.

3.13. Show that the relation between the bispinors

$$\Psi=\begin{pmatrix}\psi_{A}\\ \phi^{\dot{A}}\end{pmatrix}\qquad\text{and}\qquad\widetilde{\Psi}=\begin{pmatrix}\widetilde{\psi}_{A}\\ \widetilde{\phi}^{\dot{A}}\end{pmatrix},$$

given by (3.128), can be expressed in the form

$$\widetilde{\Psi}=\frac{1}{\sqrt{2}}(Y^{ab}\gamma_{5}\gamma_{a}\nabla_{b}\Psi+\tfrac{2}{3}K^{a}\gamma_{a}\Psi).$$

3.14. Show that if Y_{ab} is a Killing–Yano tensor and v^a is the velocity vector of an affinely parameterized geodesic, then $Y_{ab}Y_{c}{}^{b}v^{a}v^{c}$ is constant along the geodesic (this amounts to saying that $K_{ac}\equiv Y_{ab}Y_{c}{}^{b}$ is a two-index Killing tensor, $\nabla_{(a}K_{bc)}=0$). Show that $Y_{ab}v^{b}$ is covariantly constant along the geodesic.

3.15. Show that if K is a Killing vector, the Lie derivative of a bispinor Ψ with respect to K is given by

$$\pounds_K \Psi = K^a \nabla_a \Psi + \tfrac{1}{4} (\nabla_a K_b)\, \gamma^a \gamma^b \Psi$$

with

$$\Psi = \begin{pmatrix} \psi_A \\ \phi^{\dot{A}} \end{pmatrix}.$$

Chapter 4
Further Applications

In this chapter some further examples of the applications of the two-component spinor formalism are given. In contrast to the preceding chapter, manifolds with Euclidean or ultrahyperbolic signature are also considered here. In Section 4.1 the self-dual Yang–Mills fields in a possibly curved manifold are considered. In Section 4.2, a reduced expression for the metric of a Riemannian manifold with one algebraically special Weyl spinor (C_{ABCD} or $C_{\dot{A}\dot{B}\dot{C}\dot{D}}$) is derived; this class of manifolds includes the $\mathscr{H}$ spaces, which were originally studied by Newman (see, e.g., Newman 1976, Hansen et al. 1978) and Penrose (see, e.g., Penrose 1976, Penrose and Ward 1980, Ward and Wells 1990). Section 4.3 deals with Killing bispinors and eigenbispinors of the Dirac operator, which can be defined in Riemannian manifolds of any dimension or signature. The discussion is restricted to four-dimensional manifolds, which allows one to obtain stronger results than those derived by means of the formalism based on Clifford algebras, applicable to manifolds of arbitrary dimension. The sections of this chapter are independent of each other.

With the only exception of the null tetrad $\overset{\text{c}}{\partial}_{A\dot{B}}$ introduced in Section 4.2, the null tetrads employed in this chapter satisfy the condition (2.100) appropriate for the signature of the metric.

4.1 Self-Dual Yang–Mills Fields

A Yang–Mills field is locally represented by a one-index vector field (or one-form) A_a with values in the Lie algebra $\mathfrak{g}$ of some Lie group G. (More precisely, A_a is the local expression of a connection in a principal fiber bundle with structure group G (see, e.g., Ward and Wells 1990).) The field strength F_{ab} is then given by

$$F_{ab} = \nabla_a A_b - \nabla_b A_a + [A_a, A_b] \tag{4.1}$$

and satisfies the identities [which follow from (4.1)]

G.F.T. del Castillo, *Spinors in Four-Dimensional Spaces*, Progress in Mathematical Physics 59, DOI 10.1007/978-0-8176-4984-5_4,

$$\nabla_a{}^*F^{ab} + [A_a, {}^*F^{ab}] = 0, \tag{4.2}$$

where ${}^*F_{ab}$ denotes the dual of F_{ab} (${}^*F_{ab} = \frac{1}{2}\varepsilon_{abcd}F^{cd}$). The Yang–Mills equations are

$$\nabla_a F^{ab} + [A_a, F^{ab}] = 0. \tag{4.3}$$

The Yang–Mills equations are form-invariant under the gauge transformations

$$A_a \mapsto U^{-1}A_a U + U^{-1}\partial_a U, \tag{4.4}$$

where U is a function with values in the (gauge) group G. Under this transformation, the field strength transforms according to

$$F_{ab} \mapsto U^{-1}F_{ab}U.$$

The field strength F_{ab} vanishes if and only if there exists locally a function U with values in G such that $A_a = U^{-1}\partial_a U$.

By virtue of the identities (4.2), the Yang–Mills equations are trivially satisfied if ${}^*F_{ab} = \lambda F_{ab}$, for some constant factor λ. Since ${}^*({}^*F_{ab}) = (-1)^q F_{ab}$ [see (1.66)], it follows that $\lambda^2 = (-1)^q$; that is, ${}^*F_{ab}$ is proportional to F_{ab} only if the field is self-dual (${}^*F_{ab} = \mathrm{i}^q F_{ab}$) or anti-self-dual (${}^*F_{ab} = -(\mathrm{i}^q)F_{ab}$). However, when the metric has Lorentzian signature, a self-dual or anti-self-dual Yang–Mills field would satisfy ${}^*F_{ab} = \pm\mathrm{i}F_{ab}$, which cannot be fulfilled if A_a takes values in the (real) Lie algebra of G; but when the metric of M has Euclidean signature it is possible to have nontrivial self-dual or anti-self-dual Yang–Mills fields, because in that case, the field strength would satisfy the condition ${}^*F_{ab} = \pm F_{ab}$.

For instance, in the case of a self-dual Yang–Mills field we have $F_A{}^{\dot{R}}{}_{B\dot{R}} = 0$, where $F_{A\dot{A}B\dot{B}}$ is the spinor equivalent of F_{ab} [see (1.50) and (1.65)]; hence, making use of (4.1), one obtains the condition

$$\nabla_{(A}{}^{\dot{R}}A_{B)\dot{R}} + A_{(A}{}^{\dot{R}}A_{B)\dot{R}} = 0, \tag{4.5}$$

which involves only first derivatives of $A_{B\dot{B}}$ [cf. (3.10)]. As in the case of the abelian version of the self-duality conditions (4.5), given by (3.10), equations (4.5) can be partially integrated, reducing them to a single equation for a potential, if the curvature is equal to zero or if there exists a principal spinor l_A of the Weyl spinor C_{ABCD} satisfying the condition (3.52),

$$l^A l^B \nabla_{A\dot{C}} l_B = 0.$$

As shown in Section 3.3.1, in a manifold with Lorentzian signature, a one-index spinor satisfying this condition defines a shear-free congruence of null geodesics ($l_A \bar{l}_{\dot{A}}$ is the spinor equivalent of a vector field tangent to the congruence). However, when the signature of the metric is Euclidean, we cannot define a vector field with l_A alone. Nevertheless, the consequences of the existence of such a spinor field derived in Section 3.3.1 remain valid (e.g., Propositions 3.2–3.5).

The proof is almost identical to that given for Proposition 3.6. With respect to a null tetrad such that $l_A = \delta_A^2$, we have $\Gamma_{111\dot{A}} = 0$, and from (4.5), with $A = B = 1$, we obtain

$$0 = (\partial_1{}^{\dot{R}} - \Gamma_{121}{}^{\dot{R}} - \Gamma^{\dot{R}\dot{S}}{}_{\dot{S}1})A_{1\dot{R}} + A_1{}^{\dot{R}}A_{1\dot{R}} \tag{4.6}$$

[cf. (3.78)]. This last equation means that the components $A_{1\dot{R}}$ of the gauge field can be eliminated locally by a gauge transformation (4.4). That is, equation (4.6) is the integrability condition for the local existence of a function U with values in G such that $A_{1\dot{R}} = U^{-1}\partial_{1\dot{R}}U$. In fact, applying $\partial_1{}^{\dot{R}}$ to both sides of this last equation, we have

$$\begin{aligned}
\partial_1{}^{\dot{R}}A_{1\dot{R}} &= (\partial_1{}^{\dot{R}}U^{-1})\partial_{1\dot{R}}U + U^{-1}\partial_1{}^{\dot{R}}\partial_{1\dot{R}}U \\
&= -U^{-1}(\partial_1{}^{\dot{R}}U)U^{-1}\partial_{1\dot{R}}U + U^{-1}(-\Gamma^S{}_{11}{}^{\dot{R}}\partial_{S\dot{R}}U - \Gamma^{\dot{S}}{}_{\dot{R}}{}^{\dot{R}}{}_1\partial_{1\dot{S}}U) \\
&= -A_1{}^{\dot{R}}A_{1\dot{R}} + \Gamma_{121}{}^{\dot{R}}A_{1\dot{R}} + \Gamma^{\dot{S}\dot{R}}{}_{\dot{R}1}A_{1\dot{S}},
\end{aligned}$$

which coincides with (4.6).

Thus, with $A_{1\dot{R}}$ being equal to zero, taking $A = 1$, $B = 2$ in (4.5) we obtain the condition

$$0 = (\partial_1{}^{\dot{R}} + \Gamma_{121}{}^{\dot{R}} + \Gamma_{112}{}^{\dot{R}} - \Gamma^{\dot{R}\dot{S}}{}_{\dot{S}1})A_{2\dot{R}},$$

which has the form of (3.79); therefore, there exists locally a matrix-valued function χ such that

$$A_{2\dot{B}} = (\partial_{1\dot{B}} + 2\Gamma_{121\dot{B}} + \Gamma_{112\dot{B}})\chi \tag{4.7}$$

[see (3.80)].

Finally, taking $A = B = 2$ in (4.5) and substituting the expressions $A_{1\dot{R}} = 0$ and that for $A_{2\dot{R}}$ given by (4.7), one obtains the single nonlinear condition

$$\begin{aligned}
0 &= (\partial_2{}^{\dot{R}} + \Gamma_{122}{}^{\dot{R}} - \Gamma^{\dot{R}\dot{S}}{}_{\dot{S}2})A_{2\dot{R}} + A_2{}^{\dot{R}}A_{2\dot{R}}, \\
&= (\partial_2{}^{\dot{R}} + \Gamma_{122}{}^{\dot{R}} - \Gamma^{\dot{R}\dot{S}}{}_{\dot{S}2})(\partial_{1\dot{R}} + 2\Gamma_{121\dot{R}} + \Gamma_{112\dot{R}})\chi \\
&\quad + \left[(\partial_1{}^{\dot{R}} + 2\Gamma_{121}{}^{\dot{R}} + \Gamma_{112}{}^{\dot{R}})\chi\right](\partial_{1\dot{R}} + 2\Gamma_{121\dot{R}} + \Gamma_{112\dot{R}})\chi.
\end{aligned} \tag{4.8}$$

When M is flat, we can make use of the null tetrad $\partial_{A\dot{A}} = (1/\sqrt{2})\,\sigma^a{}_{A\dot{A}}\,\partial/\partial x^a$ induced by Cartesian coordinates x^a for which all the spin coefficients vanish. Then, equation (4.8) reduces to

$$0 = \frac{\partial}{\partial x^{2\dot{1}}}\frac{\partial \chi}{\partial x^{1\dot{2}}} - \frac{\partial}{\partial x^{2\dot{2}}}\frac{\partial \chi}{\partial x^{1\dot{1}}} + \left[\frac{\partial \chi}{\partial x^{1\dot{1}}}, \frac{\partial \chi}{\partial x^{1\dot{2}}}\right], \tag{4.9}$$

where $x^{A\dot{A}} \equiv (1/\sqrt{2})\,\sigma_a{}^{A\dot{A}}x^a$ [see (2.149)]. As pointed out above, if the metric of M has Lorentzian signature, there are no nontrivial self-dual Yang–Mills fields if A_a takes values in a real Lie algebra. However, one can have nontrivial self-dual fields by considering, for instance, $\mathrm{GL}(n,\mathbb{C})$ as the gauge group G. In that case, (4.9) corresponds to the equation for Newman's K-matrix (Newman 1978, Mason and Woodhouse 1996).

4.2 $\mathscr{H}$ and $\mathscr{H}\mathscr{H}$ Spaces

As shown in Section 3.3.1, in a manifold with Lorentzian signature, a one-index spinor field l_A satisfying the condition

$$l^A l^B \nabla_{A\dot{C}} l_B = 0 \tag{4.10}$$

defines a shear-free congruence of null geodesics. According to the Frobenius theorem, these equations are also the conditions for the complete integrability of the system of differential equations

$$l_A \theta^{A\dot{B}} = 0. \tag{4.11}$$

That is, equations (4.10) are necessary and sufficient for the local existence of two (possibly complex-valued) functions $q^{\dot{1}}$, $q^{\dot{2}}$ such that

$$l_A \theta^{A\dot{B}} = L^{\dot{B}}{}_{\dot{C}} \mathrm{d}q^{\dot{C}}, \tag{4.12}$$

where the $L^{\dot{B}}{}_{\dot{C}}$ are complex-valued functions whose values at each point form a nonsingular 2×2 matrix (see also Flaherty 1980). In fact, the system (4.11) is completely integrable if and only if

$$\mathrm{d}(l_A \theta^{A\dot{B}}) \wedge l_B \theta^{B\dot{1}} \wedge l_C \theta^{C\dot{2}} = 0.$$

Making use of (2.90), the Cartan first structure equations (2.15), (2.55), and Exercise 2.5, one can readily verify that this last condition amounts to (4.10).

For example, in the case of the Schwarzschild metric (2.106), the spin coefficients for the null tetrad (2.108) satisfy $\Gamma_{11\dot{1}\dot{A}} = 0$ [see (2.109)], and therefore, the one-index spinor field $l_A = \delta^2_A$ satisfies the condition (4.10). Making use of the explicit expressions (2.107), we find that the system of differential equations $l_A \theta^{A\dot{B}} = 0$ is equivalent to

$$\mathrm{d}\theta + \mathrm{i}\sin\theta\, \mathrm{d}\varphi = 0,$$
$$c\mathrm{d}t + \left(1 - \frac{r_g}{r}\right)^{-1} \mathrm{d}r = 0.$$

This system is, indeed, completely integrable (actually, in this particular case, each of these one-forms is integrable); one can readily see that

$$\mathrm{d}\theta + \mathrm{i}\sin\theta\, \mathrm{d}\varphi = 2\mathrm{e}^{-\mathrm{i}\varphi} \cos^2 \tfrac{1}{2}\theta\, \mathrm{d}\big(\mathrm{e}^{\mathrm{i}\varphi} \tan \tfrac{1}{2}\theta\big),$$
$$c\mathrm{d}t + \left(1 - \frac{r_g}{r}\right)^{-1} \mathrm{d}r = \mathrm{d}\big(ct + r + r_g \ln|r - r_g|\big).$$

Thus, the functions $q^{\dot{A}}$ appearing in (4.12) can be taken as $\mathrm{e}^{\mathrm{i}\varphi} \tan \frac{1}{2}\theta$ and $ct + r + r_g \ln|r - r_g|$, or any pair of functionally independent functions of them (see below).

Assuming that l_A is a multiple principal spinor of C_{ABCD}, with the aid of the functions $q^{\dot{A}}$ we construct a second null tetrad, $\overset{\mathrm{c}}{\theta}{}^{A\dot{B}}$, starting with

$$\overset{\mathrm{c}}{\theta}{}^{2\dot{A}} \equiv -\phi^{-2} \mathrm{d}q^{\dot{A}}, \tag{4.13}$$

where ϕ is defined by

$$l^B \nabla_{A\dot{A}} l_B = l_A l^B \partial_{B\dot{A}} \ln \phi \tag{4.14}$$

[see (3.73)]. Thus, from (4.12) we obtain

$$\overset{\mathrm{c}}{\theta}{}^{2\dot{A}} = -\phi^{-2} (L^{-1})^{\dot{A}}{}_{\dot{B}} l_A \theta^{A\dot{B}}, \tag{4.15}$$

where $(L^{-1})^{\dot{A}}{}_{\dot{B}}$ are the entries of the inverse of the matrix $(L^{\dot{A}}{}_{\dot{B}})$. Then, the one-forms $\overset{\mathrm{c}}{\theta}{}^{1\dot{A}}$ must be given by

$$\overset{\mathrm{c}}{\theta}{}^{1\dot{A}} = \phi^2 L_{\dot{B}}{}^{\dot{A}} m_C \theta^{C\dot{B}} + \eta\, \overset{\mathrm{c}}{\theta}{}^{2\dot{A}}, \tag{4.16}$$

where m_A is a one-index spinor field such that $m^A l_A = 1$ and η is an arbitrary function, so that $-\varepsilon_{AB}\varepsilon_{\dot{A}\dot{B}} \theta^{A\dot{A}} \theta^{B\dot{B}} = -\varepsilon_{AB}\varepsilon_{\dot{A}\dot{B}}\, \overset{\mathrm{c}}{\theta}{}^{A\dot{A}}\, \overset{\mathrm{c}}{\theta}{}^{B\dot{B}}$ [see (2.99)]. (Note that the spinor field m^A is not uniquely defined by the condition $m^A l_A = 1$; the function η is related to this ambiguity.) Thus, the vector fields $\overset{\mathrm{c}}{\partial}{}_{A\dot{B}}$ are

$$\begin{aligned} \overset{\mathrm{c}}{\partial}{}_{1\dot{A}} &= -\phi^{-2} (L^{-1})_{\dot{A}}{}^{\dot{B}} l^C \partial_{C\dot{B}}, \\ \overset{\mathrm{c}}{\partial}{}_{2\dot{A}} &= -\phi^2 L^{\dot{C}}{}_{\dot{A}} m^D \partial_{D\dot{C}} - \eta\, \overset{\mathrm{c}}{\partial}{}_{1\dot{A}} \,. \end{aligned} \tag{4.17}$$

A somewhat lengthy computation, making use of (1.106), (2.23), (4.10), the fact that $\mathrm{dd}q^{\dot{C}} = \mathrm{d}[(L^{-1})^{\dot{C}}{}_{\dot{B}} l_A \theta^{A\dot{B}}] = 0$, and the Cartan first structure equations (2.15), yields

$$\begin{aligned} [\overset{\mathrm{c}}{\partial}{}_{1\dot{1}}, \overset{\mathrm{c}}{\partial}{}_{1\dot{2}}] &= \overset{\mathrm{c}}{\partial}{}_1{}^{\dot{R}}\, \overset{\mathrm{c}}{\partial}{}_{1\dot{R}} \\ &= \phi^{-4} \left[\det(L^{\dot{R}}{}_{\dot{S}})\right]^{-1} \Big[-2l^C (\partial_C{}^{\dot{R}} \ln\phi) l^S \partial_{S\dot{R}} + \tfrac{3}{2} (\nabla^{R\dot{D}} l_R) l^S \partial_{S\dot{D}} \\ &\qquad - \tfrac{3}{2} l_R (\nabla^{R\dot{D}} l^S) \partial_{S\dot{D}} + \tfrac{1}{2} l^R (\nabla^{S\dot{D}} l_R) \partial_{S\dot{D}} \Big] \\ &= -2\phi^{-4} \left[\det(L^{\dot{R}}{}_{\dot{S}})\right]^{-1} \left(l^C \partial_C{}^{\dot{R}} \ln\phi - m^C l^B \nabla_C{}^{\dot{R}} l_B \right) l^S \partial_{S\dot{R}}, \end{aligned}$$

which is equal to zero as a consequence of (4.14). We can now introduce a second pair of (possibly complex-valued) functions, $p^{\dot{A}}$, defined by

$$\overset{\mathrm{c}}{\partial}{}_{1\dot{A}} p^{\dot{B}} = \delta^{\dot{B}}_{\dot{A}}. \tag{4.18}$$

The integrability conditions for these equations are trivially satisfied since the vector fields $\overset{\mathrm{c}}{\partial}_{1\dot{A}}$ commute. Then, the differential of $p^{\dot{A}}$ is [see (2.90)]

$$\mathrm{d}p^{\dot{A}} = -(\overset{\mathrm{c}}{\partial}_{C\dot{B}}p^{\dot{A}})\,\overset{\mathrm{c}}{\theta}{}^{C\dot{B}} = -\,\overset{\mathrm{c}}{\theta}{}^{1\dot{A}} - (\overset{\mathrm{c}}{\partial}_{2\dot{B}}p^{\dot{A}})\,\overset{\mathrm{c}}{\theta}{}^{2\dot{B}},$$

and using (4.13) we obtain

$$\overset{\mathrm{c}}{\theta}{}^{1\dot{A}} = -\mathrm{d}p^{\dot{A}} + \phi^{-2}(\overset{\mathrm{c}}{\partial}_{2\dot{B}}p^{\dot{A}})\,\mathrm{d}q^{\dot{B}}. \tag{4.19}$$

On the other hand, from (4.17) we have

$$\overset{\mathrm{c}}{\partial}_{2\dot{A}}p^{\dot{A}} = -\phi^2 L^{\dot{C}}{}_{\dot{A}}m^D\partial_{D\dot{C}}p^{\dot{A}} - 2\eta.$$

Therefore, choosing $\eta = -\frac{1}{2}\phi^2 L^{\dot{C}}{}_{\dot{A}}m^D\partial_{D\dot{C}}p^{\dot{A}}$, we obtain $\overset{\mathrm{c}}{\partial}_{2\dot{A}}p^{\dot{A}} = 0$, which means that

$$Q^{\dot{A}\dot{B}} \equiv -\phi^{-2}\,\overset{\mathrm{c}}{\partial}_2{}^{\dot{A}}p^{\dot{B}} \tag{4.20}$$

is symmetric on the indices $\dot{A}$, $\dot{B}$. Then, (4.19) gives

$$\overset{\mathrm{c}}{\theta}{}^{1\dot{A}} = -\mathrm{d}p^{\dot{A}} - Q^{\dot{A}}{}_{\dot{B}}\mathrm{d}q^{\dot{B}}. \tag{4.21}$$

In this manner, we have proved the validity of the following (Plebański and Robinson 1976, Finley and Plebański 1976, Torres del Castillo 1983) result.

Proposition 4.1. *The existence of a repeated principal spinor of the Weyl spinor C_{ABCD} that satisfies (4.10) implies the local existence of (possibly complex) coordinates $q^{\dot{A}}$, $p^{\dot{A}}$ such that the metric of the manifold can be expressed in the form*

$$g = 2\phi^{-2}\mathrm{d}q^{\dot{A}}(\mathrm{d}p_{\dot{A}} + Q_{\dot{A}\dot{B}}\mathrm{d}q^{\dot{B}}). \tag{4.22}$$

It may be remarked that in the preceding derivation no assumption is made concerning the signature of the metric (see Example 4.1 below). The manifolds described in this last proposition are called $\mathscr{H}\mathscr{H}$ spaces.

In the case of the Schwarzschild metric, already considered at the beginning of this section, it is convenient to choose the coordinates $q^{\dot{A}}$ in such a way that

$$\mathrm{d}q^{\dot{1}} = \csc\theta\,\mathrm{d}\theta + \mathrm{i}\mathrm{d}\varphi, \qquad \mathrm{d}q^{\dot{2}} = c\mathrm{d}t + \left(1 - \frac{r_g}{r}\right)^{-1}\mathrm{d}r. \tag{4.23}$$

Then, by comparing (2.107) with (4.12), taking $l_A = \delta_A^2$ as above, we find that

$$(L^{\dot{A}}{}_{\dot{B}}) = \mathrm{diag}\,(-r\sin\theta/\sqrt{2}, 1).$$

The conformal factor ϕ is determined by $\Gamma_{112\dot{A}} = \partial_{1\dot{A}}\ln\phi$ (i.e., $\rho = D\ln\phi$, $\tau = \delta\ln\phi$) [see (3.74)]; hence, we can take $\phi = r^{-1}$ [see (2.109) and (2.108)]. Then, according to (4.17), the vector fields $\overset{\mathrm{c}}{\partial}_{1\dot{A}}$ of the new tetrad are $\overset{\mathrm{c}}{\partial}_{1\dot{1}} = -r^2\partial_{1\dot{1}}$, $\overset{\mathrm{c}}{\partial}_{1\dot{2}} =$

$\sqrt{2}\,r\csc\theta\,\partial_{1\dot{2}}$, and one can readily verify that a possible choice for the coordinates $p^{\dot{A}}$ is

$$p^{\dot{1}} = r^{-1}, \qquad p^{\dot{2}} = -\cos\theta. \tag{4.24}$$

Taking $m^A = \delta^A_2$, the vector fields $\overset{c}{\partial}_{2\dot{A}}$ are given by [see (4.17)]

$$\overset{c}{\partial}_{2\dot{1}} = \frac{\sin\theta}{\sqrt{2}\,r}\partial_{2\dot{1}} - \eta\,\overset{c}{\partial}_{1\dot{1}}, \qquad \overset{c}{\partial}_{2\dot{2}} = -\frac{1}{r^2}\partial_{2\dot{2}} - \eta\,\overset{c}{\partial}_{1\dot{2}}\,.$$

Then, making use of the expressions derived above, one finds that $\eta = 0$ and

$$Q_{\dot{1}\dot{1}} = -\tfrac{1}{2}\sin^2\theta, \qquad Q_{\dot{1}\dot{2}} = 0, \qquad Q_{\dot{2}\dot{2}} = -\frac{r - r_g}{2r^3}. \tag{4.25}$$

It may be pointed out that if we replace the relations (4.23) by

$$\mathrm{d}q^{\dot{1}} = \csc\theta\,\mathrm{d}\theta + c\mathrm{d}t, \qquad \mathrm{d}q^{\dot{2}} = \mathrm{i}\mathrm{d}\varphi + \left(1 - \frac{r_g}{r}\right)^{-1}\mathrm{d}r,$$

maintaining (4.24), (4.25), and $\phi = r^{-1}$, then (4.22) gives another type-D solution of the Einstein vacuum field equations (with Lorentzian signature, assuming that r, θ, φ, and t are real), known as the Ehlers–Kundt $B1$ metric (Flaherty 1980).

Example 4.1. The standard metric of S^4 in complex coordinates.

An explicit example of a metric with Euclidean signature that satisfies the conditions of Proposition 4.1 and therefore can be written in the form (4.22) is given by the standard metric of S^4 (see Example 2.1, pp. 84–86). As shown in Section 2.2, the conformal curvature of this metric vanishes, $C_{ABCD} = 0$, $C_{\dot{A}\dot{B}\dot{C}\dot{D}} = 0$, and the spin coefficients for the null tetrad (2.102) satisfy $\Gamma_{11\dot{1}\dot{A}} = 0$; hence, the one-index spinor $l_A = \delta^2_A$ satisfies (4.10) and, trivially, $l^A l^B l^C C_{ABCD} = 0$. The one-forms $l_A\theta^{A\dot{B}} = \theta^{2\dot{B}}$ must form an integrable system; in fact, $\theta^{2\dot{A}} = L^{\dot{A}}{}_{\dot{B}}\mathrm{d}q^{\dot{B}}$, with

$$\mathrm{d}q^{\dot{1}} = \csc\theta\,\mathrm{d}\theta + \mathrm{i}\mathrm{d}\varphi, \qquad \mathrm{d}q^{\dot{2}} = \csc\psi\,\mathrm{d}\psi + \mathrm{i}\mathrm{d}\chi,$$

and

$$(L^{\dot{A}}{}_{\dot{B}}) = \mathrm{diag}(-\sin\psi\sin\chi\sin\theta/\sqrt{2}, \sin\psi/\sqrt{2}).$$

The conformal factor ϕ is determined by the conditions $\Gamma_{112\dot{A}} = \partial_{1\dot{A}}\ln\phi$. Hence, one finds that ϕ can be taken as

$$\phi = \csc\psi\csc\chi,$$

and from (4.17) we obtain

$$\overset{c}{\partial}_{1\dot{1}} = -\sin\psi\sin^2\chi\left(\frac{\partial}{\partial\psi} + \frac{\mathrm{i}}{\sin\psi}\frac{\partial}{\partial\chi}\right),$$

$$\overset{c}{\partial}_{1\dot{2}} = \frac{1}{\sin\theta}\left(\frac{\partial}{\partial\theta} + \frac{\mathrm{i}}{\sin\theta}\frac{\partial}{\partial\varphi}\right).$$

By inspection, one readily finds that the coordinates $p^{\dot{A}}$ can be chosen as

$$p^{\dot{1}} = -\mathrm{i}\cot\chi, \qquad p^{\dot{2}} = -\cos\theta.$$

Then, one finally finds that the functions $Q_{\dot{A}\dot{B}}$ are given by

$$Q_{\dot{1}\dot{1}} = -\tfrac{1}{2}\sin^2\theta, \qquad Q_{\dot{1}\dot{2}} = 0, \qquad Q_{\dot{2}\dot{2}} = \tfrac{1}{2}\csc^2\chi,$$

which are second-degree polynomials in the $p^{\dot{A}}$.

Going back to the general case, from (4.13) and (4.21) it follows that the null tetrad $\partial'_{A\dot{B}}$, in terms of the coordinates $q^{\dot{A}}$, $p^{\dot{A}}$, is given by

$$\begin{aligned} \overset{\mathrm{c}}{\partial}_{1\dot{A}} &= \frac{\partial}{\partial p^{\dot{A}}}, \\ \overset{\mathrm{c}}{\partial}_{2\dot{A}} &= \phi^2 \left(\frac{\partial}{\partial q^{\dot{A}}} - Q_{\dot{A}}{}^{\dot{B}} \frac{\partial}{\partial p^{\dot{B}}} \right). \end{aligned} \tag{4.26}$$

Note that this tetrad need not satisfy one of the conditions (2.100). In fact, equations (4.17) can be written in the form $\overset{\mathrm{c}}{\partial}_{A\dot{B}} = K_A{}^C K_{\dot{B}}{}^{\dot{D}} \partial_{C\dot{D}}$, with

$$\begin{aligned} K_1{}^C &= -\phi^{-2}[\det(L^{\dot{R}}{}_{\dot{S}})]^{-1/2} l^C, \\ K_2{}^C &= \phi^2[\det(L^{\dot{R}}{}_{\dot{S}})]^{1/2} m^C + \eta\phi^{-2}[\det(L^{\dot{R}}{}_{\dot{S}})]^{-1/2} l^C, \end{aligned}$$

and

$$K_{\dot{A}}{}^{\dot{B}} = [\det(L^{\dot{R}}{}_{\dot{S}})]^{1/2} (L^{-1})_{\dot{A}}{}^{\dot{B}}.$$

The matrices $(K^A{}_B)$ and $(K^{\dot{A}}{}_{\dot{B}})$ belong to SL(2,$\mathbb{C}$), but they do not have to satisfy one of the relations (1.185).

The spin coefficients of the metric (4.22) with respect to the null tetrad (4.26) can be readily obtained by making use of (2.97) and (2.98). Since

$$\overset{\mathrm{c}}{S}{}^{\dot{A}\dot{B}} = \overset{\mathrm{c}}{\theta}{}^{1(\dot{A}} \wedge \overset{\mathrm{c}}{\theta}{}^{|2|\dot{B})} = \phi^{-2}(\mathrm{d}p^{(\dot{A}} - Q^{\dot{C}(\dot{A}}\mathrm{d}q_{\dot{C}}) \wedge \mathrm{d}q^{\dot{B})}$$

[see (2.57)], we have

$$\begin{aligned} \mathrm{d}\,\overset{\mathrm{c}}{S}{}^{\dot{A}\dot{B}} &= -2\phi^{-1}\mathrm{d}\phi \wedge \overset{\mathrm{c}}{S}{}^{\dot{A}\dot{B}} - \phi^{-2}\mathrm{d}Q^{\dot{C}(\dot{A}} \wedge \mathrm{d}q_{\dot{C}} \wedge \mathrm{d}q^{\dot{B})} \\ &= 2\phi^{-1}(\overset{\mathrm{c}}{\partial}_{R\dot{C}}\,\phi)\, \overset{\mathrm{c}}{\theta}{}^{R\dot{C}} \wedge \overset{\mathrm{c}}{S}{}^{\dot{A}\dot{B}} + \phi^{-2}(\overset{\mathrm{c}}{\partial}_{R\dot{D}}\, Q^{\dot{C}(\dot{A}})\, \overset{\mathrm{c}}{\theta}{}^{|R\dot{D}|} \wedge \mathrm{d}q_{\dot{C}} \wedge \mathrm{d}q^{\dot{B})} \\ &= 2\phi^{-1}(\overset{\mathrm{c}}{\partial}_{R\dot{C}}\,\phi)\, \overset{\mathrm{c}}{\theta}{}^{R\dot{C}} \wedge \overset{\mathrm{c}}{S}{}^{\dot{A}\dot{B}} + \phi^{-2}(\overset{\mathrm{c}}{\partial}_{R\dot{D}}\, Q^{\dot{C}(\dot{A}})\, \overset{\mathrm{c}}{\theta}{}^{|R\dot{D}|} \wedge \varepsilon_{\dot{C}}{}^{\dot{B})} \phi^4\, \overset{\mathrm{c}}{S}{}^{\dot{2}\dot{2}}, \end{aligned}$$

where we have made use of the fact that $\overset{\mathrm{c}}{S}{}^{\dot{2}\dot{2}} = \overset{\mathrm{c}}{\theta}{}^{2\dot{1}} \wedge \overset{\mathrm{c}}{\theta}{}^{2\dot{2}} = \phi^{-4}\mathrm{d}q^{\dot{1}} \wedge \mathrm{d}q^{\dot{2}}$. The exterior products appearing in the last equation are expressed in terms of the three-forms $\overset{\mathrm{c}}{\breve{\theta}}{}^{A\dot{B}}$ with the aid of (2.71), and we obtain

$$\mathrm{d}\overset{\mathrm{c}}{S}{}^{\dot{A}\dot{B}} = \left[-2\phi^{-1}(\overset{\mathrm{c}}{\partial}_R{}^{(\dot{A}}\phi)\varepsilon^{\dot{B})}{}_{\dot{D}} - \phi^2(\overset{\mathrm{c}}{\partial}_{1\dot{D}}\,Q^{\dot{A}\dot{B}})\varepsilon^2{}_R\right]\,\overset{\mathrm{c}}{\breve{\theta}}{}^{R\dot{D}}. \tag{4.27}$$

Then, making use of (2.98), we find that

$$\overset{\mathrm{c}}{\Gamma}_{\dot{A}\dot{B}\dot{C}D} = \phi^{-1}\,\overset{\mathrm{c}}{\partial}_{D(\dot{A}}\,\phi\,\varepsilon_{\dot{B})\dot{C}} - \phi^2(\overset{\mathrm{c}}{\partial}_{1(\dot{A}}\,Q_{\dot{B})\dot{C}})\varepsilon^2{}_D, \tag{4.28}$$

and hence

$$\overset{\mathrm{c}}{\mathbf{\Gamma}}_{\dot{A}\dot{B}} = \phi^{-1}\partial_{(\dot{A}}\phi\,(\mathrm{d}p_{\dot{B})} + Q_{\dot{B})\dot{C}}\mathrm{d}q^{\dot{C}}) + \phi^{-1}D_{(\dot{A}}\phi\,\mathrm{d}q_{\dot{B})} - \partial_{(\dot{A}}Q_{\dot{B})\dot{C}}\,\mathrm{d}q^{\dot{C}}, \tag{4.29}$$

where we have made use of the definitions

$$\begin{aligned} \partial_{\dot{A}} &\equiv \frac{\partial}{\partial p^{\dot{A}}}, \\ D_{\dot{A}} &\equiv \left(\frac{\partial}{\partial q^{\dot{A}}} - Q_{\dot{A}}{}^{\dot{B}}\frac{\partial}{\partial p^{\dot{B}}}\right). \end{aligned} \tag{4.30}$$

(Note that $\partial^{\dot{A}} = -\partial/\partial p_{\dot{A}}$.) In a similar manner, computing separately $\mathrm{d}\overset{\mathrm{c}}{S}{}^{11}$, $\mathrm{d}\overset{\mathrm{c}}{S}{}^{12}$, and $\mathrm{d}\overset{\mathrm{c}}{S}{}^{22}$, we obtain

$$\begin{aligned} \overset{\mathrm{c}}{\mathbf{\Gamma}}_{11} &= \phi^{-3}\partial_{\dot{A}}\phi\,\mathrm{d}q^{\dot{A}}, \\ \overset{\mathrm{c}}{\mathbf{\Gamma}}_{12} &= -\tfrac{3}{2}\phi^{-1}\partial_{\dot{A}}\phi\,(\mathrm{d}p^{\dot{A}} + Q^{\dot{A}}{}_{\dot{B}}\mathrm{d}q^{\dot{B}}) - \tfrac{1}{2}(\phi^{-1}D_{\dot{A}}\phi + \partial^{\dot{B}}Q_{\dot{B}\dot{A}})\,\mathrm{d}q^{\dot{A}}, \\ \overset{\mathrm{c}}{\mathbf{\Gamma}}_{22} &= -\phi D_{\dot{A}}\phi\,(\mathrm{d}p^{\dot{A}} + Q^{\dot{A}}{}_{\dot{B}}\mathrm{d}q^{\dot{B}}) - \phi^2 D^{\dot{B}}Q_{\dot{B}\dot{A}}\,\mathrm{d}q^{\dot{A}}. \end{aligned} \tag{4.31}$$

The curvature spinors can now be calculated, e.g., with the aid of (2.91)–(2.94); the Weyl spinors are

$$\overset{\mathrm{c}}{C}_{\dot{A}\dot{B}\dot{C}\dot{D}} = -\phi^2\partial_{(\dot{A}}\partial_{\dot{B}}Q_{\dot{C}\dot{D})}, \tag{4.32}$$

and

$$\begin{aligned} \overset{\mathrm{c}}{C}_{1111} &= 0, \\ \overset{\mathrm{c}}{C}_{1112} &= 0, \\ \overset{\mathrm{c}}{C}_{1122} &= -\tfrac{1}{6}\phi^2\partial_{\dot{A}}\partial_{\dot{B}}Q^{\dot{A}\dot{B}}, \\ \overset{\mathrm{c}}{C}_{1222} &= -\tfrac{1}{2}\phi^4 D_{\dot{A}}\partial_{\dot{B}}Q^{\dot{A}\dot{B}}, \\ \overset{\mathrm{c}}{C}_{2222} &= -\phi^6\left[D_{\dot{A}}D_{\dot{B}}Q^{\dot{A}\dot{B}} - (D_{\dot{B}}Q^{\dot{A}\dot{B}})\partial^{\dot{C}}Q_{\dot{A}\dot{C}}\right], \end{aligned} \tag{4.33}$$

the spinor equivalent of the traceless part of the Ricci tensor

$$\begin{aligned} \overset{\mathrm{c}}{C}_{11\dot{A}\dot{B}} &= \phi^{-1}\partial_{\dot{A}}\partial_{\dot{B}}\phi, \\ \overset{\mathrm{c}}{C}_{12\dot{A}\dot{B}} &= -\tfrac{1}{2}\phi^2\partial^{\dot{C}}\partial_{(\dot{A}}Q_{\dot{B})\dot{C}} + \phi\partial_{(\dot{A}}D_{\dot{B})}\phi, \\ \overset{\mathrm{c}}{C}_{22\dot{A}\dot{B}} &= -\phi^5\partial_{(\dot{A}}(\phi^{-1}D^{\dot{C}}Q_{\dot{B})\dot{C}}) + \phi^3(D^{\dot{C}}\phi)\partial_{\dot{C}}Q_{\dot{A}\dot{B}} + \phi^3 D_{(\dot{A}}D_{\dot{B})}\phi, \end{aligned} \tag{4.34}$$

and the scalar curvature

$$R = 2\phi^2 \partial_{\dot{A}} \partial_{\dot{B}} Q^{\dot{A}\dot{B}} + 12\phi^3 \partial_{\dot{A}} (\phi^{-2} D^{\dot{A}} \phi). \tag{4.35}$$

Integration of Field Equations

Expression (4.22) allows us to partially integrate several sets of equations of geometric or physical interest. As an example of the reduction that can be achieved by combining the two-component spinor formalism and the form of the metric (4.22) (which is itself adapted to the spinor formalism) we shall give an alternative derivation of two results obtained in Section 3.3.1, showing that locally the solution of the source-free Maxwell equations, or of the Weyl equation, can be expressed in terms of a single scalar function that satisfies a second-order linear partial differential equation.

The source-free Maxwell equations $\nabla_{A\dot{B}} f^{\dot{A}\dot{B}} = 0$ are equivalent to $\mathrm{d}(\overset{\mathrm{c}}{f}_{\dot{A}\dot{B}} \overset{\mathrm{c}}{S}{}^{\dot{A}\dot{B}}) = 0$ (see Exercise 2.5). Thus making use of (2.71), (2.90), and (4.27), we have

$$\phi^2 \overset{\mathrm{c}}{\partial}_{1\dot{B}} (\phi^{-2} \overset{\mathrm{c}}{f}{}^{\dot{A}\dot{B}}) = 0 \tag{4.36}$$

and

$$\phi^2 \overset{\mathrm{c}}{\partial}_{2\dot{B}} (\phi^{-2} \overset{\mathrm{c}}{f}{}^{\dot{A}\dot{B}}) + \phi^2 \overset{\mathrm{c}}{f}{}^{\dot{B}\dot{C}} \overset{\mathrm{c}}{\partial}_1{}^{\dot{A}} Q_{\dot{B}\dot{C}} = 0. \tag{4.37}$$

Owing to (4.26), equation (4.36) is equivalent to $\partial_{\dot{B}}(\phi^{-2} \overset{\mathrm{c}}{f}{}^{\dot{A}\dot{B}}) = 0$. The solution of this last equation can be obtained making use of the following proposition.

Proposition 4.2. *Let $M_{\dot{A}\dot{B}}$ be a set of functions satisfying the conditions $M_{\dot{A}\dot{B}} = M_{\dot{B}\dot{A}}$ and $\partial_{\dot{A}} M^{\dot{A}\dot{B}} = 0$. Then there exists locally a function H such that $M_{\dot{A}\dot{B}} = \partial_{\dot{A}} \partial_{\dot{B}} H$.*

Proof. Conditions $\partial_{\dot{A}} M^{\dot{A}\dot{B}} = 0$ are equivalent to

$$\frac{\partial M_{\dot{1}}{}^{\dot{B}}}{\partial p^{\dot{2}}} = \frac{\partial M_{\dot{2}}{}^{\dot{B}}}{\partial p^{\dot{1}}},$$

which imply the local existence of functions $F^{\dot{B}}$ such that

$$M_{\dot{1}}{}^{\dot{B}} = \frac{\partial F^{\dot{B}}}{\partial p^{\dot{1}}}, \qquad M_{\dot{2}}{}^{\dot{B}} = \frac{\partial F^{\dot{B}}}{\partial p^{\dot{2}}}.$$

Since $M_{\dot{1}\dot{2}} = M_{\dot{2}\dot{1}}$, we have $\partial F_{\dot{2}} / \partial p^{\dot{1}} = \partial F_{\dot{1}} / \partial p^{\dot{2}}$, which in turn implies the local existence of a function H such that $F_{\dot{A}} = \partial H / \partial p^{\dot{A}}$; hence, $M_{\dot{A}\dot{B}} = \partial^2 H / \partial p^{\dot{A}} \partial p^{\dot{B}}$. □

Thus, there exists locally a function H such that

$$\overset{\mathrm{c}}{f}_{\dot{A}\dot{B}} = \phi^2 \partial_{\dot{A}} \partial_{\dot{B}} H. \tag{4.38}$$

Substituting this expression into (4.37), making use of (4.26) and (4.30), we obtain

$$
\begin{aligned}
0 &= D_{\dot{B}}\partial^{\dot{A}}\partial^{\dot{B}}H + (\partial^{\dot{B}}\partial^{\dot{C}}H)\partial^{\dot{A}}Q_{\dot{B}\dot{C}} \\
&= \partial^{\dot{A}}\left(\frac{\partial}{\partial q^{\dot{B}}}\partial^{\dot{B}}H + Q_{\dot{B}\dot{C}}\partial^{\dot{B}}\partial^{\dot{C}}H\right) \\
&= \partial^{\dot{A}}\left(D_{\dot{B}}\partial^{\dot{B}}H\right),
\end{aligned}
$$

which means that $F \equiv D_{\dot{B}}\partial^{\dot{B}}H$ is a function of $q^{\dot{A}}$ only. The function F can always be written in the form $F = \partial h^{\dot{A}}/\partial q^{\dot{A}}$, where the $h^{\dot{A}}$ are also functions of $q^{\dot{A}}$ only. Letting $\tilde{H} \equiv H - h_{\dot{A}}p^{\dot{A}}$, we find that [see (4.30)]

$$
D_{\dot{B}}\partial^{\dot{B}}\tilde{H} = D_{\dot{B}}(\partial^{\dot{B}}H - h^{\dot{B}}) = F - D_{\dot{B}}h^{\dot{B}} = F - \frac{\partial h^{\dot{B}}}{\partial q^{\dot{B}}} = 0.
$$

On the other hand, $\overset{\mathrm{c}}{f}_{\dot{A}\dot{B}} = \phi^2\partial_{\dot{A}}\partial_{\dot{B}}H = \phi^2\partial_{\dot{A}}\partial_{\dot{B}}\tilde{H}$. Thus (dropping the tilde), we conclude that locally, any solution of the source-free Maxwell equations can be written in the form $\overset{\mathrm{c}}{f}_{\dot{A}\dot{B}} = \phi^2\partial_{\dot{A}}\partial_{\dot{B}}H$, where H is a function such that $D_{\dot{B}}\partial^{\dot{B}}H = 0$.

In order to make use of this result it is not necessary to express the metric of the background space-time in the form (4.22), since with respect to an arbitrary null tetrad, this solution is given by

$$
f_{\dot{A}\dot{B}} = \nabla^{B}{}_{(\dot{A}}[\phi^{-2}\nabla^{S}{}_{\dot{B})}(\phi^2 l_S l_B \psi)] \tag{4.39}
$$

(Torres del Castillo 1984b), where ψ is some function [cf. (3.83)].

With the aid of (4.28), one finds that with respect to the null tetrad (4.26), the Weyl equation $\nabla_{A\dot{B}}\eta^{\dot{B}} = 0$ is given by

$$
\begin{aligned}
\phi^{3/2}\,\overset{\mathrm{c}}{\partial}_{1\dot{B}}\,(\phi^{-3/2}\,\overset{\mathrm{c}}{\eta}{}^{\dot{B}}) &= 0, \\
\phi^{3/2}\,\overset{\mathrm{c}}{\partial}_{2\dot{B}}\,(\phi^{-3/2}\,\overset{\mathrm{c}}{\eta}{}^{\dot{B}}) + \tfrac{1}{2}\phi^2(\overset{\mathrm{c}}{\partial}_1{}^{\dot{C}}Q_{\dot{C}\dot{B}})\,\overset{\mathrm{c}}{\eta}{}^{\dot{B}} &= 0.
\end{aligned} \tag{4.40}
$$

The first of these equations is locally equivalent to the existence of a function H such that $\overset{\mathrm{c}}{\eta}_{\dot{B}} = \phi^{3/2}\partial_{\dot{B}}H$, and substituting into the second equation one finds that the function H has to satisfy the equation

$$
D_{\dot{B}}\partial^{\dot{B}}H + \tfrac{1}{2}(\partial^{\dot{C}}Q_{\dot{C}\dot{B}})\partial^{\dot{B}}H = 0.
$$

With respect to an arbitrary null tetrad we have

$$
\eta_{\dot{B}} = \phi^{-1}\nabla^{C}{}_{\dot{B}}(\phi l_C \psi)
$$

[cf. (3.88)].

Left-Flat Spaces

The so-called $\mathscr{H}$ spaces are "half-flat," which means that the curvature two-forms $\mathscr{R}_{AB}$ (or $\mathscr{R}_{\dot{A}\dot{B}}$) vanish. In the same form as the vanishing of the curvature two-forms Ω_{ab} implies the local existence of a tetrad for which the connection one-forms vanish [see (2.51)], the vanishing of the two-forms $\mathscr{R}_{AB}$, for instance, is locally equivalent to the existence of a null tetrad such that $\boldsymbol{\Gamma}_{AB} = 0$ [see (2.68)]; therefore, according to Proposition 4.1, the metric of such a manifold can be expressed in the form

$$g = 2\mathrm{d}q^{\dot{A}}(\mathrm{d}p_{\dot{A}} + Q_{\dot{A}\dot{B}}\mathrm{d}q^{\dot{B}}) \tag{4.41}$$

in terms of some local coordinates $q^{\dot{A}}$, $p^{\dot{A}}$ [since $\Gamma_{ABC\dot{D}} = 0$, ϕ can be taken equal to 1; see (4.14)]. As we shall show below, the coordinates $q^{\dot{A}}$, $p^{\dot{A}}$ can be chosen in such a way that $\overset{\mathrm{c}}{\boldsymbol{\Gamma}}_{AB} = 0$. We start by noticing that $\partial_{\dot{A}}\partial^{\dot{C}}Q_{\dot{B}\dot{C}} = \partial_{(\dot{A}}\partial^{\dot{C}}Q_{\dot{B})\dot{C}} + \frac{1}{2}\varepsilon_{\dot{A}\dot{B}}\partial^{\dot{R}}\partial^{\dot{C}}Q_{\dot{R}\dot{C}} = 0$, as a consequence of $\overset{\mathrm{c}}{C}_{12\dot{A}\dot{B}} = 0$ and $\overset{\mathrm{c}}{C}_{1122} = 0$, taking into account that $\phi = 1$ [see (4.34) and (4.33)]. Hence, $\partial^{\dot{C}}Q_{\dot{B}\dot{C}}$ are functions of $q^{\dot{R}}$ only.

On the other hand, from $\overset{\mathrm{c}}{C}_{1222} = 0$, we have $\partial_{\dot{A}}D_{\dot{B}}Q^{\dot{A}\dot{B}} = 0$, which turns out to be equivalent to $D_{\dot{B}}\partial_{\dot{A}}Q^{\dot{A}\dot{B}} = 0$ [see (4.30)], or, since $\partial_{\dot{A}}Q^{\dot{A}\dot{B}}$ depend on the $q^{\dot{R}}$ only, $\partial(\partial_{\dot{A}}Q^{\dot{A}\dot{B}})/\partial q^{\dot{B}} = 0$. This last equation is locally equivalent to the existence of a function $\Lambda(q^{\dot{R}})$ such that

$$\partial^{\dot{A}}Q_{\dot{A}\dot{B}} = \partial\Lambda/\partial q^{\dot{B}} \tag{4.42}$$

(cf. Proposition 4.2).

We can take advantage of the fact that the coordinates $q^{\dot{A}}$ are not uniquely defined by (4.12); in place of $q^{\dot{A}}$ we can employ any pair of independent functions $q'^{\dot{R}}$ of the $q^{\dot{A}}$, which induce a new null tetrad $\overset{\mathrm{c}}{\theta}{}'^{A\dot{B}}$ with [see (4.13)]

$$\overset{\mathrm{c}}{\theta}{}'^{2\dot{A}} = -\phi^{-2}\mathrm{d}q'^{\dot{A}} = -\phi^{-2}\frac{\partial q'^{\dot{A}}}{\partial q^{\dot{B}}}\mathrm{d}q^{\dot{B}} = T^{\dot{A}}{}_{\dot{B}}\,\overset{\mathrm{c}}{\theta}{}^{2\dot{B}},$$

where

$$T^{\dot{A}}{}_{\dot{B}} \equiv \frac{\partial q'^{\dot{A}}}{\partial q^{\dot{B}}}.$$

(There is also some freedom in the choice of the conformal factor ϕ [see(4.14)]. However, in the present case, we want to maintain the value $\phi = 1$.) Hence,

$$\overset{\mathrm{c}}{\partial}{}'_{1\dot{A}} = -T_{\dot{A}}{}^{\dot{B}}\,\overset{\mathrm{c}}{\partial}_{1\dot{B}}\,.$$

The new coordinates $p'^{\dot{A}}$ must satisfy the conditions $\overset{\mathrm{c}}{\partial}{}'_{1\dot{A}}p'^{\dot{B}} = \delta^{\dot{B}}_{\dot{A}}$ [see (4.18)], and therefore,

$$p'^{\dot{A}} = -(T^{-1})_{\dot{B}}{}^{\dot{A}}p^{\dot{B}} + \sigma^{\dot{A}}, \tag{4.43}$$

where the $\sigma^{\dot{A}}$ are functions of $q^{\dot{B}}$ only.

Since the form of the metric (4.41) must be invariant under this change of coordinates, the functions $Q'_{\dot{A}\dot{B}}$ corresponding to the new coordinates $q'^{\dot{A}}$, $p'^{\dot{A}}$ must be given by

$$Q'^{\dot{A}\dot{B}} = (T^{-1})_{\dot{C}}{}^{\dot{A}}(T^{-1})_{\dot{D}}{}^{\dot{B}}Q^{\dot{C}\dot{D}} - (T^{-1})^{\dot{C}(\dot{A}}\frac{\partial p'^{\dot{B})}}{\partial q^{\dot{C}}}. \tag{4.44}$$

Using the fact that $\overset{c}{\partial}{}'_{1\dot{A}} = \partial/\partial p'^{\dot{A}} \equiv \partial'_{\dot{A}}$ [see (4.26)] and combining the previous results, one finds that

$$\partial'^{\dot{A}}Q'_{\dot{A}\dot{B}} = (T^{-1})^{\dot{D}}{}_{\dot{B}}\left[\partial^{\dot{C}}Q_{\dot{C}\dot{D}} - \frac{\partial}{\partial q^{\dot{D}}}\ln\det(T^{\dot{R}}{}_{\dot{S}})\right].$$

Hence, by means of a change of coordinates such that $\ln\det(T^{\dot{R}}{}_{\dot{S}}) = \Lambda$, we get $\partial'^{\dot{A}}Q'_{\dot{A}\dot{B}} = 0$ [see (4.42)].

Now we note that $\partial_{\dot{A}}D^{\dot{C}}Q_{\dot{B}\dot{C}} = \partial_{(\dot{A}}D^{\dot{C}}Q_{\dot{B})\dot{C}} + \frac{1}{2}\varepsilon_{\dot{A}\dot{B}}\partial^{\dot{R}}D^{\dot{C}}Q_{\dot{R}\dot{C}} = 0$, as a consequence of $\overset{c}{C}_{22\dot{A}\dot{B}} = 0$, $\overset{c}{C}_{1222} = 0$, and $\phi = 1$. Hence, $D^{\dot{C}}Q_{\dot{B}\dot{C}}$ are also functions of $q^{\dot{A}}$ only. Assuming that the coordinates $q'^{\dot{A}}$, $p'^{\dot{A}}$ have been chosen in such a way that $\partial^{\dot{A}}Q_{\dot{A}\dot{B}} = 0$, from the condition $\overset{c}{C}_{2222} = 0$ we have $D^{\dot{B}}D^{\dot{C}}Q_{\dot{B}\dot{C}} = 0$, which is locally equivalent to the existence of a function $\Xi(q^{\dot{A}})$ such that

$$D^{\dot{C}}Q_{\dot{B}\dot{C}} = \partial\Xi/\partial q^{\dot{B}}. \tag{4.45}$$

Under the coordinate transformation $q'^{\dot{A}} = q^{\dot{A}}$, $p'^{\dot{A}} = p^{\dot{A}} + \sigma^{\dot{A}}$ [see (4.43) and (4.44)], we have

$$Q'_{\dot{A}\dot{B}} = Q_{\dot{A}\dot{B}} - \frac{\partial}{\partial q^{(\dot{A}}}\sigma_{\dot{B})}.$$

Then one finds that

$$D'^{\dot{C}}Q'_{\dot{B}\dot{C}} = D^{\dot{C}}Q_{\dot{B}\dot{C}} + \frac{\partial}{\partial q^{\dot{B}}}\frac{1}{2}\frac{\partial\sigma^{\dot{A}}}{\partial q^{\dot{A}}}.$$

Therefore, by means of a change of coordinates $q'^{\dot{A}} = q^{\dot{A}}$, $p'^{\dot{A}} = p^{\dot{A}} + \sigma^{\dot{A}}$ with $\partial\sigma^{\dot{A}}/\partial q^{\dot{A}} = -2\Xi$, we get $D'^{\dot{C}}Q'_{\dot{B}\dot{C}} = 0$, thus showing that there exist local coordinates $q^{\dot{A}}$, $p^{\dot{A}}$ such that the metric can be written in the form (4.41) with

$$\partial^{\dot{A}}Q_{\dot{A}\dot{B}} = 0, \qquad D^{\dot{A}}Q_{\dot{A}\dot{B}} = 0 \tag{4.46}$$

(which amounts to $\overset{c}{\boldsymbol{\Gamma}}_{AB} = 0$ [see (4.31)]). According to Proposition 4.2, $\partial^{\dot{A}}Q_{\dot{A}\dot{B}} = 0$ implies the local existence of a function Θ such that

$$Q_{\dot{A}\dot{B}} = \partial_{\dot{A}}\partial_{\dot{B}}\Theta. \tag{4.47}$$

Substituting (4.47) into the second set of equations in (4.46), we find that [see (4.26)]

$$0=\left(\frac{\partial}{\partial q^{\dot{A}}}-Q_{\dot{A}}{}^{\dot{C}}\frac{\partial}{\partial p^{\dot{C}}}\right)Q^{\dot{A}\dot{B}}=\partial^{\dot{B}}\left[\frac{\partial}{\partial q^{\dot{A}}}\partial^{\dot{A}}\Theta-\frac{1}{2}\left(\partial_{\dot{A}}\partial^{\dot{C}}\Theta\right)\partial_{\dot{C}}\partial^{\dot{A}}\Theta\right]. \qquad (4.48)$$

Equation (4.48) means that the expression between brackets is a function of $q^{\dot{A}}$ only, which can always be written in the form $\partial h^{\dot{A}}/\partial q^{\dot{A}}$, where the $h^{\dot{A}}$ are functions of the $q^{\dot{R}}$ only. Then, the substitution $\Theta \mapsto \Theta + h_{\dot{A}}p^{\dot{A}}$, which leaves $Q_{\dot{A}\dot{B}}$ unchanged [see (4.47)], gives

$$\frac{\partial}{\partial q^{\dot{A}}}\partial^{\dot{A}}\Theta+\frac{1}{2}(\partial_{\dot{A}}\partial_{\dot{B}}\Theta)\partial^{\dot{A}}\partial^{\dot{B}}\Theta=0. \qquad (4.49)$$

This is the so-called second heavenly equation (Plebański 1975, Boyer et al. 1980), and as we have shown, the metric of any left-flat space can be locally expressed in the form

$$g=2\mathrm{d}q^{\dot{A}}(\mathrm{d}p_{\dot{A}}+\partial_{\dot{A}}\partial_{\dot{B}}\Theta\,\mathrm{d}q^{\dot{B}}), \qquad (4.50)$$

where Θ is a solution of (4.49).

According to (4.32), the only components of the curvature that can be different from zero are given by $C'_{\dot{A}\dot{B}\dot{C}\dot{D}} = -\partial_{\dot{A}}\partial_{\dot{B}}\partial_{\dot{C}}\partial_{\dot{D}}\Theta$.

4.3 Killing Bispinors. The Dirac Operator

Apart from the definition of a Killing spinor given in Section 3.4 for a manifold with Lorentzian signature, there exists another concept of a Killing spinor in the literature, which we shall call *Killing bispinor* (defined for a manifold of any signature). For a four-dimensional manifold, a pair of one-index spinor fields $(\psi_A, \phi_{\dot{A}})$ forms a Killing bispinor with Killing number λ if

$$\nabla_{A\dot{A}}\psi_B=\lambda\varepsilon_{AB}\phi_{\dot{A}}, \qquad \nabla_{A\dot{A}}\phi_{\dot{B}}=\lambda\varepsilon_{\dot{A}\dot{B}}\psi_A, \qquad (4.51)$$

for some constant λ (cf. Baum 2000, Friedrich 2000). An analogous concept can be defined in Riemannian manifolds of any dimension (Friedrich 2000, for the three-dimensional case see also Torres del Castillo 2003). As in the case of the $D(k,0)$ Killing spinors, the existence of a nontrivial solution of (4.51) imposes strong conditions on the curvature of the manifold. In fact, combining the equations (4.51), one obtains

$$\begin{aligned}\nabla_C{}^{\dot{A}}\nabla_{A\dot{A}}\psi_B&=2\lambda^2\varepsilon_{AB}\psi_C,\\ \nabla^A{}_{\dot{C}}\nabla_{A\dot{A}}\psi_B&=-\lambda^2\varepsilon_{\dot{C}\dot{A}}\psi_B,\end{aligned} \qquad (4.52)$$

and hence

$$\begin{aligned}\Box_{CA}\psi_B&=\lambda^2(\varepsilon_{AB}\psi_C+\varepsilon_{CB}\psi_A),\\ \Box_{\dot{C}\dot{A}}\psi_B&=0,\end{aligned}$$

and comparing with the Ricci identities (2.122) and (2.124), we see that $C_{ABCD} = 0$, $C_{AB\dot{A}\dot{B}} = 0$, and

$$R = -24\lambda^2. \tag{4.53}$$

In a similar way one finds that $C_{\dot{A}\dot{B}\dot{C}\dot{D}} = 0$. Equation (4.53) shows that the value of λ is determined up to sign by the scalar curvature and that λ must be real or pure imaginary.

Equations (4.51) imply that if $(\psi_A, \phi_{\dot{A}})$ is a Killing bispinor with Killing number λ, then $(-\psi_A, \phi_{\dot{A}})$ is a Killing bispinor with Killing number $-\lambda$. Each Killing bispinor gives rise to a (null, possibly complex) Killing vector $K_{A\dot{A}} = \psi_A \phi_{\dot{A}}$, since

$$\nabla_{A\dot{A}}(\psi_B \phi_{\dot{B}}) = \lambda \psi_A \psi_B \varepsilon_{\dot{A}\dot{B}} + \lambda \phi_{\dot{A}} \phi_{\dot{B}} \varepsilon_{AB} \tag{4.54}$$

[cf. (2.140)]. Equations (4.52) also yield

$$\nabla^{A\dot{A}} \nabla_{A\dot{A}} \psi_B = -2\lambda^2 \psi_B. \tag{4.55}$$

In what follows we shall restrict ourselves to the case that the signature of the metric is Euclidean. In this case equations (4.51) are equivalent to

$$\nabla_{A\dot{A}} \widehat{\psi}_B = -\overline{\lambda} \varepsilon_{AB} \widehat{\phi}_{\dot{A}}, \qquad \nabla_{A\dot{A}} \widehat{\phi}_{\dot{B}} = -\overline{\lambda} \varepsilon_{\dot{A}\dot{B}} \widehat{\psi}_A. \tag{4.56}$$

Thus, when λ is real ($R \leqslant 0$), $(\psi_A, \phi_{\dot{A}})$ and $(-\widehat{\psi}_A, \widehat{\phi}_{\dot{A}})$ are two linearly independent Killing bispinors with the same Killing number λ, while if λ is pure imaginary ($R \geqslant 0$), $(\psi_A, \phi_{\dot{A}})$ and $(\widehat{\psi}_A, \widehat{\phi}_{\dot{A}})$ are two linearly independent Killing bispinors with the same Killing number λ.

From (4.54) and its mate it follows that $\psi_A \phi_{\dot{A}} - \widehat{\psi}_A \widehat{\phi}_{\dot{A}}$ and $\mathrm{i}(\psi_A \phi_{\dot{A}} + \widehat{\psi}_A \widehat{\phi}_{\dot{A}})$ are the spinor equivalents of two *real* Killing vectors, and from the relation

$$\nabla_{A\dot{A}}(\psi_B \widehat{\phi}_{\dot{B}} \pm \widehat{\psi}_B \phi_{\dot{B}}) = \varepsilon_{AB}(\lambda \phi_{\dot{A}} \widehat{\phi}_{\dot{B}} \mp \overline{\lambda} \phi_{\dot{B}} \widehat{\phi}_{\dot{A}}) \pm \varepsilon_{\dot{A}\dot{B}}(\lambda \psi_A \widehat{\psi}_B \mp \overline{\lambda} \psi_B \widehat{\psi}_A)$$

it follows that if λ is real, then $\mathrm{i}(\psi_A \widehat{\phi}_{\dot{A}} - \widehat{\psi}_A \phi_{\dot{A}})$ is the spinor equivalent of a real Killing vector, while if λ is pure imaginary, $\psi_A \widehat{\phi}_{\dot{A}} + \widehat{\psi}_A \phi_{\dot{A}}$ is the spinor equivalent of a real Killing vector. Thus, a Killing bispinor gives rise to *three* linearly independent Killing vectors (cf. Friedrich 2000, Sec. 5.2).

Making use of (4.51) and (4.56) one can verify that if λ is real, then $\psi^A \widehat{\psi}_A - \phi^{\dot{A}} \widehat{\phi}_{\dot{A}}$ is a (real) constant, and if λ is pure imaginary, then $\psi^A \widehat{\psi}_A + \phi^{\dot{A}} \widehat{\phi}_{\dot{A}}$ is a constant.

Eigenbispinors of the Dirac Operator

We shall say that the pair of nonvanishing one-index spinor fields $(\psi_A, \phi_{\dot{A}})$ is an eigenbispinor of the Dirac operator if

$$\nabla^A{}_{\dot{A}} \psi_A = \lambda \phi_{\dot{A}}, \qquad \nabla_A{}^{\dot{A}} \phi_{\dot{A}} = \lambda \psi_A, \tag{4.57}$$

for some constant λ [cf. (3.34)]. Note that a Killing bispinor is an eigenbispinor of the Dirac operator, but the converse is not true. If we further assume that M is compact, then λ must be pure imaginary, since when the metric of M is positive definite, we have

$$\begin{aligned}\lambda \int_M \psi^A \widehat{\psi}_A \, \mathrm{d}v &= \int_M \widehat{\psi}_A \nabla^{A\dot{A}} \phi_{\dot{A}} \, \mathrm{d}v \\ &= \int_M \left[\nabla^{A\dot{A}} (\widehat{\psi}_A \phi_{\dot{A}}) - \phi_{\dot{A}} \nabla^{A\dot{A}} \widehat{\psi}_A\right] \mathrm{d}v \\ &= \int_M \phi_{\dot{A}} \overline{\lambda} \widehat{\phi}^{\dot{A}} \, \mathrm{d}v,\end{aligned}$$

i.e.,

$$\lambda \int_M \psi^A \widehat{\psi}_A \, \mathrm{d}v = -\overline{\lambda} \int_M \phi^{\dot{A}} \widehat{\phi}_{\dot{A}} \, \mathrm{d}v, \tag{4.58}$$

and similarly one finds that $\lambda \int_M \phi^{\dot{A}} \widehat{\phi}_{\dot{A}} \, \mathrm{d}v = -\overline{\lambda} \int_M \psi^A \widehat{\psi}_A \, \mathrm{d}v$, which implies that

$$(\lambda + \overline{\lambda}) \int_M (\psi^A \widehat{\psi}_A + \phi^{\dot{A}} \widehat{\phi}_{\dot{A}}) \, \mathrm{d}v = 0.$$

Hence, $\overline{\lambda} = -\lambda$, and if λ does not vanish, (4.58) yields

$$\int_M \psi^A \widehat{\psi}_A \, \mathrm{d}v = \int_M \phi^{\dot{A}} \widehat{\phi}_{\dot{A}} \, \mathrm{d}v. \tag{4.59}$$

Making use of the Ricci identities, from equations (4.57) and their mates one finds that if λ is equal to zero then the curvature of M vanishes.

A lower bound for $|\lambda|$ can be obtained by noting that for any spinor field χ^A, the identity

$$\begin{aligned}\nabla_B{}^{\dot{C}} \nabla_{A\dot{C}} \chi^A &= \nabla_{(B}{}^{\dot{C}} \nabla_{A)\dot{C}} \chi^A + \tfrac{1}{2} \varepsilon_{BA} \nabla^{S\dot{C}} \nabla_{S\dot{C}} \chi^A \\ &= \square_{BA} \chi^A + \tfrac{1}{2} \nabla^{A\dot{C}} \nabla_{A\dot{C}} \chi_B \\ &= \tfrac{1}{8} R \chi_B + \tfrac{1}{2} \nabla^{A\dot{C}} \nabla_{A\dot{C}} \chi_B\end{aligned} \tag{4.60}$$

holds. Therefore, if $(\psi_A, \phi_{\dot{A}})$ is an eigenbispinor of the Dirac operator, then $\nabla^{A\dot{C}} \nabla_{A\dot{C}} \psi_B = -2\lambda^2 \psi_B - \frac{1}{4} R \psi_B$ and

$$\begin{aligned}\int_M (2\lambda^2 + \tfrac{1}{4} R) \psi^B \widehat{\psi}_B \, \mathrm{d}v &= \int_M \widehat{\psi}^B \nabla^{A\dot{C}} \nabla_{A\dot{C}} \psi_B \, \mathrm{d}v \\ &= \int_M [\nabla^{A\dot{C}} (\widehat{\psi}^B \nabla_{A\dot{C}} \psi_B) - (\nabla^{A\dot{C}} \widehat{\psi}^B) \nabla_{A\dot{C}} \psi_B] \, \mathrm{d}v \\ &= -\int_M (\nabla^{A\dot{C}} \psi^B) (\nabla_{A\dot{C}} \psi_B)^{\frown} \, \mathrm{d}v.\end{aligned} \tag{4.61}$$

The integrand in the last expression is greater than or equal to zero, and therefore

$$2|\lambda|^2 \int_M \psi^B \widehat{\psi}_B \,\mathrm{d}v \geqslant \tfrac{1}{4} \int_M R \psi^B \widehat{\psi}_B \,\mathrm{d}v \geqslant \tfrac{1}{4} R_0 \int_M \psi^B \widehat{\psi}_B \,\mathrm{d}v,$$

where R_0 denotes the minimum of the scalar curvature on M; thus, $|\lambda|^2 \geqslant \frac{1}{8} R_0$.

In order to find a smaller lower bound for $|\lambda|$ we introduce the spinor fields

$$\xi_{AB\dot{C}} \equiv \nabla_{A\dot{C}} \psi_B - \tfrac{1}{2}\lambda \varepsilon_{AB} \phi_{\dot{C}}, \qquad \eta_{\dot{A}\dot{B}C} \equiv \nabla_{C\dot{A}} \phi_{\dot{B}} - \tfrac{1}{2}\lambda \varepsilon_{\dot{A}\dot{B}} \psi_C$$

[cf. (4.51)], so that (4.57) are equivalent to $\xi_{AB\dot{C}} = \xi_{(AB)\dot{C}}$ and $\eta_{\dot{A}\dot{B}C} = \eta_{(\dot{A}\dot{B})C}$. Then

$$\begin{aligned} (\nabla^{A\dot{C}} \psi^B)(\nabla_{A\dot{C}} \psi_B)\widehat{\ } &= (\xi^{AB\dot{C}} + \tfrac{1}{2}\lambda \varepsilon^{AB} \phi^{\dot{C}})(\widehat{\xi}_{AB\dot{C}} + \tfrac{1}{2}\overline{\lambda} \varepsilon_{AB} \widehat{\phi}_{\dot{C}}) \\ &= \xi^{AB\dot{C}} \widehat{\xi}_{AB\dot{C}} + \tfrac{1}{2}|\lambda|^2 \phi^{\dot{C}} \widehat{\phi}_{\dot{C}}, \end{aligned}$$

and making use of (4.61) and (4.59), we have

$$\begin{aligned} 0 &\leqslant \int_M \xi^{AB\dot{C}} \widehat{\xi}_{AB\dot{C}} \,\mathrm{d}v \\ &= \int_M \big[(\nabla^{A\dot{C}} \psi^B)(\nabla_{A\dot{C}} \psi_B)\widehat{\ } - \tfrac{1}{2}|\lambda|^2 \phi^{\dot{C}} \widehat{\phi}_{\dot{C}}\big] \,\mathrm{d}v \\ &= \int_M \big[-(2\lambda^2 + \tfrac{1}{4}R)\psi^B \widehat{\psi}_B - \tfrac{1}{2}|\lambda|^2 \phi^{\dot{C}} \widehat{\phi}_{\dot{C}}\big] \,\mathrm{d}v \\ &= \int_M \big(\tfrac{3}{2}|\lambda|^2 - \tfrac{1}{4}R\big) \psi^B \widehat{\psi}_B \,\mathrm{d}v. \end{aligned}$$

Thus $|\lambda|^2 \geqslant \frac{1}{6} R_0$. Furthermore, if $|\lambda|^2 = \frac{1}{6} R_0$, then R is constant and $\xi_{AB\dot{C}} = 0$, which means that $\nabla_{A\dot{C}} \psi_B = \frac{1}{2}\lambda \varepsilon_{AB} \phi_{\dot{C}}$. In a similar manner one finds that $\eta_{\dot{A}\dot{B}C} = 0$, which is equivalent to $\nabla_{C\dot{A}} \phi_{\dot{B}} = \frac{1}{2}\lambda \varepsilon_{\dot{A}\dot{B}} \psi_C$, and therefore, $(\psi_A, \phi_{\dot{A}})$ is a Killing bispinor with Killing number $\frac{1}{2}\lambda$.

Exercises

4.1. Show that the solution of the zero-rest-mass field equations (3.42) with respect to the null tetrad (4.26) can be expressed in terms of a scalar potential [cf. (4.38)].

4.2. Find the commutators of the three linearly independent Killing vector fields induced by a Killing bispinor when the signature of the metric is Euclidean.

Appendix A
Bases Induced by Coordinate Systems

In the traditional tensor formalism, the vector or tensor fields and the connections are given through their components with respect to the bases induced by coordinate systems, instead of rigid bases as the orthonormal or the null tetrads. Given the components $t_{\mu\nu\ldots\rho}$ of a tensor field with respect to the basis induced by a coordinate system x^μ, the components of its spinor equivalent are defined by

$$t_{A\dot{A}B\dot{B}\ldots D\dot{D}} \equiv \frac{1}{\sqrt{2}}\sigma^\mu{}_{A\dot{A}}\frac{1}{\sqrt{2}}\sigma^\nu{}_{B\dot{B}}\cdots\frac{1}{\sqrt{2}}\sigma^\rho{}_{D\dot{D}}\, t_{\mu\nu\ldots\rho}, \tag{A.1}$$

where the Infeld–van der Waerden symbols $\sigma^\mu{}_{A\dot{B}}$ are complex-valued *functions* such that

$$g_{\mu\nu}\sigma^\mu{}_{A\dot{B}}\sigma^\nu{}_{C\dot{D}} = -2\varepsilon_{AC}\varepsilon_{\dot{B}\dot{D}} \tag{A.2}$$

and the $g_{\mu\nu}$ are the components of the metric tensor with respect to the coordinate system x^μ. (By contrast, the Infeld–van der Waerden symbols associated with a rigid tetrad are constant.) We use here almost the same notation as in the preceding chapters for the Infeld–van der Waerden symbols, only with a Greek letter instead of a Latin letter for the first index.

The Infeld–van der Waerden symbols can be explicitly obtained for a metric given in terms of a coordinate system, by finding first a set of one-forms $\theta^{A\dot{B}}$ such that the metric tensor is expressed as $g = -2\theta^{1\dot{1}}\theta^{2\dot{2}} + 2\theta^{1\dot{2}}\theta^{2\dot{1}}$ [see (2.99)], and then reading off the coefficients in $\theta^{A\dot{B}} = (1/\sqrt{2})\sigma_\mu{}^{A\dot{B}}\,\mathrm{d}x^\mu$ [cf. (2.14)]. Finally, $\sigma^\nu{}_{A\dot{B}} = g^{\mu\nu}\varepsilon_{AC}\varepsilon_{\dot{B}\dot{D}}\sigma_\nu{}^{C\dot{D}}$.

The components of the covariant derivative of a tensor field involve the Christoffel symbols, which are given by the well-known formula

$$\Gamma^\mu_{\nu\rho} = \tfrac{1}{2}g^{\mu\lambda}\left(\frac{\partial g_{\nu\lambda}}{\partial x^\rho} + \frac{\partial g_{\rho\lambda}}{\partial x^\nu} - \frac{\partial g_{\nu\rho}}{\partial x^\lambda}\right). \tag{A.3}$$

For instance, the components of the covariant derivative of a vector field are

$$\nabla_\mu t^\nu = \frac{\partial t^\nu}{\partial x^\mu} + \Gamma^\nu_{\rho\mu}t^\rho,$$

which are also denoted by $t^{\nu}{}_{;\mu}$. Therefore, using the fact that

$$t^{\rho} = -\frac{1}{\sqrt{2}}\sigma^{\rho}{}_{E\dot{F}}\, t^{E\dot{F}},$$

the components of the spinor equivalent of $\nabla_{\mu}t^{\nu}$ are given by

$$\begin{aligned}
\nabla_{A\dot{B}}\, t^{C\dot{D}} &= \frac{1}{2}\sigma^{\mu}{}_{A\dot{B}}\sigma_{\nu}{}^{C\dot{D}}\left(\frac{\partial t^{\nu}}{\partial x^{\mu}} + \Gamma^{\nu}_{\rho\mu}\, t^{\rho}\right) \\
&= -\frac{1}{2\sqrt{2}}\sigma^{\mu}{}_{A\dot{B}}\sigma_{\nu}{}^{C\dot{D}}\left(\sigma^{\nu}{}_{E\dot{F}}\frac{\partial t^{E\dot{F}}}{\partial x^{\mu}} + t^{E\dot{F}}\frac{\partial \sigma^{\nu}{}_{E\dot{F}}}{\partial x^{\mu}} + \Gamma^{\nu}_{\rho\mu}\sigma^{\rho}{}_{E\dot{F}}\, t^{E\dot{F}}\right) \\
&= \partial_{A\dot{B}}\, t^{C\dot{D}} - \Gamma^{C\dot{D}}{}_{E\dot{F}A\dot{B}}\, t^{E\dot{F}},
\end{aligned}$$

where the vector fields

$$\partial_{A\dot{B}} \equiv \frac{1}{\sqrt{2}}\sigma^{\mu}{}_{A\dot{B}}\frac{\partial}{\partial x^{\mu}} \tag{A.4}$$

form a null tetrad and

$$\Gamma^{C\dot{D}}{}_{E\dot{F}A\dot{B}} = \frac{1}{2\sqrt{2}}\sigma^{\mu}{}_{A\dot{B}}\sigma_{\nu}{}^{C\dot{D}}\left(\frac{\partial \sigma^{\nu}{}_{E\dot{F}}}{\partial x^{\mu}} + \Gamma^{\nu}_{\rho\mu}\sigma^{\rho}{}_{E\dot{F}}\right). \tag{A.5}$$

(An example in which one can identify the functions $\sigma^{\mu}{}_{A\dot{B}}$ is given by (2.108).)

Since in the present case the components of the metric tensor need not be constant, one has to be careful with the order in which the partial derivatives and the raising or lowering of indices are applied. Using the fact that $\partial g_{\nu\lambda}/\partial x^{\mu} = \Gamma^{\kappa}_{\nu\mu}g_{\kappa\lambda} + \Gamma^{\kappa}_{\lambda\mu}g_{\nu\kappa}$, which is equivalent to (A.3), from (A.2) we obtain

$$\begin{aligned}
\sigma_{\nu}{}^{C\dot{D}}\frac{\partial \sigma^{\nu}{}_{E\dot{F}}}{\partial x^{\mu}} &= -\sigma^{\nu}{}_{E\dot{F}}\frac{\partial \sigma_{\nu}{}^{C\dot{D}}}{\partial x^{\mu}} \\
&= -\sigma^{\nu}{}_{E\dot{F}}\frac{\partial}{\partial x^{\mu}}(g_{\nu\lambda}\sigma^{\lambda C\dot{D}}) \\
&= -\sigma_{\lambda E\dot{F}}\frac{\partial \sigma^{\lambda C\dot{D}}}{\partial x^{\mu}} - \sigma^{\nu}{}_{E\dot{F}}\sigma^{\lambda C\dot{D}}(\Gamma^{\kappa}_{\nu\mu}g_{\kappa\lambda} + \Gamma^{\kappa}_{\lambda\mu}g_{\nu\kappa}),
\end{aligned}$$

and substituting into (A.5), we find that $\Gamma^{C\dot{D}}{}_{E\dot{F}A\dot{B}} = -\Gamma_{E\dot{F}}{}^{C\dot{D}}{}_{A\dot{B}}$, which implies that $\Gamma_{C\dot{D}E\dot{F}A\dot{B}} = \Gamma_{CEA\dot{B}}\varepsilon_{\dot{D}\dot{F}} + \Gamma_{\dot{D}\dot{F}\dot{B}A}\varepsilon_{CE}$. Hence, the spin coefficients for the null tetrad (A.4) are related to the Christoffel symbols by

$$\begin{aligned}
\Gamma_{CEA\dot{B}} &= \frac{1}{4\sqrt{2}}\sigma^{\mu}{}_{A\dot{B}}\sigma_{\nu C}{}^{\dot{D}}\left(\frac{\partial \sigma^{\nu}{}_{E\dot{D}}}{\partial x^{\mu}} + \Gamma^{\nu}_{\rho\mu}\sigma^{\rho}{}_{E\dot{D}}\right), \\
\Gamma_{\dot{D}\dot{F}\dot{B}A} &= \frac{1}{4\sqrt{2}}\sigma^{\mu}{}_{A\dot{B}}\sigma_{\nu}{}^{C}{}_{\dot{D}}\left(\frac{\partial \sigma^{\nu}{}_{C\dot{F}}}{\partial x^{\mu}} + \Gamma^{\nu}_{\rho\mu}\sigma^{\rho}{}_{C\dot{F}}\right)
\end{aligned} \tag{A.6}$$

[cf. (2.18)].

If we consider two conformally related metrics $g = \phi^2 g'$ as in Section 2.3, for a given null tetrad $\partial_{A\dot{B}}$ associated with the metric g, we can define a null tetrad $\partial'_{A\dot{B}}$ for the metric g', by $\partial'_{A\dot{B}} = \phi\, \partial_{A\dot{B}}$ [see (2.132)]. Expressing the vector fields $\partial'_{A\dot{B}}$ in the form (A.4), in terms of the basis induced by the coordinates x^μ, we have

$$\sigma'^{\mu}{}_{A\dot{B}} = \phi\, \sigma^{\mu}{}_{A\dot{B}},$$

or equivalently,

$$\sigma'_{\mu A\dot{B}} = \phi^{-1} \sigma_{\mu A\dot{B}}.$$

(The tensor indices of $\sigma'^{\mu}{}_{A\dot{B}}$ are lowered by means of $g'_{\mu\nu} = \phi^{-2} g_{\mu\nu}$.) With the aid of the coefficients $\sigma^{\mu}{}_{A\dot{B}}$ and $\sigma'^{\mu}{}_{A\dot{B}}$ we can find the relation between the tensor components of objects corresponding to the metrics g and g', with respect to a given coordinate system. For example, making use of (2.136) and the fact that $C_{AB\dot{C}\dot{D}}$ is the spinor equivalent of $\frac{1}{2}(R_{\mu\nu} - \frac{1}{4} R g_{\mu\nu})$, we obtain

$$R'_{\mu\nu} - \tfrac{1}{4} R' g'_{\mu\nu} = R_{\mu\nu} - \tfrac{1}{4} R g_{\mu\nu} + 2\phi^{-1} (\nabla_\mu \nabla_\nu \phi - \tfrac{1}{4} g_{\mu\nu} \nabla^\rho \nabla_\rho \phi),$$

where $R'_{\mu\nu}$ and R' are the components of the Ricci tensor and the scalar curvature, respectively, of the metric g'. Hence,

$$R'_{\mu\nu} = R_{\mu\nu} + 2\phi^{-1} \nabla_\mu \nabla_\nu \phi + \phi^{-1} g_{\mu\nu} \nabla^\rho \nabla_\rho \phi - 3\phi^{-2} g_{\mu\nu} (\nabla^\rho \phi)(\nabla_\rho \phi). \tag{A.7}$$

In a similar way, making use of the fact that $C_{ABCD}\varepsilon_{\dot{A}\dot{B}}\varepsilon_{\dot{C}\dot{D}} + C_{\dot{A}\dot{B}\dot{C}\dot{D}}\varepsilon_{AB}\varepsilon_{CD}$ is the spinor equivalent of the Weyl tensor, from (2.135) and (2.138) we find that

$$C'_{\mu\nu\rho\sigma} = \phi^{-2} C_{\mu\nu\rho\sigma}. \tag{A.8}$$

References

1. Ares de Parga, G., Chavoya, O., and López-Bonilla, J.L. (1989). Lanczos potential, *J. Math. Phys.* **30**, 1294.
2. Ashtekar, A. (1987). New Hamiltonian formulation of general relativity, *Phys. Rev.* D **36**, 1587.
3. Ashtekar, A. (1991). *Lectures on Non-perturbative Canonical Gravity* (World Scientific, Singapore).
4. Bampi, F. and Caviglia, G. (1983). Third-order tensor potentials for the Riemann and Weyl tensors, *Gen. Rel. Grav.* **15**, 375.
5. Baum, H. (2000). Twistor and Killing spinors in Lorentzian geometry, *Séminaires et Congrès, Société Mathématique de France* **4**, 35.
6. Berestetskii, V.B., Lifshitz, E.M., and Pitaevskii, L.P. (1982). *Quantum Electrodynamics*, 2nd ed. (Pergamon, Oxford).
7. Bergqvist, G. and Lankinen, P. (2004). Unique characterization of the Bel–Robinson tensor, *Class. Quantum Grav.* **21**, 3499.
8. Boyer, C.P., Finley, III, J.D., and Plebański, J.F. (1980). Complex general relativity, $\mathfrak{H}$ and $\mathfrak{H}\mathfrak{H}$ spaces – a survey of one approach, in *General Relativity and Gravitation*, Vol. 2, ed. A. Held (Plenum, New York), pp. 241–281.
9. Buchdahl, H.A. (1958). On the compatibility of relativistic wave equations for particles of higher spin in the presence of a gravitational field, *Nuovo Cim.* **10**, 96.
10. Cartan, E. (1966). *The Theory of Spinors* (Hermann, Paris) (Dover, New York, reprinted 1981).
11. Carter, B. and McLenaghan, R.G. (1979). Generalized total angular momentum operator for the Dirac equation in curved space-time, *Phys. Rev.* D **19**, 1093.
12. Catenacci, R., Marzuoli, A., and Salmistraro F. (1980). A note on Killing vectors in algebraically special vacuum space-times, *Gen. Rel. Grav.* **12**, 575.
13. Chandrasekhar, S. (1983). *The Mathematical Theory of Black Holes* (Clarendon, Oxford).
14. Chevalley, C. (1996). *The Algebraic Theory of Spinors and Clifford Algebras* (Springer, Berlin).
15. Cohen, J.M. and Kegeles, L.S. (1974). Electromagnetic fields in curved spaces: A constructive procedure, *Phys. Rev.* D **10**, 1070.
16. Conlon, L. (2001). *Differentiable Manifolds*, 2nd ed. (Birkhäuser, Boston).
17. Davydov, A.S. (1988). *Quantum Mechanics*, 2nd ed. (Pergamon, Oxford).
18. Debever, R. (1958). La super-énergie en relativité générale. *Bull. Soc. Math. Belg.* **10**, 112.
19. Debney, G.C. (1971a). Invariant approach to a space-time symmetry, *J. Math. Phys.* **12**, 1088.
20. Debney, G.C. (1971b). On vacuum space-times admitting a null Killing bivector, *J. Math. Phys.* **12**, 2372.

21. Dunajski, M. and West, S. (2008). Anti-self-dual conformal structures in neutral signature, in *Recent Developments in Pseudo-Riemannian Geometry*, ESI Lectures in Mathematics and Physics, ed. D. Alekseevsky and H. Baum (European Mathematical Society Publishing House, Zürich), pp. 113–148.
22. Ernst, F.J. (1968a). New formulation of the axially symmetric gravitational field problem, *Phys. Rev.* **167**, 1175.
23. Ernst, F.J. (1968b). New formulation of the axially symmetric gravitational field problem. II, *Phys. Rev.* **168**, 1415.
24. Finley III, J.D. and Plebański, J.F. (1976). The intrinsic spinorial structure of hyperheavens, *J. Math. Phys.* **17**, 2207.
25. Flaherty, Jr., E.J. (1980). Complex variables in relativity, in *General Relativity and Gravitation*, Vol. 2, ed. A. Held (Plenum, New York), pp. 207–239.
26. Friedrich, Th. (2000). *Dirac Operators in Riemannian Geometry* (American Mathematical Society, Providence, Rhode Island).
27. Geroch, R. (1968). Spinor structure of space-times in general relativity. I, *J. Math. Phys.* **9**, 1739.
28. Geroch, R. (1970). Spinor structure of space-times in general relativity. II, *J. Math. Phys.* **11**, 343.
29. Geroch, R.P. (1971). A method for generating solutions of Einstein's equations, *J. Math. Phys.* **12**, 918.
30. Geroch, R.P., Held, A., and Penrose, R. (1973). A space-time calculus based on pairs of null directions, *J. Math. Phys.* **14**, 874.
31. Goldberg, J.N. and Sachs, R.K. (1962). A theorem on Petrov types, *Acta Phys. Polon., Suppl.* **22**, 13. Reprinted in *Gen. Rel. Grav.* **41**, 433 (2009).
32. Goldblatt, E. (1994a). A Newman-Penrose formalism for gravitational instantons, *Gen. Rel. Grav.* **26**, 979.
33. Goldblatt, E. (1994b). Symmetries of type D^+D^- gravitational instantons, *J. Math. Phys.* **35**, 3029.
34. Hall, G.S. (2004). *Symmetries and Curvature Structure in General Relativity* (World Scientific, Singapore).
35. Hansen, R.O., Newman, E.T., Penrose, R., and Tod, K.P. (1978). The metric and curvature properties of $\mathscr{H}$-space, *Proc. R. Soc. Lond.* A **363**, 445.
36. Heusler, M. (1996). *Black Hole Uniqueness Theorems* (Cambridge University Press, Cambridge).
37. Hirsch, M. and Smale, S. (1974). *Differential Equations, Dynamical Systems, and Linear Algebra* (Academic, New York).
38. Huggett, S.A. and Tod, K.P. (1994). *An Introduction to Twistor Theory*, 2nd ed. (Cambridge University Press, Cambridge).
39. Hughston, L.P., Penrose, R., Sommers, P., and Walker, M. (1972). On a quadratic first integral for the charged particle orbits in the charged Kerr solution, *Commun. Math. Phys.* **27**, 303.
40. Hughston, L.P. and Sommers, P. (1973). The symmetries of Kerr black holes, *Commun. Math. Phys.* **33**, 129.
41. Infeld, L. and van der Waerden, B.L. (1933). Die Wellengleichung des Elektrons in der allgemeinen Relativitätstheorie, *Sitzber. preuss. Akad. Wiss. Physik-math. Kl.* **9**, 380.
42. Israel, W. (1979) *Differential Forms in General Relativity*, 2nd ed. (Dublin Institute for Advanced Studies, Dublin).
43. Jackson, J.D. (1975). *Classical Electrodynamics*, 2nd ed. (Wiley, New York).
44. Jeffryes, B.P. (1984). Space-times with two-index Killing spinors, *Proc. R. Soc. Lond.* A **392**, 323.
45. Kamran, N. (1987). Séparation des variables pour les potentiels de Debye dans toutes les solutions de type D des équations d'Einstein, *C.R. Acad. Sci. Paris*, **304**, Ser. I, 299.
46. Kamran, N. and McLenaghan, R.G. (1983). Separation of variables and quantum numbers for Weyl neutrino fields on curved spacetime, *Lett. Math. Phys.* **7**, 381.
47. Kamran, N. and McLenaghan, R.G. (1984a). Symmetry operators for neutrino and Dirac fields on curved spacetime, *Phys. Rev.* D **30**, 357.

48. Kamran, N. and McLenaghan, R.G. (1984b). Separation of variables and symmetry operators for the neutrino and Dirac equations in the space-times admitting a two-parameter abelian orthogonally transitive isometry group and a pair of shearfree geodesic null congruences, *J. Math. Phys.* **25**, 1019.
49. Kobayashi, S. and Nomizu, K. (1963). *Foundations of Differential Geometry*, Vol. 1, (Wiley-Interscience, New York).
50. Lanczos, C. (1962). The splitting of the Riemann tensor, *Rev. Mod. Phys.* **34**, 379.
51. Lanczos, C. (1970). *The Variational Principles of Mechanics*, 4th ed. (University of Toronto Press, Toronto) (Dover, New York, reprinted 1986).
52. Lawson, H.B. and Michelsohn, M.-L. (1989). *Spin Geometry* (Princeton University Press, Princeton, N.J.).
53. Lichnerowicz, A. (1964). Propagateurs, commutateurs et anticommutateurs en relativité générale, in *Relativity, Groups and Topology*, ed. B.S. DeWitt and C. DeWitt (Gordon and Breach, New York), pp. 821–861.
54. Lounesto, P. (1997). *Clifford Algebras and Spinors* (Cambridge University Press, Cambridge).
55. Maher, Jr., W.F. and Zund, J.D. (1968). A spinor approach to the Lanczos spin-tensor, *Nuovo Cim.* **57**, 638.
56. Mariot, L. (1954). Le champ électromagnétique singulier, *C. R. Acad. Sci. Paris* **238**, 2055.
57. Mason, L.J. and Woodhouse, N.M.J. (1996). *Integrability, Self-Duality, and Twistor Theory* (Oxford University Press, Oxford).
58. McLehaghan, R.G., Smith, S.N., and Walker, D.M. (2000). Symmetry operators for spin-1/2 relativistic wave equations on curved space-time, *Proc. R. Soc. Lond.* A **456**, 2629.
59. Merzbacher, E. (1998). *Quantum Mechanics*, 3rd ed. (Wiley, New York).
60. Messiah, A. (1962). *Quantum Mechanics*, Vol. II, (North Holland, Amsterdam).
61. Misner, C.W., Thorne, K.S., and Wheeler, J.A. (1973). *Gravitation* (Freeman, New York), Chap. 41.
62. Mustafa, E. and Cohen, J.M. (1987). Hertz and Debye potentials and electromagnetic fields in general relativity, *Class. Quantum Grav.* **4**, 1623.
63. Newman, E.T. (1976). Heaven and its properties, *Gen. Rel. Grav.* **7**, 107.
64. Newman, E.T. (1978). Source-free Yang-Mills theories, *Phys. Rev.* D **18**, 2901.
65. Newman, E.T. and Penrose, R. (1962). An approach to gravitational radiation by a method of spin coefficients, *J. Math. Phys.* **3**, 566; (1963) **4**, 998.
66. Newman, E.T. and Penrose, R. (1966). Note on the Bondi–Metzner–Sachs group, *J. Math. Phys.* **7**, 863.
67. O'Donnell, P. (2003). *Introduction to 2-Spinors in General Relativity* (World Scientific, Singapore).
68. Papapetrou, A. (1953). Eine rotationssymmetrische Lösung in der allgemeinen Relativitätstheorie, *Ann. Physik* **12**, 309.
69. Penrose, R. (1960). A spinor approach to general relativity, *Ann. Phys.* **10**, 171.
70. Penrose, R. (1965). Zero rest-mass fields including gravitation: asymptotic behaviour, *Proc. R. Soc. Lond.* A **284**, 159.
71. Penrose, R. (1976). Nonlinear gravitons and curved twistor theory, *Gen. Rel. Grav.* **7**, 31.
72. Penrose, R. and Rindler, W. (1984). *Spinors and Space-Time*, Vol. 1, (Cambridge University Press, Cambridge).
73. Penrose, R. and Ward, R.S. (1980). Twistors for flat and curved space-time, in *General Relativity and Gravitation*, Vol. 2, ed. A. Held (Plenum, New York), pp. 283–328.
74. Petersen, P. (1998). *Riemannian Geometry* (Springer, New York).
75. Pirani, F.A.E. (1965). In *Lectures on General Relativity, 1964 Brandeis Summer Institute*, Vol. 1, ed. S. Deser and K.W. Ford (Prentice-Hall, Englewood Cliffs, NJ).
76. Plebański, J.F. (1965). The "vectorial" optics of fields with arbitrary spin, rest-mass zero, *Acta Phys. Polon.* **27**, 361.
77. Plebański, J.F. (1974). *Spinors, Tetrads and Forms*, unpublished monograph, Centro de Investigación y de Estudios Avanzados del IPN, México, D.F., México.

78. Plebański, J.F. (1975). Some solutions of complex Einstein equations, *J. Math. Phys.* **16**, 2395.
79. Plebański, J.F. and Robinson, I. (1976). Left-degenerate vacuum metrics, *Phys. Rev. Lett.* **37**, 493.
80. Porteous, I.R. (1995). *Clifford Algebras and the Classical Groups* (Cambridge University Press, Cambridge).
81. Rindler, W. (1977). *Essential Relativity, Special, General, and Cosmological*, 2nd ed. (Springer, New York).
82. Robinson, I. (1961). Null electromagnetic fields, *J. Math. Phys.* **2**, 290.
83. Sen, A. (1981). On the existence of neutrino "zero-modes" in vacuum spacetimes, *J. Math. Phys.* **22**, 1781.
84. Sen, A. (1982). Quantum theory of spin-3/2 field in Einstein spaces, *Int. J. Theor. Phys.* **21**, 1.
85. Shaw, W.T. (1983a). Spinor fields at spacelike infinity, *Gen. Rel. Grav.* **15**, 1163.
86. Shaw, W.T. (1983b). Twistor theory and the energy–momentum and angular momentum of the gravitational field at spatial infinity, *Proc. R. Soc. Lond.* A **390**, 191.
87. Sommers, P. (1980). Space spinors, *J. Math. Phys.* **21**, 2567.
88. Stephani, H., Kramer, D., MacCallum, M.A.H., Hoenselaers, C., and Herlt, E. (2003). *Exact Solutions to Einstein's Equations*, 2nd ed. (Cambridge University Press, Cambridge).
89. Stewart, J. (1990). *Advanced General Relativity* (Cambridge University Press, Cambridge).
90. Taxiarchis, P. (1985). Space-Times admitting Penrose–Floyd tensors, *Gen. Rel. Grav.* **17**, 149.
91. Torres del Castillo, G.F. (1983). Null strings and Bianchi identities, *J. Math. Phys.* **24**, 590.
92. Torres del Castillo, G.F. (1984a). A note on $D(k,0)$ Killing spinors, *Commun. Math. Phys.* **92**, 485.
93. Torres del Castillo, G.F. (1984b). Null strings and Hertz potentials, *J. Math. Phys.* **25**, 342.
94. Torres del Castillo, G.F. (1985). Killing spinors and massless spinor fields, *Proc. R. Soc. Lond.* A **400**, 119.
95. Torres del Castillo, G.F. (1986). Killing spinors and gravitational perturbations, *J. Math. Phys.* **27**, 1583.
96. Torres del Castillo, G.F. (1988). The Teukolsky–Starobinsky identities in type D vacuum backgrounds with cosmological constant, *J. Math. Phys.* **29**, 2078.
97. Torres del Castillo, G.F. (1999). Debye potentials for self-dual fields, *Gen. Rel. Grav.* **31**, 205.
98. Torres del Castillo, G.F. (2003). *3-D Spinors, Spin-Weighted Functions and Their Applications* (Birkhäuser, Boston).
99. Wald, R.M. (1978). Construction of solutions of gravitational, electromagnetic, or other perturbation equations from solutions of decoupled equations, *Phys. Rev. Lett.* **41**, 203.
100. Wald, R.M. (1984). *General Relativity* (University of Chicago Press, Chicago).
101. Walker, M. and Penrose, R. (1970). On quadratic first integrals of the geodesic equations for type $\{22\}$ spacetimes, *Commun. Math. Phys.* **18**, 265.
102. Ward, R.S. and Wells, Jr., R.O. (1990). *Twistor Geometry and Field Theory* (Cambridge University Press, Cambridge).
103. Zund, J.D. (1975). The theory of the Lanczos spinor, *Ann. Mat. Pura Appl., Ser.* IV, **104**, 239.

Index

FSC
www.fsc.org
MIX
Papier | Fördert gute Waldnutzung
FSC® C083411

Kolibri 360 GmbH
Ferdinand-Jühlke-Straße 7
99095 Erfurt, Deutschland
produktsicherheit@kolibri360.de